# Symposium:
# SULFUR IN NUTRITION

# SYMPOSIUM: SULFUR IN NUTRITION

Oregon State University, 1969



*Editor* O. H. Muth, D.V.M.

*Professor, Department of Veterinary Medicine*
*Oregon State University*
*Corvallis, Oregon*

*Associate Editor* J. E. Oldfield, Ph.D.

*Head, Department of Animal Science*
*Oregon State University*
*Corvallis, Oregon*

WESTPORT, CONNECTICUT

**THE AVI PUBLISHING COMPANY, INC.**

1970

THE AVI PUBLISHING COMPANY, INC.
*Westport, Connecticut*

REGISTERED AT STATIONERS' HALL
LONDON, ENGLAND, 1970



*Library of Congress Catalog Card Number: 70-115500*
*ISBN 0-87055-073-X*



*Printed in the United States of America*

# Contributors to the Volume

W. H. ALLAWAY, U.S. Dept. Agr., A.R.S., Plant, Soil and Nutrition Laboratory, Ithaca, New York

H. J. ALMQUIST, Pacific Vegetable Oil Corporation, Kelseyville, California

D. D. DZIEWIATKOWSKI, School of Dentistry, University of Michigan, Ann Arbor, Michigan

J. D. FINKELSTEIN, Biochemistry Research Laboratory and Hepatology Unit, Veterans Administration Hospital, Washington, D.C.

U. S. GARRIGUS, Department of Animal Science, University of Illinois, Urbana, Illinois

J. L. KICE, Department of Chemistry, Oregon State University, Corvallis, Oregon

G. MATRONE, Dept. of Biochemistry, North Carolina State University, Raleigh, North Carolina

P. F. MAY, C.S.I.R.O., Division of Mathematical Statistics, Pastoral Research Laboratory, Armidale, N.S.W., Australia

I. W. McDONALD, C.S.I.R.O., Division of Animal Physiology, Ian Clunies Ross Animal Research Laboratory, Parramatta, Australia

R. J. MOIR, Institute of Agriculture, University of Western Australia, Nedlands, Western Australia

D. P. MOORE, Department of Soils, Oregon State University, Corvallis, Oregon

S. H. MUDD, Laboratory of General and Comparative Biochemistry, National Institute of Mental Health, U.S. Department of Health, Education and Welfare, Public Health Service, National Institutes of Health, Bethesda, Maryland

L. W. PARKS, Department of Microbiology, Oregon State University, Corvallis, Oregon

H. D. PECK, JR., Department of Biochemistry, University of Georgia, Athens, Georgia

I. K. SMITH, Department of Agronomy, Cornell University, Ithaca, New York

A. STRACHER, Department of Biochemistry, State University of New York, Downstate Medical Center, Brooklyn, New York

M. E. SWENDSEID, Department of Nutrition and Biological Chemistry, University of California, Los Angeles, California

J. F. THOMPSON, U.S. Dept. Agr., A.R.S., Plant, Soil and Nutrition Laboratory, Ithaca, New York

A. R. TILL, C.S.I.R.O., Division of Animal Physiology, Pastoral Research Laboratory, Armidale, N.S.W., Australia

E. J. UNDERWOOD, Institute of Agriculture, University of Western Australia, Nedlands, Western Australia

M. WANG, Division of Environmental and Nutritional Sciences, School of Public Health, University of California, Los Angeles, California

P. D. WHANGER, Department of Agricultural Chemistry, Oregon State University, Corvallis, Oregon

# Preface

This book records the proceedings of a Symposium, *Sulfur in Nutrition* held at Corvallis, Oregon September 11–12, 1969, co-sponsored by The Sulphur Institute of Washington, D.C. and The Nutrition Research Institute of Oregon State University. The Symposium was conceived and organized by two committees: one consisting of W. H. Allaway, U.S. Garrigus, Gennard Matrone, and J. E. Oldfield who did the preliminary planning; the other including A. F. Anglemier, G. H. Arscott, D. P. Moore, O. H. Muth, J. E. Oldfield, P. D. Whanger, and Elisabeth Yearick, all members of the faculty of Oregon State University who completed the detailed arrangements on site that were necessary for executing the program.

The Symposium was designed to bring together scientists active in sulfur research to review the basic knowledge and recent findings relating to the nutritional role of sulfur. Presentation progressed through discussions of the chemistry and biochemistry of sulfur to their implications for the metabolism of this most useful element by microorganisms, higher plants, ruminant and monogastric animals, and man. It was hoped that such an interaction would be useful in documenting the current status of sulfur in nutrition, in identifying problem areas of deficiencies of knowledge and subsequently in stimulating additional productive research.

This book, a more tangible product of the Symposium, brings together reviews of research which should prove a useful reference for students and scientists interested in sulfur metabolism, as well as for those who work with sulfur in agriculture or industry.

Acknowledgement is made of the generous funding of The Sulphur Institute and the enthusiastic cooperation of The Nutrition Research Institute that made this project possible, and of the efforts of the program participants and the members of the two committees involved in the planning and execution of the Symposium. Thanks are also given to Mrs. J. R. Harr for her very considerable assistance in editing the manuscripts and to Dr. D. K. Tressler of the AVI Publishing Company for his interest and cooperation in all phases of publication of this book.

J. E. Oldfield

O. H. Muth

*October 3, 1969*

# Contents

## Section V. Sulfur in Human Nutrition



## Section VI. Conclusion

SECTION I

# Introduction

CHAPTER 1

## The Scope of the Symposium: Outline of Current Problems Related to Sulfur in Nutrition

W. H. Allaway

A shortage of sulfur-containing amino acids is currently one of the more important world problems in human nutrition. At the same time, excesses of inorganic sulfur compounds in smokes or in mine drainage waters are blamed for serious deterioration of the quality of the environment. It is evident, therefore that a knowledge of the chemical pathways involved in the formation of the important organic compounds of sulfur from the inorganic compounds of this element is of key importance to our efforts to understand and improve life. One of the objectives of this Symposium is to examine the present state of knowledge concerning these pathways. Other objectives include consideration of the essential functions of sulfur compounds in living cells and the impacts of sulfur deficiencies upon humans, animals, and plants.

The topics included in this Symposium are an important segment of the global cycle of sulfur. The movements of this cycle, and estimates of the amounts of sulfur present in the different pools, are outlined by Bowen (1966). Sulfur is one of the more abundant elements in the earth, the oceans, and the atmosphere. The sulfur in the earth is present as sulfides of metals, hydrogen sulfide in gas and oil, elemental sulfur, organic sulfides in coal, as calcium sulfate or gypsum, and in other forms. Elemental sulfur, sulfur compounds from gas and oil, and the sulfide ores of metals are major sources of sulfur for commercial production.

As sulfur enters the biosphere, it is generally oxidized to sulfates. Any of the less oxidized forms of sulfur will normally be converted to sulfates within a short time in the rooting zone of plants grown on aerated soils. Sulfur enters living materials, and therefore enters the scope of this Symposium primarily by the movement of sulfate ions into plant roots, or into bacteria. It should be kept in mind, however, that $SO_2$ from the atmosphere may enter plants through the leaves, and also that rice, the most important food crop in the world, may take in reduced forms of sulfur.

Within the plant, the sulfate ion enters into a series of reactions involving stepwise reduction to sulfides and the incorporation of the sulfur into cysteine. The synthesis of sulfur amino acids from sulfates in plants is one of the key reactions in biology, comparable in importance to the reduction of carbon in photosynthesis. The ability to perform the reduction of sulfate and the synthesis of sulfur amino acids is an important marker of the boundary between the plant and the animal kingdoms. Plants perform this series of reactions and animals do not. Plants also synthesize methionine from cysteine by way of cystathionine; whereas in animals, sulfur moves in the opposite direction on this same pathway.

Another essential feature in the movement of sulfur through the biological phase of its cycle is the incorporation of cysteine and methionine into polypeptide chains during protein synthesis, and the formation of disulfide bonds between cysteine residues in polypeptide chains to impart essential spatial configurations to the protein. The presence of methionine, cysteine, and disulfide bonds in the proteins is critical to the functioning of these proteins as enzymes and as structural units in all living cells. The ratio of sulfur amino acids to the other amino acids in proteins has a marked effect upon the nutritional quality of these proteins when they are consumed by the next organism in the biological cycle.

In addition to the proteins, there are a number of lower molecular weight organic sulfur compounds that are required for the metabolic processes of living cells. S-adenosylmethionine, lipoic acid, coenzyme A, glutathione, thiamine, and biotin are a few examples of this group of compounds. The processes for which these compounds are essential include transmethylation, acetylation, $CO_2$ transfer, and the maintenance of proper redox potentials in the cell.

As the metabolism of sulfur compounds in the animal body progresses, these compounds are eventually oxidized. Animals excrete oxidized compounds of sulfur, including sulfates. The excreted sulfur compounds may reenter the biological cycle or return to the geological reserves of this element.

A brief treatment such as this cannot hope to even mention all of the important roles of sulfur compounds in living material. Indeed, the entire Symposium must, of necessity, limit its considerations to just a few of the biochemical and nutritional roles of these compounds. It is clear, however, that sulfur compounds are essential to so many life processes that under conditions of an absolute deficiency of sulfur, there is no living material.

When one becomes aware of the importance of sulfur compounds to all living cells, it seems, at first, somewhat surprising that this element has not been the object of greater public concern, and of more massive scientific study. A closer look at the situation reveals that in the past we have accidentally partially minimized the possibility of severe deficiencies of sulfur. This has been done through the deposition of sulfur from smoke and industrial fumes upon crops and agricultural soils, and through the use of superphosphate fertilizers that contained calcium sulfate. Modern technology, however, is working to remove both of these sources of sulfur from agriculture. Efforts to curb air pollution and shifts to sulfur-free sources of energy are already reducing the deposition of airborne sulfur compounds upon agricultural areas, and this trend will almost certainly continue. Developments in fertilizer technology have resulted in the widespread use of phosphorus fertilizers that contain higher levels of phosphorus, but no sulfur. Ammonium sulfate has lost part of its share of the nitrogen fertilizer market to competition with sulfur-free nitrogen compounds such as anhydrous ammonia and ammonium nitrate. So, sulfur deficiencies that limit plant growth are becoming increasingly common, and research workers in the area of soil fertility and plant nutrition are directing more effort to the study of sulfur in soils and its availability to plants. This trend is evidenced by the fact that in 1940, the President of the American Society of Agronomy, in his presidential address, considered sulfur as a neglected plant nutrient (Alway 1940), but in 1966 a Symposium concerned primarily with the soil chemistry and plant uptake of sulfur could, and did, include reviews of impressive progress in this field (Hanna 1966).

Some of the basic mechanisms of the process of protein synthesis in plants have operated to protect farm animals from extreme sulfur deficiencies. If a plant grows and synthesizes any proteins, its genetic makeup dictates that the proteins formed must contain fairly definite proportions of the sulfur amino acids. So, if an animal eats protein from plant sources, it is assured of at least some intake of the sulfur amino acids. Even though the sulfur content of the diet may be less than optimum in many cases, the effects of sulfur deficiency are not so dramatic as the effects of deficiencies of some of the other elements (Rendig and Weir 1957). If the diet contains a wide ratio of nitrogen to sulfur, the animal adjusts to this ratio by wasting nitrogen, and a decreased efficiency of feed protein utilization is the principal effect of sulfur deficiency.

But the current practice of substituting urea for plant proteins in the diets of ruminant animals can remove the safeguards against sulfur deficiency that are built into the process of protein synthesis in plants. Plant proteins frequently have ratios of N to S ranging from 10 to 1, to about 18 to 1. The optimum ratio for ruminant animals may be less than 10 to 1 (Moir *et al.* 1967–1968). When stockmen add urea to low-protein, high-energy diets the nitrogen-sulfur ratio may become much wider than this optimum. Unsatisfactory growth of cattle and sheep may result when these diets are fed, unless the diets are also supplemented with sulfur. This will be discussed in some of the papers in this Symposium.

In addition to the growing problems of sulfur nutrition of plants and ruminant animals, the problem of supplying adequate amounts of the essential sulfur amino acids to monogastric animals and humans is becoming increasingly severe. As the population expands toward the limits of the food supply and we strive for better nutrition for the peoples of developing countries, plants become increasingly important as sources of human dietary protein. Many of the more available plant proteins, such as soybean meal, are limited in nutritional value because they are deficient in the essential sulfur amino acids. The problem is of sufficient importance that some nutritionists (Miller and Donoso 1963) have suggested that the nitrogen-to-sulfur ratio of food proteins can be used as a measure of the nutritional value of these proteins. As new developments, such as the "miracle rice" and the Mexican wheats, help to meet the needs of people for calories, and high-lysine corn comes to play an important role in removing lysine as a factor limiting the quality of plant proteins, the need for supplying the essential sulfur amino acids will occupy an increasingly prominent position in the world food problem.

This Symposium, therefore, is being held at a time when studies of sulfur in the nutrition of plants, animals, and people are coming to be of increased importance to a great number of people. It is perhaps appropriate to consider some of the goals for research on the biological, biochemical, and nutritional aspects of sulfur. Certainly, one of the goals would be to increase the concentration of sulfur amino acids in food plants. It may also be possible to develop specialized supplements, possibly proteins from single-celled organisms that are especially rich in sulfur amino acids, for use with some plant proteins. Another goal would be to improve the efficiency of conversion of plants to animal food products that contain higher levels of sulfur amino acids. We also must seek a complete understanding of the metabolic pathways and essential functions of sulfur compounds in living cells, and to identify the factors that inhibit or block the proper functioning of these compounds. Only through this understanding can we hope to correct some of the errors of sulfur metabolism. This understanding must, in turn, be based on knowledge of the nature of the bonds between sulfur and other atoms in the compounds found in living cells; of how sulfhydryl groups react with other active groups and with the different heavy metals; and of how the sulfhydryl groups and disulfide bonds respond to subtle changes in the redox potential of the cell. And these can only be considered as a very incomplete list of goals for research on sulfur in nutrition.

This Symposium will not provide the answers to many of these important questions. There is a long road of difficult research still ahead. But, neither is this Symposium a starting point for this research. There has already been substantial progress along this road. This Symposium is rather to be considered an advance base or staging area where we can jointly profit from experience gained in traversing the road so far, and indicate promising routes to additional progress.

## BIBLIOGRAPHY

ALWAY, F. J. 1940. A nutrient element slighted in agricultural research. Agron. J. *32*, 913–921.

BOWEN, H. J. M. 1966. The Trace Elements in Biochemistry. Academic Press, New York.

HANNA, W. J. (Editor). 1966. Symposium on sulfur in agriculture. Soil Sci. *101*, 229–345.

MILLER, D. S., and DONOSO, G. 1963. Relation between the sulfur/nitrogen ratio and protein value of diets. J. Sci. Food Agr. *14*, 345–349.

MOIR, R. J., SOMERS, M., and BRAY, A. C. 1967–1968. Utilization of dietary sulfur and nitrogen by ruminants. Sulphur Inst. J. *3*, 15–18.

RENDIG, V. V., and WEIR, W. C. 1957. Evaluation by lamb feeding tests of alfalfa hay grown on a low sulfur soil. J. Animal Sci. *16*, 451–461.

SECTION II

# Chemistry and Biochemistry of Sulfur and Its Compounds

CHAPTER 2

John L. Kice

## The Chemistry of the Sulfur-Sulfur Bond

### INTRODUCTION

As any biochemist knows, the sulfur–sulfur bond is such a ubiquitous feature of the structure of proteins, enzymes, antibodies, etc., that an understanding of both its physical properties and chemical behavior is obviously of great importance. The organic chemist finds the S–S bond an intriguing functional group because, while it is easily isolable, it is at the same time very labile chemically and is subject to cleavage by a wide variety of reagents. It therefore possesses a rather unique character and chemistry.

Most chemists and biochemists think of sulfur–sulfur bonds only with reference to disulfides, RS—SR, not realizing that sulfur atoms of higher oxidation state can also be joined by a sulfur–sulfur bond. Some of the various types of compounds possessing a single sulfur–sulfur bond which are known are shown below:

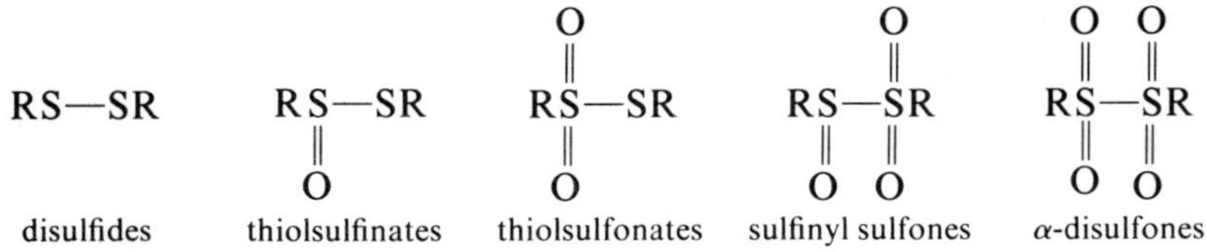

disulfides   thiolsulfinates   thiolsulfonates   sulfinyl sulfones   α-disulfones

| $R_2\overset{+}{S}—SR$ | $RS—SO_3^-$ | $RS—SCN$ |
|---|---|---|
| dialkylalkylthio-sulfonium salts | Bunte salts (S-alkyl thiosulfates) | sulfenyl thiocyanates |

Although, compared to disulfides, most of these compounds are probably not of great importance in biological or biochemical systems, Bunte salts, $RSSO_3^-$, are produced by the treatment of the disulfide links in a protein with sulfite (Milligan and Swan 1962), and thiolsulfinates, RS(O)SR, are formed on mild oxidation of disulfides with various oxidizing agents [equation (1)] (Small *et al.* 1947; Marangelli *et al.* 1960). Since, under a given set of reaction conditions, thiolsulfinates undergo cleavage of their

$$RS—SR \xrightarrow{[O]} \underset{\underset{O}{\|}}{RS}—SR \qquad (1)$$

sulfur–sulfur bond orders of magnitude more rapidly than the corresponding disulfide, it is entirely possible for a thiolsulfinate to be an intermediate in a reaction sequence leading to cleavage of a disulfide link, and initiated by oxidation of one of the sulfur atoms of that disulfide link, without the concentration of thiolsulfinate ever building up to the point where its presence would be detectable by any usual means.

Given the nature of the orientation of this Symposium we shall confine most of our discussion of the chemistry of sulfur–sulfur bonds to the chemistry of those in disulfides. However, because of the considerations just outlined we will also say a few words about the chemistry of the S–S bonds in thiolsulfinates and Bunte salts.

## PHYSICAL CHARACTERISTICS AND PROPERTIES OF THE SULFUR–SULFUR BOND IN A DISULFIDE

### Bond Length and Bond Angle

The length of the S–S bond in a disulfide is typically between 2.04 and 2.06 Å (Hordvik 1966; Pryor 1962). The R—S⟋S bond angle is somewhat variable but usually lies somewhere in the range 100° to 108° (Pryor 1962). Shorter S–S bond lengths than 2.04 Å mean an increase in the degree of overlap of unshared pairs of 2p electrons on one sulfur with the 3d orbitals of the other sulfur and an increase in the multiple bond character of the S–S bond (Hordvik 1966), i.e.

$$[R—\ddot{\underset{..}{S}}—\ddot{\underset{..}{S}}—R \leftrightarrow R—\overset{-}{\ddot{\underset{..}{S}}}=\overset{+}{\underset{..}{S}}—R]$$

## Dihedral Angle and Torsional Strain

The preferred dihedral angle, $\phi$, between the 2 substituents attached to the S–S bond of a disulfide is about 90° (Hordvik 1966), presumably because this particular conformation minimizes unfavorable repulsions be-

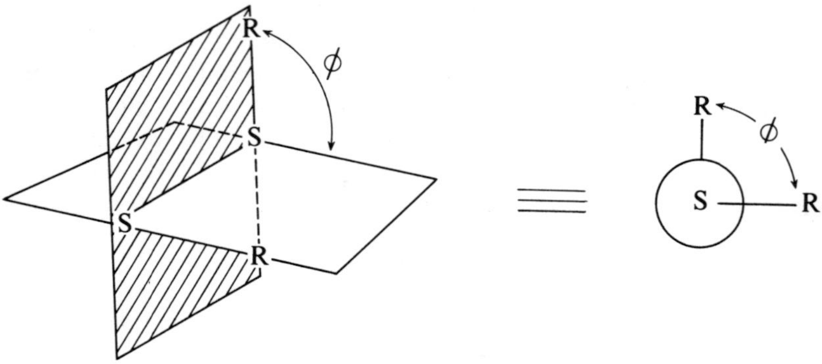


tween the unshared electron pairs on adjacent sulfur atoms. Geometric constraints which cause this angle to be significantly less than 90° lead to what organic chemists term torsional strain, because smaller dihedral angles result in increased repulsions between the nonbonding electron pairs on the two atoms. This torsional strain is particularly severe in cyclic 5-membered ring disulfides (1,2-dithiacyclopentanes) such as lipoic acid (I) where the dihedral angle is only about 27° (Hordvik 1966).

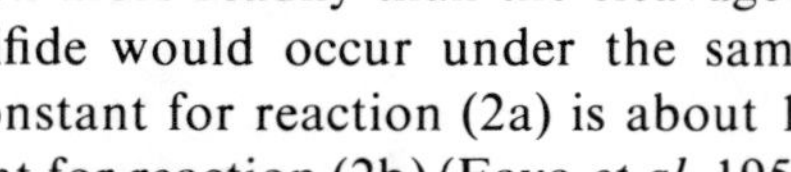


lipoic acid (I)

Such torsional strain in 1,2-dithiacyclopentanes (I) is important because it causes them to undergo reactions leading to ring opening via cleavage of the S–S bond much more readily than the cleavage of the S–S bond of an open chain disulfide would occur under the same conditions. For example, the rate constant for reaction (2a) is about 10,000 times greater than the rate constant for reaction (2b) (Fava *et al.* 1957).

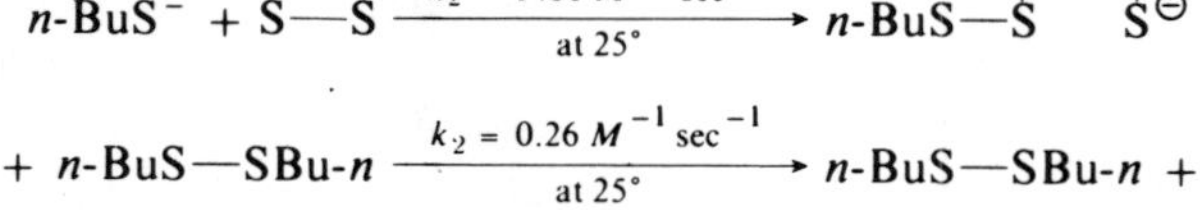


(2a)

$$n\text{-BuS}^- + n\text{-BuS—SBu-}n \xrightarrow[\text{at } 25^\circ]{k_2 = 0.26\ M^{-1}\ \text{sec}^{-1}} n\text{-BuS—SBu-}n + n\text{-BuS}^- \quad (2b)$$

One important coenzymatic function of lipoic acid is in the oxidative decarboxylation of $\alpha$-ketoacids, where its role is to carry out the interesting oxidation and acyl group transfer shown in equation (3) (Reed 1958). With an ordinary disulfide one would not observe cleavage of the S–S bond under such conditions. Much, if not all, of the driving force for the initial step of equation (3) must come from the torsional strain in the 1,2-dithiacyclopentane ring of (I).

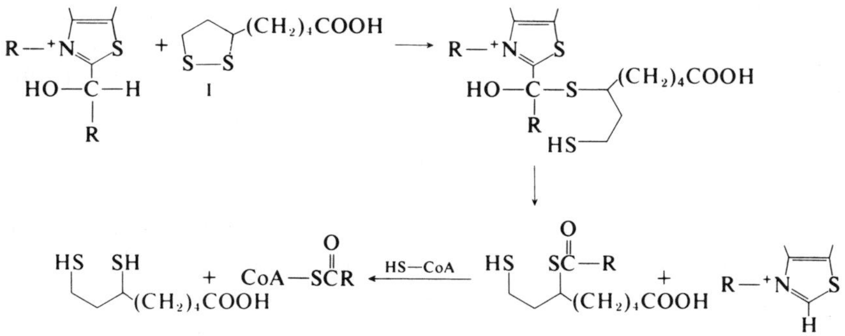


(3)

**FREE RADICAL REACTIONS INVOLVING SULFUR–SULFUR BONDS**

Sulfur–sulfur bonds are subject to both homolytic (reactions involving free radical intermediates) and heterolytic (reactions involving ionic intermediates) cleavage. Homolytic cleavage of a sulfur–sulfur bond, which will be discussed in this section, can occur either by direct dissociation [equation (4)] (Pryor 1962), or as a result of attack of a free radical on one of the sulfurs [equation (5)] (Pryor 1962; Pryor and Guard 1964).

$$RS—SR \rightarrow 2RS\cdot \tag{4}$$

$$R'\cdot + RS—SR \rightarrow R'SR + RS\cdot \tag{5}$$

**Thermal Homolytic Dissociation of Sulfur–Sulfur Bonds**

The reversible coloration which certain disulfides exhibit on being heated was at one time thought to be due to their appreciable reversible dissociation at elevated temperatures into a pair of RS· radicals [equation (4)]. However, subsequent work (Lecher 1954; Brand and Davidson 1956; Davis and Perrin 1960) has shown that this phenomenon of thermochromism has nothing to do with homolytic dissociation of the disulfide. Although we now therefore realize that heating a disulfide does not result in any significant equilibrium concentration of RS· radicals, there is evidence that at sufficiently elevated temperatures disulfides do undergo homolytic dissociation according to equation (4) (Davis and Perrin 1960; Ferrington and Tobolsky 1955, 1958; Kern 1955). Quantitative studies (Davis and Perrin 1960; Minch 1965) of this phenomenon, however, indicate that, except in special cases where the structure of R is such as to provide particularly effective resonance stabilization of the RS· radical, the rate of dissociation of RSSR is very slow ($k_d < 10^{-8}$ sec$^{-1}$) even at 100°C. It therefore seems very unlikely that direct radical dissociation of cystine disulfide bridges will ever be important under biological conditions.

One should note, however, that thiolsulfinates, RS(O)SR, which, as we mentioned earlier, are formed on oxidation of disulfides with a variety of reagents, undergo homolytic dissociation [equation (6)] much more readily than disulfides. In the one case so far studied quantitatively

$$\underset{\underset{\displaystyle O}{\|}}{\mathrm{RS}}\!-\!\mathrm{SR} \rightarrow \mathrm{R\dot{S}O} + \mathrm{RS}\cdot \qquad (6)$$

(Koch and Fava 1968) the thiolsulfinate dissociated homolytically at 100°C at least 1,000 times faster than the corresponding disulfide.

Perhaps even more important is the fact that $\alpha$-disulfoxides (II) apparently decompose extremely rapidly into a pair of RSO· radicals [equation (7)] even at temperatures below room temperature (Barnard and

$$\underset{\mathrm{II}}{\underset{\underset{\displaystyle O}{\|}}{\mathrm{RS}}\!-\!\underset{\underset{\displaystyle O}{\|}}{\mathrm{SR}}} \xrightarrow[\text{fast}]{\text{very}} 2\mathrm{RSO}\cdot \qquad (7)$$

Percy 1960; Marangelli *et al.* 1960). Since $\alpha$-disulfoxides are thought to be the initial product of the further oxidation of thiolsulfinates by the same reagents that oxidize disulfides to thiolsulfinates, one can see that one could conceivably get homolytic dissociation of a disulfide S–S bond in a biological system via a reaction sequence involving two oxidation steps followed by equation (7), i.e.,

$$\mathrm{RS}\!-\!\mathrm{SR} \xrightarrow{[\mathrm{O}]} \underset{\underset{\displaystyle O}{\|}}{\mathrm{RS}}\!-\!\mathrm{SR} \xrightarrow{[\mathrm{O}]} \underset{\underset{\displaystyle O}{\|}}{\mathrm{RS}}\!-\!\underset{\underset{\displaystyle O}{\|}}{\mathrm{SR}} \xrightarrow[\text{fast}]{\text{very}} 2\mathrm{RSO}\cdot$$

**Radical Displacement Reactions of Disulfides**

Although homolytic dissociation of the S–S bond in disulfides is slow even at elevated temperatures, disulfides do react readily with other free radicals at ordinary temperatures in the manner shown in equation (5). Most studies of such reactions to date have involved investigation of the effect of disulfides on free radical polymerization of a vinyl monomer like styrene (Pryor 1962) or methyl methacrylate (Otsu *et al.* 1964), since in such a polymerization the addition of a simple *n*-alkyl disulfide leads, via the reaction sequence of equations (8) and (9), to a marked reduction in the average molecular weight of the polymer being formed without any decrease in the rate of polymerization. (Compounds which act in this way

$$\sim CH_2\underset{\displaystyle R'}{\underset{|}{\dot{C}H}} + RSSR \xrightarrow{k_{tr}} \sim CH_2\underset{\displaystyle R'}{\underset{|}{C}}HSR + RS\cdot \tag{8}$$

$$RS\cdot + CH_2{=}CHR' \rightarrow RSCH_2\underset{\displaystyle R'}{\underset{|}{\dot{C}H}} \xrightarrow{CH_2=CHR'} RSCH_2\underset{\displaystyle R'}{\underset{|}{C}}HCH_2\underset{\displaystyle R'}{\underset{|}{C}}H\cdot \tag{9}$$

on a free radical polymerization are called chain transfer agents.) The reactivity of disulfides in equation (8), $k_{tr}$, is generally high enough (Pryor 1962) to suggest that in biological systems radical displacement reactions of free radicals with disulfide bridges could be important under the right conditions.

Pryor and Guard (1964) have shown that the reactivity of a disulfide RSSR in a radical displacement reaction of the type shown in equation (5) depends markedly on the bulk of the R group. They decomposed the azo compound $(C_6H_5)_3CN{=}NC_6H_5$ in carbon tetrachloride solutions containing various alkyl disulfides. The decomposition of the azo compound [equation (10)] yields phenyl radicals, and if these attack the solvent [equation (11)] they produce chlorobenzene, while if they undergo a radical displacement with the disulfide they yield the sulfide $C_6H_5SR$ [equation (12)]. By determining the ratio (moles $C_6H_5SR/C_6H_5Cl$) from

$$(C_6H_5)_3C{-}N{=}N{-}C_6H_5 \longrightarrow (C_6H_5)_3C\cdot + N_2 + C_6H_5\cdot \tag{10}$$

$$C_6H_5\cdot + CCl_4 \xrightarrow{k_{Cl}} C_6H_5Cl + \cdot CCl_3 \tag{11}$$

$$C_6H_5\cdot + RS{-}SR \xrightarrow{k_S} C_6H_5SR + RS\cdot \tag{12}$$

the decomposition of a small amount of the azo compound in the presence of a large excess of $CCl_4$ and RSSR, each present at a known concentration, one can determine $k_S/k_{Cl}$, since in such a situation

$$\frac{\text{moles } C_6H_5SR}{\text{moles } C_6H_5Cl} = \frac{k_S}{k_{Cl}} \times \frac{[RSSR]}{[CCl_4]}$$

Pryor and Guard's results, which are given in Table 2.1, show that increased branching at the $\alpha$-carbon of the alkyl group R leads to a marked decrease in $k_S/k_{Cl}$, and that each successive replacement of H by $CH_3$ on that carbon results in a proportionately greater decrease in reactivity. This pattern of reactivity is typically found in those organic reactions whose

TABLE 2.1

STERIC REQUIREMENTS IN RADICAL DISPLACEMENT REACTIONS OF DISULFIDES[1]

| R in RSSR | $(k_S/k_{Cl})$ for Equations (11) and (12) |
|---|---|
| $CH_3$ | 31 |
| $CH_3CH_2$ | 17 |
| $CH_3CH_2CH_2$ | 16 |
| $(CH_3)_2CH$ | 3.3 |
| $(CH_3)_3C$ | 0.23 |

[1]Data of Pryor and Guard (1964).

rates are subject to marked retardation by increases in the steric bulk (size) of substituents near the reaction center. In particular, from comparison of the data in Table 2.1 with data for other displacement reactions of known transition state geometry, Pryor and Guard (1964) were able to conclude that the geometry of the transition state for radical displacement reactions of a disulfide is a trigonal bipyramid with the attacking radical and the departing RS group occupying the apical positions, as shown in III. Increases in the steric bulk of R are known to alter the rates of displacements having this type of transition state geometry in the manner observed.

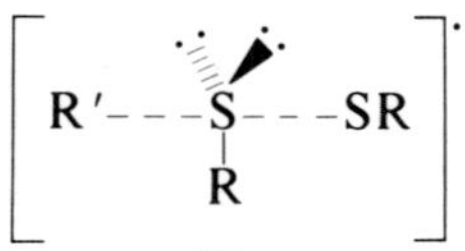


III

## NUCLEOPHILIC CLEAVAGE OF SULFUR–SULFUR BONDS

### General Considerations

By far the most common mode of cleavage of the S–S bond in a disulfide is via reactions involving attack of a nucleophile on one of the sulfurs and displacement of mercaptide ion (Parker and Kharasch 1959). A generalized representation of such a process is shown in equation (13). Because the reverse reaction [equation (14)] is frequently very rapid, one is

$$Nu^- + RS{-}SR \xrightarrow{k_{13}} RSNu + RS^- \qquad (13)$$

$$RS^- + RSNu \xrightarrow{k_{14}} RSSR + Nu^- \qquad (14)$$

often dealing with an equilibrium

$$Nu^- + RSSR \underset{k_{14}}{\overset{k_{13}}{\rightleftharpoons}} RSNu + RS^-$$

where $k_{14}$ is much larger than $k_{13}$. As a result, except in a few selected cases ($Nu^- = R'S^-$, $AlH_4^-$, $R'^-$, $BH_4^-$, $R'As(O)(OH)_2^-$ or $(R'O)_2PO^-$) such reactions generally do not proceed very far toward completion (Parker and Kharasch 1959). However, in those cases where the conjugate acid of the nucleophile, NuH, has a significantly smaller $pK_a$ than the $pK_a$ of the mercaptan, RSH, one can sometimes get otherwise unfavorable equilibria to go quite far toward completion by the expedient of carrying out the reaction at a pH at which $RS^-$ will effectively be completely converted to RSH but $Nu^-$ will still be unprotonated. This is effective because of the fact that RSH is generally at least $10^3$ less reactive as a nucleophile than $RS^-$ (Davis *et al.* 1969), i.e., $k_{15} < 10^{-3}\, k_{14}$. The equilibrium constants for the reaction of sulfite and bisulfite with various disul-

$$RSH + RSNu \xrightarrow{k_{15}} RSSR + H^+ + Nu^- \qquad (15)$$

fides (Stricks and Kohltoff 1951; Stricks *et al.* 1955) provide an excellent illustration of this phenomenon. With the disulfide derived from thioglycollic acid, for example, Stricks and Kohltoff (1951) found $K_{17}$ was over $10^3$ larger than $K_{16}$, and large enough, 1.40, so that on mixing equi-

$$(^-OOCCH_2S)_2 + SO_3^= \xrightleftharpoons{K_{16} = 4.7 \times 10^{-4}} {}^-OOCCH_2S^- + {}^-OOCCH_2S{-}SO_3^- \qquad (16)$$

$$(^-OOCCH_2S)_2 + HSO_3^- \xrightleftharpoons{K_{17} = 1.40} {}^-OOCCH_2SH + {}^-OOCCH_2S{-}SO_3^- \qquad (17)$$

molar concentrations of sulfite and disulfide about 50% of the disulfide is cleaved at equilibrium, in contrast to the less than 1% cleavage in solutions of higher pH where $^-OOCCH_2S^-$ is not protonated.

Another way in which equilibria $RSSR + Nu^- \rightleftarrows RSNu + RS^-$ can be driven to complete cleavage of the S–S bond is if either RSNu or $RS^-$ undergo some further irreversible reaction under the conditions of the experiment. One example of this is the fact that cyanide ion cleaves the S–S bond of phenyl disulfide in the presence of added 2,4-dinitrochlorobenzene, but not in its absence (Parker and Kharasch 1960). The reaction proceeds successfully in the presence of the added aryl halide because it reacts irreversibly with the mercaptide ion formed in the S–S bond cleavage [equation (18)]. Another example is the fact that cystine is cleaved by cyanide ion while simple alkyl disulfides are not (Schoberl *et al.* 1951). This turns out to be due to the fact that the thiocyanate formed as one of the products of the cleavage of the S–S bond reacts

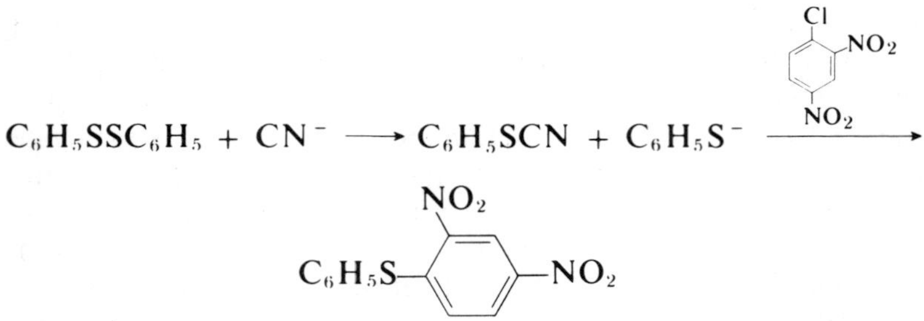

(18)

further in the manner shown in equation (19). Thiocyanates from cleavage of simple alkyl disulfides do not possess a suitably situated amino group; thus the further reaction can't happen in such cases, and, because the initial equilibrium is quite unfavorable, cleavage of the S–S bond is not observed.

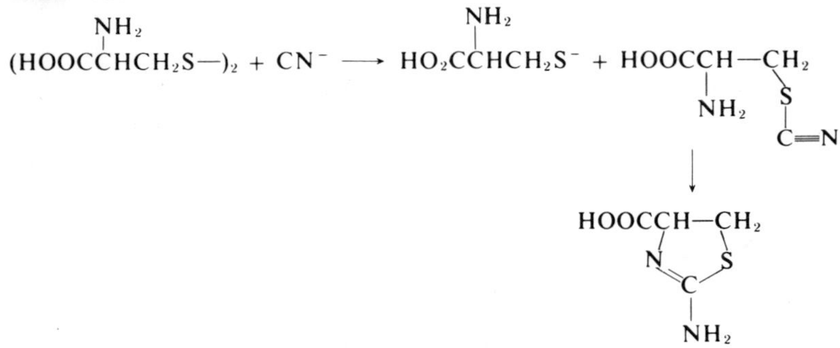

(19)

Both the forward and reverse steps of these $Nu^- + RSSR \rightleftarrows RSNu + RS^-$ equilibria, i.e., equations (13) and (14), are nucleophilic substitution reactions involving attack of the nucleophile on a divalent, or sulfenyl, sulfur. A general knowledge of the important characteristics of nucleophilic substitutions at sulfenyl sulfur is accordingly necessary if we are to understand such reactions fully.

**Mechanistic Aspects of Nucleophilic Substitution at Sulfenyl Sulfur**

A generalized representation of a nucleophilic substitution at sulfenyl sulfur is shown in equation (20). We will now discuss some of the important characteristics of this type of reaction.

$$Nu^- + R{-}S{-}X \rightarrow R{-}S{-}Nu + X^- \qquad (20)$$

**Steric Effects in Nucleophilic Substitutions at Sulfenyl Sulfur.**—The rates of nucleophilic substitutions at sulfenyl sulfur [equation (20)] are greatly reduced by increased branching at the $\alpha$-carbon of the alkyl group R in RSX. The effects are of the same general character but are considerably larger in magnitude than those (Table 2.1) observed for radical displacements at sulfur. The most extensive study of this phenomenon was by Fava and Ilceto (1958) who measured the rate of exchange of radioactive sulfite with a series of alkyl Bunte salts $RSSO_3^-$ [equation (21)] and compared the variation of rate with R with that for exchange of radioactive bromide ion with alkyl bromides $RCH_2Br$ [equation (22)] (de la Mare 1955). The data for equations (21) and (22) are given in Table 2.2.

$$*SO_3^{=} + RS{-}SO_3^- \rightarrow RS{-}*SO_3^- + SO_3^{=} \qquad (21)$$

$$*Br^- + RCH_2{-}Br \rightarrow RCH_2{-}*Br + Br^- \qquad (22)$$

They reveal that both reactions show a similar decrease in rate with increased branching in R and, more important, that both exhibit that proportionately greater decrease in rate with each successive replacement

TABLE 2.2

COMPARISON OF STERIC EFFECTS IN NUCLEOPHILIC SUBSTITUTIONS AT SULFENYL SULFUR AND TETRAHEDRAL CARBON

| R— | $(k/k_{R=Et})$ | |
|---|---|---|
| | $RSSO_3^- + *SO_3^{=}$† | $RCH_2Br + *Br^-$‡ |
| $CH_3$— | 2.0 | 1.54 |
| $CH_3CH_2$— | (1.0) | (1.0) |
| $(CH_3)_2CH$— | 0.014 | 0.05 |
| $(CH_3)_3C$— | 0.000012 | 0.000023 |

†Fava and Ilceto (1958).
‡de la Mare (1955).

of H by $CH_3$ which we said earlier was typical of a steric retardation effect. Similar changes in rate with the structure of R have been observed in each of the substitutions at sulfenyl sulfur shown in equations (23)–(25) (Fava *et al.* 1957; Ciuffarin and Fava 1968; Kice and Morkved 1964), so that the effect is clearly a general one as far as substitution at sulfenyl sulfur is concerned.

$$R{*}S^- + RS{-}SR \rightarrow RS{*}{-}SR + RS^- \qquad (23)$$

$$n\text{-}Bu\ddot{N}H_2 + RS{-}SCN \rightarrow n\text{-}BuN^+(H)_2{-}SR + SCN^- \qquad (24)$$

$$n\text{-}Bu_2S + RS{-}\overset{+}{S}(R){-}S(=O)Ar \rightarrow n\text{-}Bu_2\overset{+}{S}{-}SR + RS(=O)SAr \qquad (25)$$

The transition state for the displacement at tetrahedral carbon shown in equation (22) is known to have the geometry shown in IV. From the

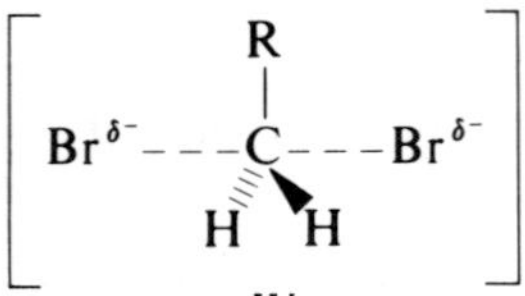

IV

close correspondence of the effect of R on the rates of equations (21) and (22), Fava and Ilceto (1958) concluded that a similar geometry (V) applies to substitutions at sulfenyl sulfur and that the entering and leaving groups

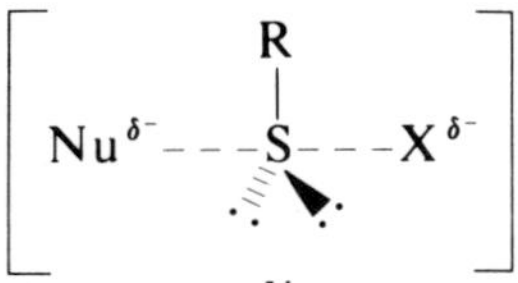

V

in such substitutions occupy in effect the apical positions of a trigonal bipyramid.

**The Question of an Intermediate in Substitutions at Sulfenyl Sulfur.**—In a nucleophilic substitution at tetrahedral carbon such as that shown in equation (22) the making of the new bond is synchronous with the breaking of the old one. Although the geometry of the transition state for substitutions at sulfenyl sulfur is apparently the same as for substitutions at tetrahedral carbon (*vide supra*), this in itself does not require that bond making and bond breaking be exactly synchronous as they are in the substitution at carbon. Indeed, the availability of *d* orbitals on sulfur means that a mechanism where making of the Nu–S bond in equation (20) has progressed much further than the breaking of the S–X bond by the time the transition state is reached must be given serious consideration. In the most extreme case one might even conceive of having an actual intermediate (VI) formed in the course of the substitution.

$$\mathrm{Nu^- + R{-}\ddot{\underset{..}{S}}{-}X \rightarrow Nu{-}\overset{\ominus}{\underset{\underset{R}{|}}{S}}{-}X \rightarrow R{-}\ddot{\underset{..}{S}}{-}Nu + X^-} \qquad (26)$$

VI

This question has been investigated by three different groups (Ciuffarin and Fava 1968; Kice and Anderson 1968; Brown and Hogg 1967) by determining the effect of *meta* and *para* substituents on the rates of substitutions at sulfenyl sulfur of the general type

$$\mathrm{Nu^- + ArS{-}X \rightarrow ArSNu + X^-}$$

and comparing the results with the effect of the same substituents on the rates of analogous substitutions at tetrahedral carbon and silicon. In the case of the substitutions at silicon bond-making is known to be well ahead of bond-breaking in the transition state, due to the use of *d* orbitals on silicon, and *meta* and *para* substituents have a much larger effect on rate than they do for the analogous substitution at carbon. Therefore, if the substitution at sulfur were to involve a mechanism with an actual intermediate, as in equation (26), it should show a response to changes in substituents in the aromatic ring akin to the substitution at silicon, and quite different from the one found for substitutions of $ArCH_2X$. In actual fact, it does not, behaving instead in every respect similarly to carbon rather than to silicon. From this one concludes that, despite the availability of *d* orbitals on sulfur, VI is not involved as an intermediate in substitutions at sulfenyl sulfur, and that such substitutions proceed with the same kind of synchronous bond making and bond breaking as one has in analogous substitutions at tetrahedral carbon.

**Rate of Nucleophilic Substitution at Sulfenyl Sulfur as Compared to Rate of Substitution at Tetrahedral Carbon.**—Nucleophilic substitution at sulfenyl sulfur is orders of magnitude faster than an analogous substitution at tetrahedral carbon. This can be seen from the fact that the rate constant, $k_S$, for the reaction shown in equation (27) is greater than $10^5\ M^{-1}$ $sec^{-1}$ (Kice and Favstritsky 1969), while the rate constant, $k_c$, for an analogous substitution ($-I$ and $-\overset{+}{S}(CH_3)_2$ are about comparable as leaving groups) at tetrahedral carbon [equation (28)] at the same temperature is

$$(CH_3)_2S + CH_3S\overset{+}{S}(CH_3)_2 \xrightarrow{k_S} (CH_3)_2\overset{+}{S}\text{—}SCH_3 + (CH_3)_2S \quad (27)$$

$$(CH_3)_2S + CH_3\text{—}I \xrightarrow{k_c} (CH_3)_2\overset{+}{S}\text{—}CH_3 + I^- \quad (28)$$

only $4.5 \times 10^{-5}\ M^{-1}\ sec^{-1}$ (Pearson *et al.* 1968). Thus the displacement at sulfenyl sulfur is about $10^{10}$ faster than the same displacement at tetrahedral carbon, a huge rate difference.

An important practical consequence of this is the following. Even in cases of $Nu^- + RSSR \rightleftarrows RSNu + RS^-$ equilibria where the equilibrium constant is very small, the rate of the forward reaction [equation (13)] may still be quite fast, the equilibrium constant being small simply because the rate of the $RS^- + RSNu$ reaction [equation (14)] is so much faster. Therefore, even though no buildup of RSNu is detectable, one can still have breaking and remaking of disulfide S–S bonds occurring quite rapidly via the reaction sequence of equation (13) followed by equation (14). The potential implications of this for systems of biological interest are obvious.

**The Relative Reactivity of Different Nucleophiles in Substitution at Sulfenyl Sulfur.**—It is important to know how different nucleophiles compare in their reactivity toward sulfenyl sulfur in a substitution of the kind shown in equation (20), and what types of nucleophiles react most readily with this center. Because of the problems caused by the reversibility of most nucleophilic displacements of disulfides [equations (13) and (14)], it is easiest to get such data for systems where the leaving group X in equation (20) is something other than a mercaptide ion. However, the results so obtained should also be applicable to predicting the relative rates at which different nucleophiles will perform displacements on disulfides [equation (13)].

One fairly extensive set of data on the relative reactivity of a series of nucleophiles toward sulfenyl sulfur comes from a study of the acid and nucleophile catalyzed racemization of optically active phenyl benzenethiolsulfinate (VII) in 60% dioxane (Kice and Large 1968). The racemization of VII was shown to proceed by a mechanism [equations (29a) to (29c)] involving rate-determining attack of the nucleophile on the sulfenyl sulfur of protonated VII.

$$(+)\text{-}C_6H_5S\underset{\underset{\text{O}}{\|}}{S}C_6H_5 + H^+ \rightleftharpoons (+)\text{-}C_6H_5S\underset{\underset{\text{OH}}{|}}{\overset{+}{S}}C_6H_5 \qquad (29a)$$

VII

$$Nu^- + (+)\text{-}C_6H_5S\text{—}\underset{\underset{\text{OH}}{|}}{\overset{+}{S}}C_6H_5 \xrightarrow[\text{rate determining}]{k_{Nu}} C_6H_5SNu + C_6H_5SOH \qquad (29b)$$

$$C_6H_5SNu + C_6H_5SOH \rightarrow (\pm)\text{-}C_6H_5S\underset{\underset{\text{O}}{\|}}{S}C_6H_5 + H^+ + Nu^- \qquad (29c)$$

The relative reactivities of the various nucleophiles studied as catalysts compared to that of chloride ion as the standard, i.e., $(k_{Nu}/k_{Cl})$, are given in the second column of Table 2.3.

Data on the relative reactivity of an entirely different group of nucleophiles toward sulfenyl sulfur have been obtained from studies of the rates of a series of nucleophilic substitution reactions of trithionate ion in aqueous methanol [equation (30)] (Fava and Pajaro 1954, 1956; Krueger and

$$Nu^- + {}^-O_3S\text{—}S\text{—}SO_3^- \xrightarrow{k_{Nu}} {}^-O_3S\text{—}S\text{—}Nu + SO_3^= \qquad (30)$$

Ritter 1970). These results are shown in the third column of Table 2.3, using cyanide ion as the nucleophile to which the reactivity of the others is compared.

TABLE 2.3

RELATIVE REACTIVITY OF NUCLEOPHILES TOWARD SULFENYL SULFUR

| Nucleophile | $(k_{Nu}/k_{Cl})$ Equation (29b)* | $(k_{Nu}/k_{CN})$ Equation (30)† |
|---|---|---|
| $Cl^-$ | (1.0) | — |
| $Br^-$ | 35 | — |
| $n$-$Bu_2S$ | $8.2 \times 10^2$ | — |
| $SCN^-$ | $5.4 \times 10^3$ | — |
| $I^-$ | $1.4 \times 10^4$ | — |
| $S_2O_3^=$ | — | 0.13‡ |
| $SO_3^=$ | — | 0.42‡ |
| $CN^-$ | — | (1.0) |
| $(C_6H_5)_3P$ | — | 12 |
| $C_6H_5S^-$ | — | 32 |
| $CH_3CH_2S^-$ | — | $3.6 \times 10^3$ |

*Kice and Large (1968).
†Fava and Pajaro 1954, 1956; Krueger and Ritter 1970.
‡Solvent is water rather than 50% methanol, so that $(k_{Nu}/k_{CN})$ given is actually $k_{Nu}(H_2O)/k_{CN}$ (50% MeOH). Value of $k_{Nu}$ (50% MeOH)/$k_{CN}$ (50% MeOH) would be somewhat smaller than this.

Pearson and Songstad (1967) have shown that the different patterns of relative nucleophile reactivity in substitutions at different centers can be nicely rationalized by application of the concepts of the Theory of Hard and Soft Acids and Bases, HSAB (Pearson 1963). According to HSAB, nucleophiles which are "soft" bases, i.e., of high polarizability and low electronegativity, like iodide ion, cyanide ion, mercaptide ions, triphenylphosphine, etc., should react particularly readily in substitutions at atoms which represent "soft" electrophilic centers. On the other hand, nucleophiles which are "hard" bases, i.e., of low polarizability and high electronegativity, like fluoride ion, alkoxide ions, hydroxide ion, etc., enjoy a special advantage in substitutions at atoms which represent "hard" electrophilic centers. Because of its two unshared pairs of outer shell electrons and relatively low positive charge on sulfur, sulfenyl sulfur would be predicted to be a rather soft electrophilic center by HSAB. The data for equation (29b), which show that nucleophile reactivity for that subsitution is in the order $I^- > SCN^- > n\text{-}Bu_2S \gg Br^- \gg Cl^-$, indicate that this is indeed the case.

Although significantly "softer" than tetrahedral carbon sulfenyl sulfur is not as soft as such extremely soft centers as $Pt^{II}$. This is seen by comparison of the reactivity order for equation (30), $C_6H_5S^- > (C_6H_5)_3P \gg CN^- > SO_3^= > S_2O_3^=$, with the order for a typical substitution at $Pt^{II}$ (Pearson *et al.* 1968), $(C_6H_5)_3P > S_2O_3^= \cong C_6H_5S^- \cong CN^- \gg SO_3^=$. Therefore, although the polarizability of a nucleophile is the dominant factor in determining its reactivity toward sulfenyl sulfur, its electronegativity and proton basicity also play some role. However, the important point to carry away from this discussion is the realization that the best nucleophiles for substitutions at sulfenyl sulfur will be those which are soft bases, like $RS^-$, $CN^-$, $R_3P$, $I^-$, $SCN^-$, etc., and that hard bases, like the amines, hydroxide or alkoxide ions, or fluoride ion, are in general relatively much poorer nucleophiles toward sulfenyl sulfur.

**Influence of the Leaving Group.**—The effect of changes in the leaving group X on the rates of nucleophilic substitutions of the type $Nu^- + RSX \rightarrow RSNu + X^-$ [equation (20)] is also worth discussing briefly.

Parker and Kharasch (1960) investigated the cleavage of various unsymmetrical disulfides RSSR′ by selected nucleophiles. In such a system, as shown in equation (31), cleavage will lead to RSNu if $R'S^-$ is more easily displaced than $RS^-$ and to R′SNu if the reverse is the case. For systems where R′ was equal to either 2,4-dinitrophenyl or *o*-nitrophenyl

$$RS^- + R'SNu \xleftarrow[\text{b}]{\text{path}} RSSR' + Nu^- \xrightarrow[\text{a}]{\text{path}} RSNu + R'S^- \quad (31)$$

and R was equal to either ethyl, benzyl or $—CH_2CH_2OH$ they found that in each case cleavage took place exclusively by path *a*. Since this is the one that yields the more thermodynamically stable of the two possible mercaptide ions, this indicates that the more stable $X^-$ is as an anion the better it is as a leaving group and the faster it will be displaced by a given nucleophile. This conclusion has also been borne out by studies of other nucleophilic substitutions at sulfenyl sulfur involving leaving groups of different stability (Fava and Bresadola 1955; Fava and Pajaro 1954; Foss 1958).

Some indication of how large an effect a large change in leaving group stability can have on rate is provided by a comparison of the rate constant for the displacement shown in equation (32), involving $CH_3S—$ as the

$$CH_3{*}S^- + CH_3S—SCH_3 \xrightarrow[\text{at } 25°]{k_M = 0.6\ M^{-1}\,\text{sec}^{-1}} CH_3{*}S—SCH_3 + CH_3S^- \quad (32)$$

leaving group (Fava *et al.* 1957) with the rate constant ($k_S$) for the displacement shown in equation (27), where $(CH_3)_2\overset{+}{S}—$ is the leaving group (Kice and Favstritsky 1969). Since $CH_3S^-$ is a much more reactive nucleophile than $(CH_3)_2S$, the fact that $k_S$ is more than $10^5$ larger than $k_M$ must be due entirely to the fact that $(CH_3)_2\overset{+}{S}—$ is a much better leaving group than $CH_3S—$.

A later section of this chapter will deal with "Electrophilic Catalysis of the Cleavage of Sulfur–Sulfur Bonds." In it we shall see that all such electrophilic catalysis really amounts to in the case of disulfides is the conversion of RS— into a much better leaving group which can be much more readily displaced by nucleophiles.

## ELIMINATION REACTION OF DISULFIDES

As mentioned earlier, because hydroxide and alkoxide ions are "hard" bases they are not particularly reactive in nucleophilic substitutions at sulfenyl sulfur. At the same time, being of higher proton basicity, they will be much more reactive than other common nucleophiles in elimination reactions of the types outlined schematically in equations (33) and (34). As a result, with $Nu^-$ equal to either $OH^-$ or $R'O^-$ the elimination reactions shown in equations (33) and (34) can, under the proper circumstances, become faster than attack of $Nu^-$ on sulfur (equation (13),

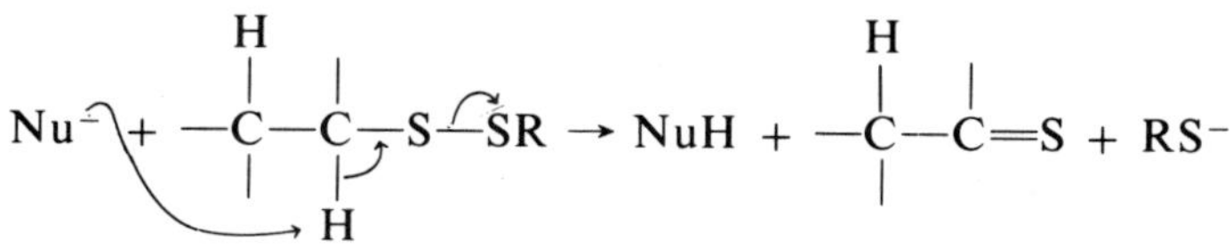
(33)

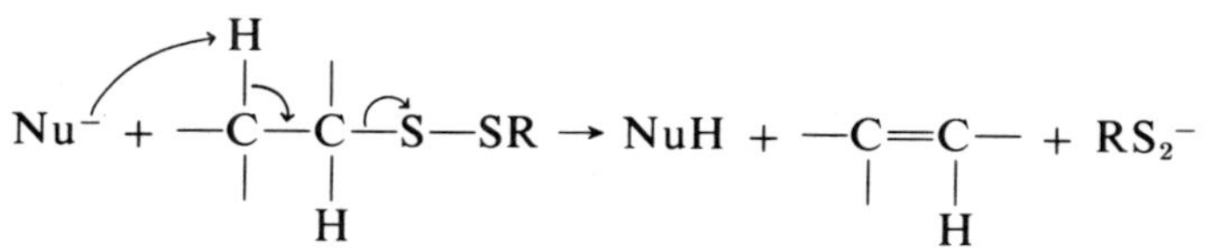
(34)

$Nu^- = OH^-$ or $R'O^-$). The earlier literature on the subject has been reviewed in detail by Danehy (1966), and Danehy and his co-workers (Danehy and Hunter 1967; Danehy and Parameswaran 1968) have conducted further studies aimed at elucidating the relative importance of the different reactions, i.e., equations (13), (33), and (34), under various conditions and with various disulfides. They have also found evidence for a fourth pathway [equation (35)] for alkaline decomposition in certain cases.

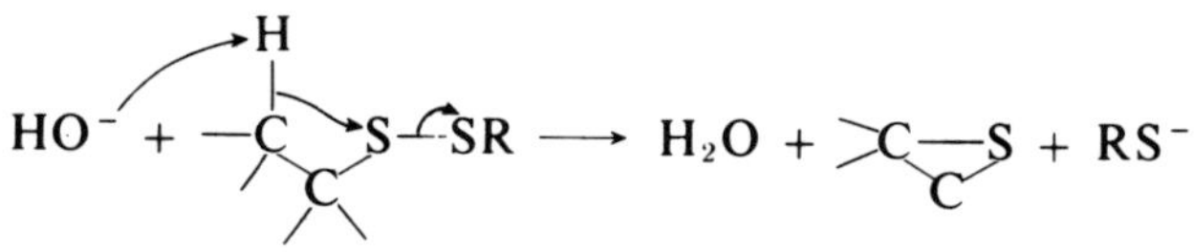
(35)

The picture that emerges from their work and the earlier studies is a complex one with a number of important points not yet adequately understood. However, the following generalization probably can be safely made at this time: the eliminations shown in equations (33) to (35) are not likely to compete favorably with equation (13), even with hydroxide or alkoxide as the attacking nucleophile, except when the carbon atom from which the proton is to be removed in equations (33) to (35) has attached directly to it some group, like carboxyl, —C≡N, or phenyl, which significantly increases the acidity of that hydrogen.

Even in those cases where cleavage of the disulfide in alkaline solution does involve either equation (33), (34), or (35), the specific nature of the subsequent steps which lead to the observed final products are often still obscure. They are, however, no more obscure than the exact sequence of reactions that leads to the final set of products in those cases where the initial cleavage does occur by equation (13), i.e.,

$$\mathrm{HO^- + RS{-}SR \rightarrow RSOH + RS^-}$$

Clearly the whole matter of the alkaline cleavage of disulfides is an area needing more investigation.

## ELECTROPHILIC CATALYSIS OF THE CLEAVAGE OF SULFUR–SULFUR BONDS

An electrophilic species (which we shall represent generally by $E^+$) can catalyze the cleavage of the sulfur–sulfur bond in a disulfide by coordinating with an unshared electron pair on one of the sulfurs, thereby converting RS— into a much better leaving group, $\mathrm{R\overset{+}{S}(E)}$—, i.e.,

$$\mathrm{RS{-}SR + E^+ \rightleftharpoons RS{-}\boxed{\overset{+}{S}R\ (E)}}\text{—(much better leaving group than RS)}$$

VIII

In earlier discussions of such catalysis (Parker and Kharasch 1959) the belief was that VIII then underwent cleavage of the S–S bond by a unimolecular dissociation into RSE and the sulfenium ion $RS^+$ [equation (36)]. However, there is now ample evidence (Ciuffarin and Fava 1968;

$$\underset{\text{VIII}}{\mathrm{RS{-}\overset{+}{S}R\ (E)}} \rightarrow \mathrm{RS^+ + RSE} \qquad (36)$$

Kice 1968) that such a picture is wrong and that the vast majority of the electrophilically-catalyzed cleavages of a sulfur–sulfur bond also involve the participation of a nucleophile as well in the rate-determining step, as shown in equation (37). They should thus actually be regarded as "nucleophilic substitutions assisted by electrophiles" (Ciuffarin and Fava

$$\mathrm{RS{-}SR + E^+ \rightleftharpoons RS{-}\overset{+}{S}R\ (E)} \qquad (37a)$$

$$\mathrm{Nu^- + RS{-}\overset{+}{S}R\ (E)} \xrightarrow[\text{determining}]{\text{rate}} \mathrm{RSNu + RSE} \qquad (37b)$$

1968) or as examples of "concomitant electrophilic and nucleophilic catalysis of the cleavage of a sulfur–sulfur bond" (Kice 1968).

There are presumably two reasons why equation (36) is unimportant for VIII compared to equation (37b). First, studies by Helmkamp *et al.* (1968) have shown that free sulfenium ions, $RS^+$, are formed only with extreme difficulty. They found, for example, that treatment of $CH_3SBr$ with silver 2,4,6-trinitrobenzenesulfonate in various aprotic, polar solvents and removal of the resulting silver bromide left solutions containing *no free sulfenium ion* $CH_3S^+$, the only species containing the $CH_3S$ group being either the covalent compound shown as IX, or, in acetonitrile as solvent, the ion shown as X. Second, we discussed in an earlier section the fact

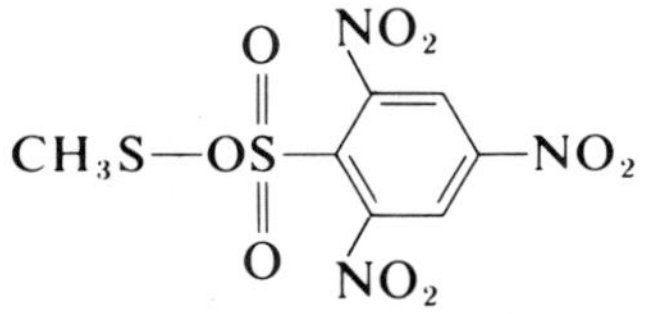

IX

$$CH_3S-\overset{+}{N}\equiv CCH_3$$

X

that nucleophilic substitution reactions at sulfenyl sulfur, such as equation (37b), often have extremely fast rates, particularly compared to the rates of analogous substitutions at tetrahedral carbon. These two factors, the generally fast rates for nucleophilic substitution at sulfenyl sulfur and the difficulty in forming sulfenium ions (which should make equation (36) very slow), explain why equation (36) can almost never compete effectively with equation (37b), even though at tetrahedral carbon processes of the type

$$\gt C-X \rightarrow \gt C^+ + X^-$$

frequently do compete with bimolecular nucleophilic substitution,

$$Nu^- + \gt C-X \rightarrow \gt C-Nu + X^-$$

Let us now discuss one specific example of the kind of concomitant electrophilic and nucleophilic catalysis of the cleavage of a disulfide that was outlined in a general form in equation (37). In acid solution the reaction between disulfides and the corresponding sulfinic acid [equation (38)] is dramatically catalyzed by the addition of small amounts of alkyl sulfides (Kice and Morkved 1964; Kice and Bowers 1962) the reaction proceeding 50 times faster in the presence of 0.005 *M* *n*-butyl sulfide than

$$4\,ArSO_2H + ArSSAr \rightarrow 3\,Ar\overset{\overset{O}{\|}}{\underset{\underset{O}{\|}}{S}}-SAr + 2\,H_2O \qquad (38)$$

it does in its absence. The mechanism of the sulfide-catalyzed reaction has been shown (Kice and Morkved 1964) to be the following:

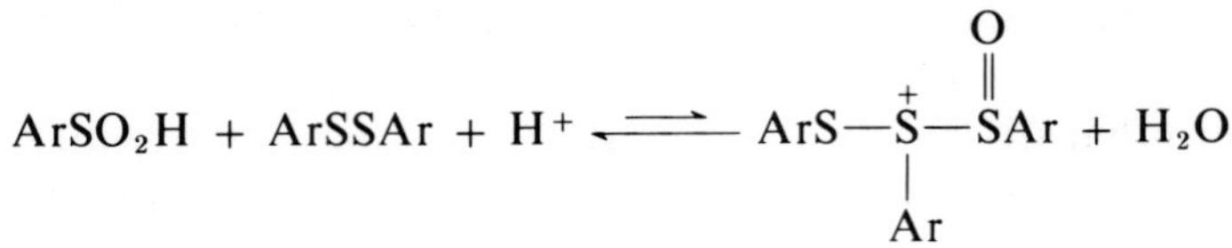

XI

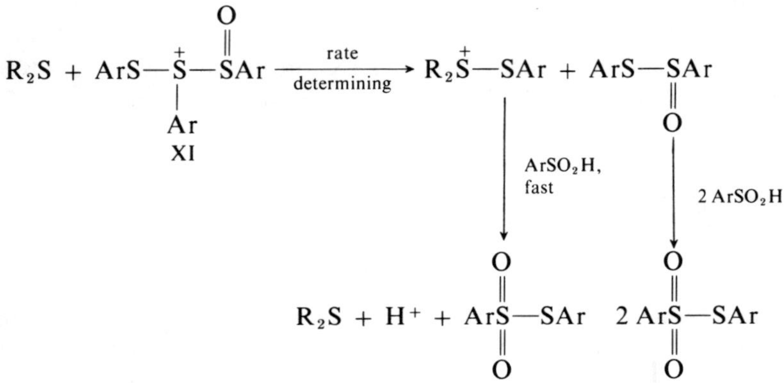

The important features of this mechanism are: (1) conversion of ArS to a much better leaving group by attachment of the electrophilic fragment $Ar\overset{+}{S}O$ to one of the sulfur atoms of the disulfide, and (2) cleavage of the original S–S bond in the disulfide by nucleophilic attack of the sulfide catalyst on the other sulfur.

When the catalyzing sulfide is absent the rate-determining step by which XI goes to products is *not* unimolecular dissociation to $ArS^+$ and ArS(O)SAr, but rather the following reaction in which a second molecule of disulfide assumes the role of the nucleophile in equation (37b) (Kice and Bowers 1962):

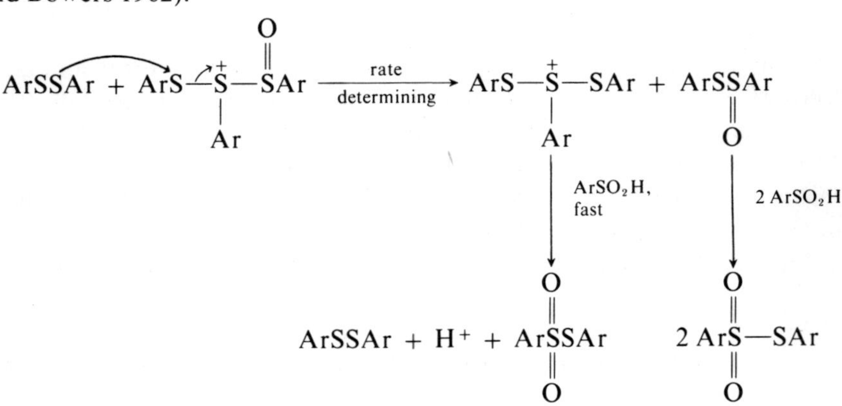

Since an aryl disulfide is a rather weak nucleophile this illustrates rather forcefully the very strong preference which electrophilically-assisted S–S bond cleavages exhibit for following the mechanism shown in equation (37) rather than that shown in equation (36).

There are, however, a few electrophilically-catalyzed cleavages of sulfur–sulfur bonds where nucleophilic assistance is not required. One of these is the acid-catalyzed hydrolysis of Bunte salts [equation (39)]. A

$$RS{-}SO_3^- + H_2O \xrightarrow{H^+} RSH + HSO_4^- \qquad (39)$$

careful study of the mechanism of this reaction (Kice *et al.* 1966) has shown it to proceed in the manner outlined in equation (40). In this mechanism only electrophilic catalysis is involved. There are presumably

$$RS{-}SO_3^- + H^+ \rightleftharpoons \underset{\text{XII}}{R\overset{+}{\underset{|\atop H}{S}}{-}SO_3^-} \xrightarrow[\text{determining}]{\text{rate}}$$

$$RSH + SO_3 \xrightarrow[\text{fast}]{H_2O} H^+ + HSO_4^- \qquad (40)$$

two reasons why such a mechanism is observed here rather than one involving both nucleophilic and electrophilic catalysis. First, nucleophilic attack on tetracoordinate sulfur (here the $-SO_3^-$ group) is much slower than attack on sulfenyl sulfur, and, second, there is the fact that unimolecular dissociation of XII into mercaptan plus $SO_3$ occurs much more readily than unimolecular dissociation of VIII, because of the much greater stability of sulfur trioxide ($SO_3$) as compared to a sulfenium ion.

## INTRAMOLECULAR BIFUNCATIONAL CATALYSIS OF THE SCISSION OF A SULFUR–SULFUR BOND

As is evident from inspection of the mechanism shown in equation (29), the acid- and nucleophile-catalyzed racemization of optically active phenyl benzenethiolsulfinate (VII) is another example of concomitant electrophilic and nucleophilic catalysis of the cleavage of a sulfur–sulfur bond. As with the acid- and sulfide-catalyzed disulfide-sulfinic acid reaction, the catalyzing electrophile and nucleophile are each initially separate molecular entities. One is therefore naturally curious whether the incorporation of nucleophile and electrophile into a single molecule with the proper geometry could lead to even more efficient catalysis of the racemization of the optically active thiolsulfinate. Since intramolecular bifunctional or polyfunctional catalysis is generally thought to be involved in many enzymatically-catalyzed reactions, the question is one of obvious biochemical interest, particularly given the ubiquity of S–S bonds in biological systems.

That, in fact, one can get efficient intramolecular bifunctional catalysis of the racemization of VII under the proper conditions has been demonstrated by recent work in our laboratory (Moguin 1969). In acetonitrile at 40°C optically active VII racemizes quite rapidly in the presence of added $\alpha$-methylthio-*m*-toluic acid (XIII) at a rate which is proportional

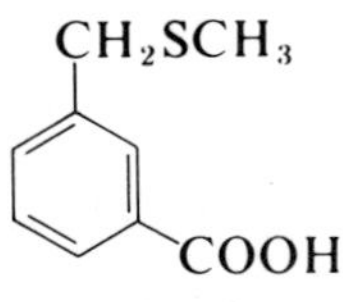

XIII

to the concentration of added XIII. In contrast, racemization is not brought about by the addition of comparable concentrations of methyl benzyl sulfide and benzoic acid, either together or separately, showing that both the —COOH group and the sulfide function are necessary for XIII to function as a catalyst under these conditions. The additional fact that intramolecular bifunctional catalysis by XIII is observed only in aprotic solvents like acetonitrile and not in protic media like 60% dioxane suggests that the first step in the process is an association of the substrate and the catalyst [equation (41a)] involving hydrogen bonding of the carboxyl group of XIII to the sulfinyl oxygen of the substrate. This is then followed by intramolecular bifunctional catalysis of the cleavage of the S–S bond of the thiolsulfinate in the manner shown in equation (41b). Recombination [equation (41c)] of the fragments produced in equation (41b) leads to racemic thiolsulfinate.

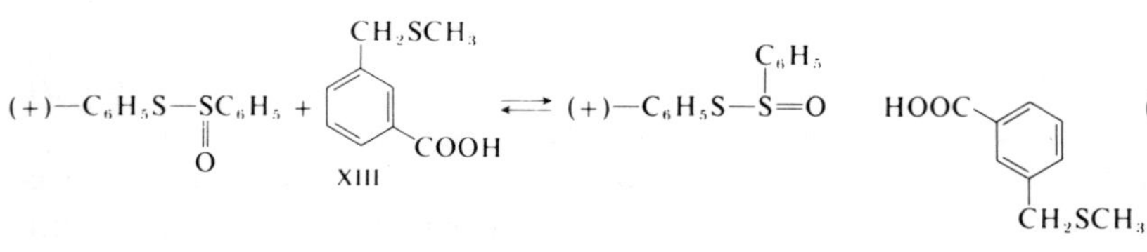

(41a)

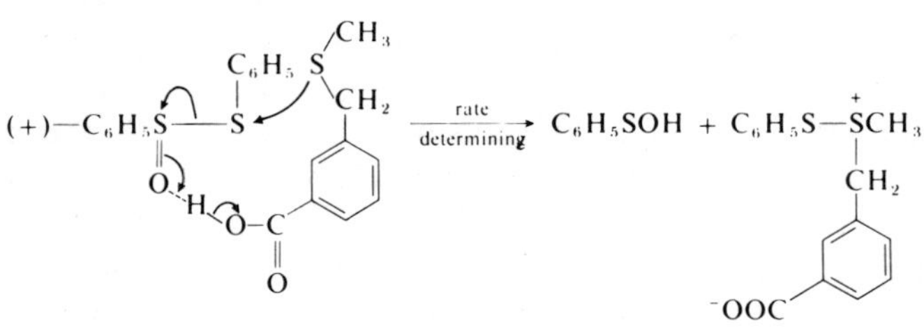

(41b)

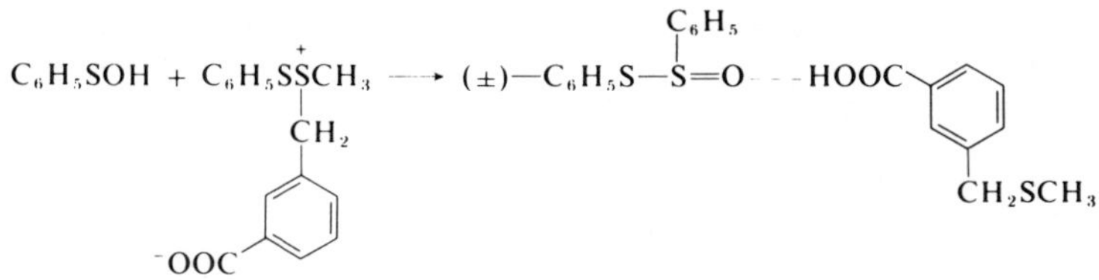


(41c)

The close analogy between the mechanism for the catalysis of the racemization by XIII and the mechanisms of typical enzyme catalyzed reactions, i.e.,

$$E + S \rightleftarrows E \cdot S \rightarrow \text{products} + E$$

is obvious and most interesting. It suggests a whole host of worthwhile additional studies of closely related systems. These are now under investigation.

## EPILOGUE

In this chapter I have tried to give a short summary of some of the key fundamentals of the chemistry of the sulfur–sulfur bond, at least as they appear to the organic sulfur chemist. Unfortunately, since most of the particularly pertinent studies in this area have been the work of organic or inorganic sulfur chemists, the examples I have discussed have dealt chiefly with disulfides and related molecules other than those derived from cysteine. However, it seems reasonable that most of the principles that we have tried to develop during this talk should be just as applicable to the chemistry of disulfides and other S–S bond compounds derived from cysteine as they are to simple alkyl disulfides and their derivatives.

## BIBLIOGRAPHY

Barnard, D., and Percy, E. J. 1960. Oxidation of thiolsulfinates to thiolsulfonates. Chem. Ind. (London) *1960*, 1332–1333.

Brand, J. C. D., and Davidson, J. R. 1956. Thermochromism of compounds containing the thiocarbonyl and disulfide functions. J. Chem. Soc. *1956*, 15–22.

Brown, C., and Hogg, D. R. 1967. Nucleophilic substitution at bivalent sulfur: the alkaline hydrolysis of esters of arylsulphenic acids. Chem. Commun. *1967*, 38–39.

Ciuffarin, E., and Fava, A. 1968. Nucleophilic substitution at sulfur. *In* Progress in Physical Organic Chemistry, Vol. 6, A. Streitweiser, Jr., and R. W. Taft (Editors). Interscience Div., John Wiley & Sons, New York.

Danehy, J. P. 1966. The alkaline decomposition of aliphatic disulfides. *In* The Chemistry of Organic Sulfur Compounds, Vol. 2, N. Kharasch and C. Y. Meyers (Editors). Pergamon Press, Oxford, England.

Danehy, J. P., and Hunter, W. E. 1967. The alkaline decomposition of organic disulfides. II. Alternative pathways as determined by structure. J. Org. Chem. *32*, 2047–2053.

DANEHY, J. P., and PARAMESWARAN, K. N. 1968. The alkaline decomposition of organic disulfides. III. Substituent effects among aromatic disulfides. J. Org. Chem. *33*, 568–572.

DAVIS, R. E., and PERRIN, C. 1960. Thermochromism of two disulfides. J. Am. Chem. Soc. *82*, 1590–1594.

DAVIS, R. E., SUBA, L., KLIMISHIN, P., and CARTER, J. 1969. Displacement reactions. XIX. The rates and heats of reactions of nucleophiles with $\beta$-propiolactone. J. Am. Chem. Soc. *91*, 104–107.

DE LA MARE, P. D. B. 1955. Mechanism of substitution at a saturated carbon atom. Part XLV. Kinetics of the interaction of bromide ions with simple alkyl bromides. J. Chem. Soc. *1955*, 3180–3187.

FAVA, A., and BRESADOLA, S. 1955. Kinetics of the catalytic rearrangement of tetrathionate. J. Am. Chem. Soc. *77*, 5792–5794.

FAVA, A., and ILCETO, A. 1958. Kinetics of displacement reactions at the sulfur atom. II. Stereochemistry. J. Am. Chem. Soc. *80*, 3478–3479.

FAVA, A., ILCETO, A., and CAMERA, E. 1957. Kinetics of thiol-disulfide exchange. J. Am. Chem. Soc. *79*, 833–838.

FAVA, A., and PAJARO, G. 1954. Kinetics of the isotopic exchange between thiosulfate and trithionate. J. Chim. Phys. *51*, 594–599. (French)

FAVA, A., and PAJARO, G. 1956. Kinetics of displacement reactions at the sulfur atom. I. Isotopic exchange between sulfite and alkylthiosulfates. J. Am. Chem. Soc. *78*, 5203–5208.

FERRINGTON, T. E., and TOBOLSKY, A. V. 1955. Organic disulfides as initiators of polymerization: Tetramethylthiuram disulfide. J. Am. Chem. Soc. *77*, 4510–4512.

FERRINGTON, T., and TOBOLSKY, A. V. 1958. Organic disulfides as initiators of polymerization. Part II. J. Am. Chem. Soc. *80*, 3215–3222.

FOSS, O. 1958. Remarks on the reactivities of the penta- and hexathionate ions. Acta Chem. Scand. *12*, 959–966.

HELMKAMP, G. K., OWSLEY, D. C., BARNES, W. M., and CASSEY, H. N. 1968. Reaction of methanesulfenyl bromide with silver 2,4,6-trinitrobenzenesulfonate in organic solvents. J. Am. Chem. Soc. *90*, 1635–1638.

HORDVIK, A. 1966. Bond-length/dihedral angle and bond-length/bond-order relationships for sulfur (II)–sulfur (II) bonds. Acta Chem. Scand. *20*, 1885–1891.

KERN, R. J. 1955. Some sulfur compounds as polymerization sensitizers. J. Am. Chem. Soc. *77*, 1382–1383.

KICE, J. L. 1968. Electrophilic and nucleophilic catalysis of the scission of the sulfur–sulfur bond. Accounts Chem. Res. *1*, 58–64.

KICE, J. L., and ANDERSON, J. M. 1968. The timing of covalency changes in nucleophilic substitutions at sulfenyl sulfur. The influence of *meta* and *para* substituents on the rate of reaction of aryl Bunte salts with cyanide ion. J. Org. Chem. *33*, 3331–3333.

KICE, J. L., ANDERSON, J. M., and PAWLOWSKI, N. E. 1966. The mechanism of the acid hydrolysis of Bunte salts (S-alkyl and S-aryl thiosulfates). J. Am. Chem. Soc. *88*, 5245–5250.

KICE, J. L., and BOWERS, K. W. 1962. Mechanisms of reactions of sulfinic acids. II. The reaction of *p*-tolyl disulfide with *p*-toluenesulfinic acid. J. Am. Chem. Soc. *84*, 2384–2389.

KICE, J. L., and FAVSTRITSKY, N. A. 1969. The extremely rapid exchange reaction between dimethyl sulfide and a dimethylmethylthiosulfonium salt. J. Am. Chem. Soc. *91*, 1751–1753.

KICE, J. L., and LARGE, G. B. 1968. The relative nucleophilicity of some common nucleophiles toward sulfenyl sulfur. The nucleophile- and acid-catalyzed racemization of optically active phenyl benzenethiolsulfinate. J. Am. Chem. Soc. *90*, 4069–4076.

KICE, J. L., and MORKVED, E. H. 1964. Mechanisms of reactions of sulfinic acids. VI. The mechanism of the disulfide-sulfinic acid reaction. J. Am. Chem. Soc. *86*, 2270–2278.

KOCH, P., and FAVA, A. 1968. The thermal racemization of aryl arenethiolsulfinates. An extraordinary rate acceleration of the inversion of sulfoxide sulfur. J. Am. Chem. Soc. *90*, 3867–3868.

KRUEGER, J. H., and RITTER, D. 1970. Nucleophilic substitution at sulfur. Kinetics of displacement reactions involving trithionate ion. J. Am. Chem. Soc. *92*, 2316–2321.

LECHER, H. Z. 1954. The radical dissociation of aryldisulfides. Science *120*, 220–222.

MARANGELLI, U., MODENA, G., and TODESCO, P. E. 1960. Researches on the oxidation of organic sulfides. VIII. On the oxidation of aromatic disulfides. Gazz. Chim. Ital. *90*, 681–693. (Italian)

MILLIGAN, B., and SWAN, J. M. 1962. Bunte salts ($RSSO_3Na$). Rev. Pure Appl. Chem. *12*, 72–94.

MINCH, M. 1965. Kinetic studies of the homolytic scission of the sulfur–sulfur bond. B.S. Thesis, Oregon State Univ., Corvallis, Oregon.

MOGUIN, A. 1969. Intramolecular bifunctional catalysis of the racemization of phenyl benzenethiolsulfinate. M.S. Thesis, Oregon State Univ., Corvallis, Oregon.

OTSU, T., KINOSHITA, Y., and IMOTO, M. 1964. The chain transfer mechanism of substituted diphenyl disulfides. Makromol. Chem. *73*, 225–230.

PARKER, A. J., and KHARASCH, N. 1959. The scission of the sulfur–sulfur bond. Chem. Rev. *59*, 583–628.

PARKER, A. J., and KHARASCH, N. 1960. Derivatives of sulfenic acids. XXXVI. The ionic scission of the sulfur–sulfur bond. Part 1. J. Am. Chem. Soc. *82*, 3071–3075.

PEARSON, R. G. 1963. Hard and soft acids and bases. J. Am. Chem. Soc. *85*, 3533–3539.

PEARSON, R. G., SOBEL, H., and SONGSTAD, J. 1968. Nucleophilic reactivity constants toward methyl iodide and trans-$Pt(py)_2Cl_2$. J. Am. Chem. Soc. *90*, 319–326.

PEARSON, R. G., and SONGSTAD, J. 1967. Application of the principles of hard and soft acids and bases to organic chemistry. J. Am. Chem. Soc. *89*, 1827–1836.

PRYOR, W. A. 1962. Mechanisms of Sulfur Reactions. McGraw-Hill Book Co., New York.

PRYOR, W. A., and GUARD, H. 1964. Walden inversion by radicals. The reaction of phenyl radicals with disulfides. J. Am. Chem. Soc. *86*, 1150–1152.

REED, L. J. 1958. Lipoic acid. *In* The Enzymes, Vol. 3, P. D. Boyer, H. Lardy, and K. Myrback (Editors). Academic Press, New York.

SCHOBERL, A., KAWOHL, M., and HAMM, R. 1951. The reaction of cystine and cystamine with potassium cyanide. A new route in thiazoline chemistry. Chem. Ber. *84*, 571–576. (German)

SMALL, L. D., BAILEY, J. H., and CAVILLITO, C. J. 1947. Alkyl thiolsulfinates. J. Am. Chem. Soc. *69*, 1710–1713.

STRICKS, W., and KOHLTOFF, I. M. 1951. Equilibrium constants for reaction of sulfite with cystine and dithioglycollic acid. J. Am. Chem. Soc. *73*, 4569–4574.

STRICKS, W., KOHLTOFF, I. M., and KAPOOR, R. C. 1955. Equilibrium constants for the reaction between sulfite and oxidized glutathione. J. Am. Chem. Soc. *77*, 2057–2061.

CHAPTER 3

Alfred Stracher

# Implication of the Sulfhydryl Residues in the Biological Functioning of Myosin[1]

Myosin is a unique protein, for in addition to being a member of the fibrous class of proteins, it is the only one of this group that possesses enzymatic properties. It is an ATPase and it is the hydrolysis of ATP which is now known to provide the chemical energy necessary for the performance of the mechanical work of muscle. In addition to its property as an ATPase, myosin interacts with the other major protein of the contractile apparatus, F-actin. Both of these properties of myosin are intimately related to its role in the molecular mechanism of muscle contraction.

It has long been recognized that the integrity of the sulfhydryl residues of myosin are necessary features in the biological functioning of the molecule. Like other sulfhydryl dependent enzymes this type of information has been obtained by chemical labeling of the enzyme with various sulfhydryl reagents such as *p*-chloromercuribenzoate (Kielley and Bradley 1956), *N*-ethylmaleimide (Kielley and Bradley 1956), iodoacetamide (Barany and Barany 1959), and studying the resultant loss in biological function.

Thus, it was early recognized (Kielley and Bradley 1956) that reaction of myosin with PCMB to a point where 50% of the -SH residues had been labeled, resulted in a sixfold activation of ATPase, whereas labeling of the remaining 50% resulted in inactivation of the enzyme. In addition, the ability to interact with F-actin was also abolished. It was concluded from these studies that the -SH groups involved in both processes were probably the same. The activation-inactivation phenomenon delineated at least two classes of reactivity of the -SH groups of myosin.

Barany and Barany (1959) were the first to show that using various sulfhydryl reagents they could isolate a myosin devoid of ATPase, but still retaining its ability to combine with F-actin. Subsequent studies from this laboratory, using a combination of disulfide exchange and labeling with iodoacetamide (Stracher 1964) confirmed these findings and in addition enabled us to localize a sulfhydryl residue, the reaction of which led to an inactivation of ATPase (Stracher 1964). It was assumed that this -SH residue was localized at the ATPase active site of the enzyme. At about

[1]Supported by National Institute of General Medical Sciences Grant Number RG 7076.

the same time, Sekine *et al.* (1962) using *N*-ethylmaleimide were able to specifically label an -SH residue involved in the activation of ATPase. Our own studies with iodoacetamide confirmed this result. Thus, two unique -SH residues in myosin could be specifically labeled with sulfhydryl reagents leading to an activation or inactivation of the enzyme. These residues were designated the $S_1$ and $S_2$ groups respectively. Inactivation was always preceded by activation.

The question as to the localization of these -SH residues in myosin and their presence at the active site was next investigated, utilizing the well-known property of myosin to be degraded by proteolytic enzymes into two molecules designated light meromyosin (LMM) and heavy meromyosin (HMM) (Szent-Gyorgyi 1953). The latter retains all of the known biological properties of the native molecule. Further digestion of HMM with trypsin leads to the production of a molecule known as subfragment-1 (molecular weight 108,000) (Mueller and Perry 1962), again retaining all of the activity of the parent molecule.

Thus, by this technique it is possible to degrade myosin (molecular weight 470,000) (Gershman *et al.* 1969) to a molecule of molecular weight 108,000 which retains full activity. By specifically labeling the $S_1$ and $S_2$ sulfhydryl residues with 1-$C^{14}$ iodoacetamide, the distribution of radioactivity in the molecular fragments of myosin was followed by the tryptic digestion of the $C^{14}$-$S_1$-blocked derivatives, which yielded HMM and subfragment-1. The results showed that 75 to 80% of the specific radioactivity of myosin (molar basis) was found to be associated with HMM suggesting that both the $S_1$ and $S_2$ sulfhydryl residues were associated with this fragment (Trotta *et al.* 1968). Further digestion of HMM with trypsin and gel filtration of the mixture on Sephadex G-200 resulted in the separation of a subfragment-1 molecule retaining only 20% of the total radioactivity originally associated with myosin, but which at the same time accounted for more than 90% of the total ATPase activity of the parent molecule. Over 65% of the radioactivity was associated with a peptide of molecular weight 2,500. Similar distribution of radioactivity was found for the $C^{14}$-$S_2$-blocked myosin, although in this case the molecule was devoid of ATPase.

Myosin may also be degraded with an insoluble papain complex to produce subfragment-1 without the intermediate formation of HMM (Kominz *et al.* 1965). Prepared in this manner the isolated subfragment-1 retained 85% of the total radioactivity. Only a small percentage ($<$ 10%) was released as small peptide material. On a molar basis subfragment-1 produced in this manner contains 42% of the specific radioactivity of the original $S_1$-blocked myosin. These data are consistent with the production of two moles of subfragment-1 from each molecule of myosin. In addition, it can be concluded that tryptic digestion leads to a removal of the $S_1$ and $S_2$ sulfhydryls from myosin, whereas papain digestion produces a subfragment-1 that retains these residues. Further confirmation of these results comes from the observation that tryptic subfragment-1 cannot be activated by iodoacetamide, whereas the papain-prepared molecule can be activated four- to sixfold, as is the case with myosin ATPase.

One can conclude from these results that since the trypsin-prepared subfragment-1 has lost the $S_1$ and $S_2$ sulfhydryls, but retains full ATPase activity, that these groups must not be present directly at the active site for ATP hydrolysis. Rather, it is more likely that they are localized in an allosteric area near the active site where they may influence the catalytic events occurring there indirectly, via, for example, a conformational change in the protein structure. These results seem to imply that there may be a flexible, dynamic character to the ATPase site of myosin, similar to the type of active site associated with allosteric enzymes.

## SULFHYDRYL RESIDUES AND THE LIGHT CHAINS OF MYOSIN

It is now well-established that the myosin molecule comprises an axial core of two heavy chains (each of molecular weight 215,000) that terminate in a globular head region containing 2 light chains (molecular weight 20,000) (Gershman *et al.* 1967). Although the $S_1$ and $S_2$ sulfhydryl residues are present in the heavy chains, the main core exhibits no residual ATPase activity following dissociation of the light chains by alkali, guanidine, heat, detergent, and high salt concentrations (Gershman *et al.* 1967). Thus, the relationship between the subunit structure of the molecule and the ATPase site is of considerable interest.

If the trypsin-prepared subfragment-1, devoid of the $S_1$ and $S_2$ sulfhydryl residues, is reacted with iodoacetamide, the ATPase activity of the molecule is abolished and the reaction follows first order kinetics. If the light chains, which are present in subfragment-1 (Trotta *et al.* 1968) are now dissociated at pH 11.0, it is found that over 90% of the radioactivity is associated with the light chains (molecular weight 20,000) after separation of these chains from residual heavy chains on Sephadex G-100. Amino acid analysis of the light chains reveals that the single sulfhydryl residue in these chains has reacted, resulting in a loss of the ATPase activity of subfragment-1 (Stracher 1969). These results suggest that a sulfhydryl residue ($S_3$) at or near the active site has been localized in this reaction.

Previous studies in this laboratory have shown that interaction of the light and heavy chains of myosin were an essential feature of the ATPase activity of the enzyme. Localization of an essential sulfhydryl residue ($S_1$ and $S_2$) on the heavy chains which dramatically alter the ATPase when reacted with iodoacetamide, further strengthen this hypothesis. Thus, of the 20 sulfhydryl residues present in myosin per subunit molecular weight, some essential role has been assigned to at least 4 of these. The $S_1$, $S_2$ and $S_3$ groups have already been discussed, whereas a fourth, Sa appears to be involved in the interaction with F-actin; however, little else is known concerning its localization in the molecule.

These sulfhydryl residues all appear to be localized in conformationally sensitive regions of the molecule. The most reactive appears to be $S_1$ and the least reactive Sa, with $S_2$ and $S_3$ being intermediate in reactivity. This can be summarized in the following diagram:

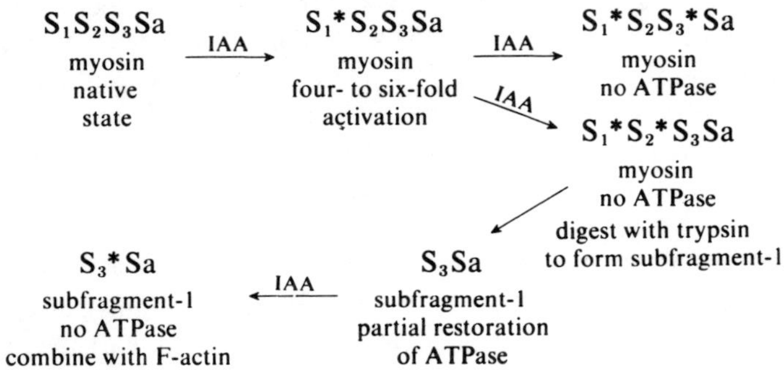


*Represents reacted with iodoacetamide

The reaction of $S_1$ with iodoacetamide is not protected by substrate (ATP), whereas $S_2$ is protected by ATP but not pyrophosphate and $S_3$ both by ATP and pyrophosphate. This result suggests that it is the purine ring moiety of ATP binding to $S_2$ but that only the phosphate portion of ATP is essential for binding to $S_3$. Since $S_2$ and $S_3$ are localized in different chains, it is feasible to assume that ATP may be "bridging" the light and the heavy chains by binding to sulfhydryl residues in each chain. This type of bridging may place the ATP molecule in the proper orientation for hydrolysis by residues localized at the catalytic site. This type of bridging would thus explain why blocking either $S_2$ or $S_3$ results in a loss of catalytic activity of myosin. Blocking of $S_2$ might possibly provide enough steric restrictions such that $S_3$, as well, is rendered inoperable. However, removal of $S_2$ as in tryptic-prepared subfragment-1, leaves $S_3$ with no such restrictions and thus enables this molecule to function normally as an ATPase.

These results provide an initial insight into the role of sulfhydryl residues in the biological functioning of myosin. It is obvious that subtle conformational changes and steric restrictions, after reaction of these groups, are sufficient to cause a loss in the ATPase activity of the molecule. The significance of the activation phase is unclear at the present time but this may be a means for regulating the activity of the enzyme during periods where greater amounts of energy are needed. Understanding the close interrelationship of these sulfhydryl residues in the control and regulation of the ATPase and F-actin combining properties of myosin will enable us to better understand the molecular mechanism of muscle contraction.

## BIBLIOGRAPHY

BARANY, M., and BARANY, K. 1959. Studies on "Active Centers" of L-Myosin. Biochim. Biophys. Acta *35*, 293.

GERSHMAN, L. C., STRACHER, A., and DREIZEN, P. 1967. Subunit interactions of myosin. *In* Symposium on Fibrous Proteins, W. Crewther (Editor), Butterworths, London. Available from Plenum Publishing Co., New York.

GERSHMAN, L. C., STRACHER, A., and DREIZEN, P. 1969. Subunit structure of myosin. III. A proposed model for rabbit skeletal myosin. J. Biol. Chem. *244*, 2726.

KIELLEY, W. W., and BRADLEY, L. B. 1956. The relationship between sulfhydryl groups and the activation of myosin ATPase. J. Biol. Chem. *218*, 653.

KOMINZ, D. R., MITCHELL, T., NIHEI, T., and KAY, C. M. 1965. The papain digestion of skeletal myosin A. Biochemistry *4*, 2373.

MUELLER, H., and PERRY, S. V. 1962. The degradation of heavy meromyosin by trypsin. Biochem. J. *85*, 431.

SEKINE, T., BARNETT, L. M., and KIELLEY, W. W. 1962. The active site of myosin ATPase. I. Localization of one of the sulfhydryl groups. J. Biol. Chem. *237*, 2769.

STRACHER, A. 1964. Disulfide-sulfhydryl interchange studies on myosin A. J. Biol. Chem. *239*, 1118.

STRACHER, A. 1969. Unpublished observations.

SZENT-GYORGYI, A. G. 1953. Meromyosins, the sub-units of myosin. Arch. Biochem. Biophys. *42*, 305.

TROTTA, P. P., DREIZEN, P., and STRACHER, A. 1968. Studies on subfragment-1, a biologically active fragment of myosin. Proc. Natl. Acad. Sci. *61*, 659.

CHAPTER 4

L. W. Parks

# Transmethylation Reactions

## INTRODUCTION

The term transmethylation was coined in the laboratory of du Vigneaud to describe the metabolic transfer *in toto* of the methyl group of one compound to another (du Vigneaud and Rachele 1965). This concept was based on reports made in 1939 describing interrelationships between methionine, homocysteine, and choline (du Vigneaud *et al.* 1939). It was fitting that on the 25th anniversary of that report, a Symposium "Transmethylation and Methionine Biosynthesis" was held. Subsequently published in book form (Shapiro and Schlenk 1965), the proceedings of that Symposium elegantly chronicled the historical development of the field and described the diversity of biochemical functions in which transmethylation reactions participate. Based on the nutritional experiments, subsequent work from du Vigneaud's laboratory (Keller *et al.* 1949; du Vigneaud *et al.* 1956) demonstrated that the methyl group of methionine could be transferred as a complete unit to various acceptors.

The apparent simplicity of the transmethylation process was challenged by the work of Borsook and Dubnoff (1948) who demonstrated that an energy source was necessary to mediate methyl group transfer. Based on additional findings that the sulfonium compounds dimethyl acetothetin and dimethyl-$\beta$-propiothetin were methyl donors to homocysteine (Dubnoff and Borsook 1948), the concept of an "active methionine" was proposed (Cantoni 1951). Subsequently, S-adenosylmethionine was synthesized in liver preparations using methionine and ATP (Cantoni 1953), by the methionine activating enzyme or S-adenosylmethionine synthetase (ATP :L-methionine S-adenosyltransferase; E.C. 2.4.2.13).

$$\text{L-methionine} + \text{ATP} \rightarrow \text{S-adenosylmethionine} + \text{P-P-P}$$

The mechanism of the reaction has been extensively studied by Mudd (Mudd 1965) and Greene (Greene 1969). The structure has been confirmed by total synthesis (Baddiley and Jamieson 1954). This is shown below.

The ubiquity of distribution and biochemical versatility of S-adenosylmethionine rapidly became evident. Techniques for the rapid isolation of

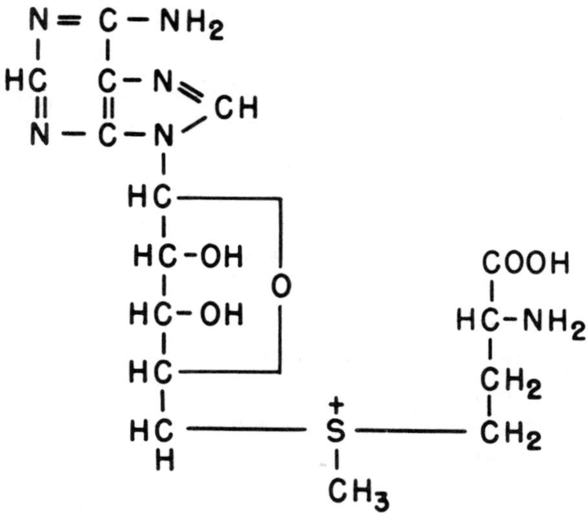


FIG. 4.1. S-ADENOSYLMETHIONINE

S-AM in quantity and high purity (Schlenk and de Palma 1957A; Schlenk and de Palma 1957B; and Schlenk *et al.* 1959) with a variety of labeled positions (Zappia *et al.* 1968; Schlenk and Zydek 1968), has greatly facilitated the work on methyl group transfer reactions. So casual has become the notion that S-AM participation and transmethylation are synonymous, that it was important to establish rigorous guidelines for determining metabolic transmethylation (Shapiro and Schlenk 1960). These are: (1) the donor and acceptor of the methyl group should be identified; (2) the nature of the reaction product and the demethylated donor molecule should be established; and (3) the methyl group should be transferred without alteration. Although these criteria appear relatively straightforward, very few methyl group transfer reactions satisfactorily meet all of these criteria. Two transmethylation reactions will be explored here. These are the S-adenosylmethionine homocysteine methyltransferase and the methylation in ergosterol biosynthesis.

### S-ADENOSYLMETHIONINE HOMOCYSTEINE METHYLTRANSFERASE

The S-adenosylmethionine homocysteine methyltransferase (S-adenosylmethionine:homocysteine S-methyltransferase; E.C. 2.1.1.10) has been extensively studied at Argonne by Shapiro (Shapiro *et al.* 1964). Probably no other transmethylation reactions has been as extensively investigated as has this one. The reaction as shown below

(S-adenosylmethionine + homocysteine →

S-adenosylhomocysteine + methionine)

has no organic cofactor requirements and equimolar products are obtained. High levels of activity are obtained in yeast. With the development of a quick, reproducible assay procedure for the enzyme based on the retention of AM by strongly cationic exchange resins (Shapiro and Yphantis 1959), Shapiro was able to effect a 500-fold purification of the enzyme from yeast. The reaction is inhibited by its products S-adenosylhomocysteine and methionine (Shapiro 1965). It is of interest that a variety of methyl sulfonium compounds will serve as substrate for the reaction. This will be discussed further below.

The metabolic function of this enzyme is of concern to biochemists, and although there have been extensive investigations in this area, the function remains obscure. The role of this enzyme in the biosynthesis of methionine is of concern. Based on feeding experiments with a large number of yeast methionine auxotrophs, a pathway for methionine formation in that organism was proposed which incorporated the S-AM homocysteine methyltransferase (Pigg *et al.* 1962). A mutant lacking the transmethylase could be fed only methionine. The proposed mechanism for methionine biosynthesis was especially attractive since reports of the regeneration of AM directly by the methylation of AH eliminated the inherent difficulty of AM formation (Duerre and Schlenk 1962). An intensive analysis of the process ensued in our laboratory with frustrating results. Firstly, no additional mutants lacking the AM homocysteine methyltransferase could be isolated. Secondly, we were unable to get any evidence *in vitro* in support of the methylation of AH. We therefore developed a cell-free system from yeast capable of converting the $\beta$-carbon of serine into the methyl group of methionine (Botsford and Parks 1967). No requirements for S-AM or S-AH could be demonstrated. It was further shown that as the levels of the cell-free synthesis of methionine varied during the culture cycle of the organism, no change in the AM-homocysteine methyltransferase occurred (Botsford and Parks 1969). We are convinced now that the AM-homocysteine methyltransferase does not participate directly in the biosynthesis of methionine. This is supported by recent observations using newer quantitative techniques that in yeast AM biosynthesis from AH involves a degradation of AH and reutilization of the scission products for AM formation (Duerre 1968; Shapiro and Ehninger 1969). In light of these findings it is difficult to imagine a biosynthetic scheme utilizing AM directly, since it appears that the only way to form AM is via the methionine: ATP generating system.

Interestingly enough AM is important in the synthesis of methionine in the $B_{12}$ pathway of bacteria. Certain enteric bacteria possess two systems for methionine formation. A $B_{12}$ pathway is formed in some organisms when the growth medium is enriched with vitamin $B_{12}$ (Guest *et al.*

1960). The second, or non$B_{12}$ pathway, differs significantly from the other in that there is no $B_{12}$ requirement; only polyglutamyl derivatives of reduced folate will participate (Woods *et al.* 1965).

Working with the vitamin $B_{12}$ induced pathway of *Escherichia coli* B, Taylor and Weissbach have recently proposed that AM and a reduced flavin are needed to regenerate methyl-$B_{12}$-enzyme after the demethylated enzyme has been oxidized to an inactive form (Taylor and Weissbach 1969A). Thus, AM does not directly participate in the sequence of reactions in methyl group formation, but is necessary for providing the methyl group to reform the methyl-$B_{12}$-enzyme. No direct role for AM is obvious for the non$B_{12}$ pathway either. It is interesting, however, that the AM-homocysteine transmethylase is stimulated by $N^5$-methyltetrahydrofolic acid. This is not peculiar to the $N^5$-methyl group since tetrahydrofolic acid also stimulates the reaction (Taylor and Weissbach 1969B).

Although the yeast and non$B_{12}$ pathways must use the diglutamyl or triglutamyl derivatives of reduced folic acid, the $B_{12}$ enzyme may use the monoglutamyl or higher multiples of the folate (Burton *et al.* 1969).

Thus the function of the AM-homocysteine methyltransferase remains a paradox. The ubiquity and high activity of the enzyme make it an attractive vehicle for methionine biosynthesis. Yet, none of the mechanisms that have been intensively studied provide a direct role for the enzyme.

## STEROL TRANSMETHYLASE

The metabolic transformations leading from acetate to sterol have been intensively investigated. The major steps to ring closure appear to be similar if not identical in plants, microorganisms and animals. Several of the phytosterols are characterized by an extra methyl or ethyl group appended to carbon-24.

It is clear that whereas the remaining 27 carbon atoms are generated from acetate, these extra ethyl or methyl groups are derived from one-carbon pools. Evidence for a transmethylation was obtained by Alexander, Gold and Schwenk who showed that methionine could provide the intact methyl group (Alexander *et al.* 1957). At this point, S-AM became available labeled and in high purity (Schlenk and de Palma 1957B). In a series of cell-free experiments I was able to prove that S-AM was the actual methyl donor (Parks 1958B).

Whereas the AM-homocysteine methyltransferase discussed above meets all of the requirements for a transmethylation reaction, the sterol methyltransferase in reality meets none. The nature of the substrate and immediate product are not known. We only know that AM will donate the methyl group. S-methyl-methionine is not an effective methyl donor here (Parks 1958B). It is clear, however, that only two of the hydrogen

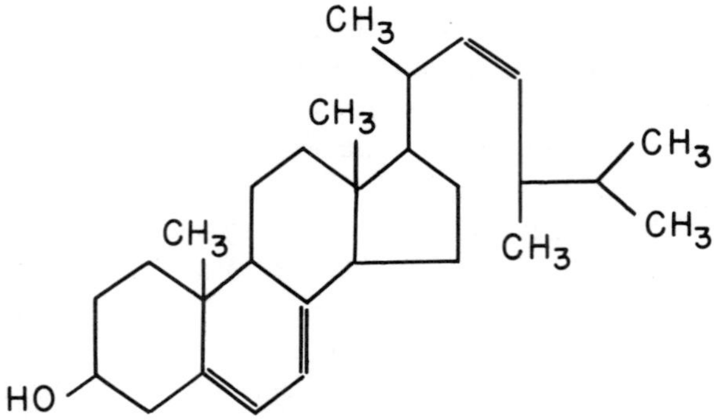


FIG. 4.2. ERGOSTEROL

atoms of the methyl group are transferred from the methionine. This was established by experiments done in Gif-sur-Yvette using methionine with three deuterium atoms on the methyl group. Mass spectrometry of the resulting ergosterol showed clearly that only two hydrogen atoms of the methyl group of methionine are retained throughout the entire methyl transfer process (Jaureguiberry *et al.* 1965).

We have shown that the immediate product of transmethylation is not ergosterol (Turner and Parks 1965). In cell-free experiments very little ergosterol is synthesized. On a silic acid chromatographic column, the labeled product is displaced from ergosterol to a more polar position. The transmethylation process is significantly stimulated by carbonate (Parks *et al.* 1965).

The mechanism of the transmethylation which would result in the loss of only one of the hydrogen atoms has been extensively discussed (Lederer 1964). The absence of a purified sterol transmethylase has made work on this problem very difficult. This is further compounded by the insolubility in aqueous media of the sterol substrates. We have been able to achieve effective aqueous solubilization of sterols using a component isolated from extracts of yeast (Adams and Parks 1967). Although considerable speculation has existed that desmosterol or lanosterol are the substrates for methylation, we have been unable to demonstrate this in our cell-free system. Thus, it remains to be established at what stage the methylation reaction occurs. It is known that (24-$^{3}$H)-lanosterol is converted to ergosterol with a migration of the H-atom from C-24 to C-25 (Akhtar *et al.* 1967). Additionally, 24-methylenelanosterol is efficiently incorporated into ergosterol (Barton *et al.* 1966). Considerable evidence implicates a $\Delta^{24}$-bond as essential for alkylation (Russell *et al.* 1967). Recently, we have found that 7,22 *bis*-dehydrocholesterol can be methylated rapidly in our cell-free system. It remains to be determined if the methylated product is ergosterol, however.

The ethyl sterols in the phytosterol group are derived from 2 transmethylation reactions with a 24-methylene intermediate (Castle *et al.* 1963). In such a system the data clearly show that when methionine-$C_{D3}$ is used as substrate, five deuterium atoms of the methyl groups are incorporated in the ethyl group (Lenfant *et al.* 1967). Although ethionine has been shown to be converted to S-adenosylethionine and capable of participation in a transethylation reaction (Parks 1958A), in the plant systems ethionine is virtually without transethylation capacity for sterol synthesis (Castle *et al.* 1963).

The importance of the extra methyl and ethyl groups in the physiology of the organism remains obscure. Indeed, little is known about the precise physiological function of sterols, such as ergosterol, in the general economy of the cells. Circumstantial evidence has been obtained which indicates a role for sterols in the respiratory mechanism of the cell (Adams and Parks 1969). With so little knowledge of this role, it is only the wildest of speculation that could predict the function of the extra methyl group at C-24 in this sterol.

## S-ADENOSYLMETHIONINE IN TRANSMETHYLATION

In a series of experiments to be published soon, Schlenk and colleagues (Zappia *et al.* 1969) have made a thorough study of the mechanism of three transmethylation reactions. By a series of clever chemical and enzymatic manipulations, they have been able to prepare a number of analogs and derivatives of S-AM. These were tested in three different transmethylase systems, homocysteine methyltransferase, histamine methyltransferase, and acetylserotonin methyltransferase catalyzing methyl group transfer to sulfur, nitrogen, and oxygen respectively.

In the methylation of histamine only S-AM and S-inosyl-L-methionine (deaminated adenosine) showed methyl transfer activity. When checked for inhibitory activity in the presence of AM, S-adenosyl-(5)-3-methylthiopropylamine (decarboxylated AM) and S-inosyl-L-methionine were effective inhibitors. AH and 5′-methylthioadenosine also were inhibitory.

Acetylserotonin methylation was accomplished only with AM. S-IM, decarboxylated AM, 5′-MTA and AH were inhibitory, the last being a very effective inhibitor.

By contrast, all of the sulfonium compounds tested acted as methyl donors to homocysteine except those lacking the 2-amino group of methionine. As mentioned earlier S-AH inhibits this enzyme.

Thus the specificity for catalytic and inhibitory activities of the sulfonium compounds is dependent upon the particular transmethylase. From the work presented by Schlenk, the S-AM-homocysteine methyltransferase stands out because of the nonspecificity of its reactants. Nonspecific *N*-methyltransferases have been reported (Axelrod 1965).

## REGULATION OF TRANSMETHYLATION REACTIONS

Although little is known about the control of transmethylase reactions, one feature is clear. In most cases S-AH is a potent inhibitor of transmethylation. This was true for all three of Schlenk's examples cited above. S-AH is a competitive inhibitor of bacterial tRNA methylases also (Hurwitz *et al.* 1964). With the diversity of regulatory mechanisms available to cells, it would be surprising if product inhibition were the main source of regulation. We have demonstrated repression of the AM homocysteine methyltransferase (Pigg *et al.* 1962). This has been confirmed and extended by K. D. Spence (unpublished experiments). Unhappily, the role of that enzyme in a biosynthetic function has not been defined. Further experimentation is needed in the control aspects of transmethylase reactions.

As the nutritional experiments of du Vigneaud initiated the labyrinthine biochemical discourse into transmethylation, truly the vehicle for this excursion was and remains S-adenosylmethionine. Launched by Cantoni, that sulfonium derivative has been brilliantly exploited by Schlenk. His insight into the chemistry and biochemistry of AM has made a study of the transmethylases one of the revealing and exciting areas of biochemistry.

## ACKNOWLEDGMENTS

The author's recent experiments reported in this paper were supported by the National Institutes of Health (AM 11927–01). I should like to thank Dr. F. Schlenk who provided the results of some of his work reported herein in advance of their publication.

## BIBLIOGRAPHY

Adams, B. G., and Parks, L. W. 1967. A water-soluble form of ergosterol and cholesterol for physiological studies. Biochem. Biophys. Res. Comm. *28*, 490–494.

Adams, B. G., and Parks, L. W. 1969. The differential effect of respiratory inhibitors on ergosterol synthesis by *Saccharomyces cerevisiae* during adaptation to oxygen. J. Bacteriol. *100*, 370–376.

Akhtar, M., Hunt, P. F., and Parvez, M. A. 1967. The transfer of hydrogen from C-24 to C-25 in ergosterol biosynthesis. Biochem. J. *103*, 616–622.

Alexander, G. J., Gold, A. M., and Schwenk, E. 1957. The methyl group of methionine as a source of $C_{28}$ in ergosterol. J. Am. Chem. Soc. *79*, 2967.

Axelrod, J. 1965. The formation and metabolism of physiologically active compounds by *N*- and *O*-methyltransferases. *In* Transmethylation and Methionine Biosynthesis, S. K. Shapiro and F. Schlenk (Editors). Univ. Chicago Press, Chicago.

Baddiley, J., and Jamieson, G. A. 1954. Synthesis of active methionine. J. Chem. Soc. (London) *1954*, 4280–4284.

BARTON, D. H. R., HARRISON, D. M., and MOSS, G. P. 1966. 24-methylenedihydrolanosterol as a precursor of steroids and triterpenoids. Chem. Commun. *1966*, 595–596.

BORSOOK, H., and DUBNOFF, J. W. 1948. On the role of the oxidation in the methylation of guanidoacetic acid. J. Biol. Chem. *171*, 363–375.

BOTSFORD, J. L., and PARKS, L. W. 1967. Role of S-adenosylmethionine in methionine biosynthesis in yeast. J. Bacteriol. *94*, 966–971.

BOTSFORD, J. L., and PARKS, L. W. 1969. Serine transhydroxymethylase in methionine biosynthesis in *Saccharomyces cerevisiae*. J. Bacteriol. *97*, 1176–1183.

BURTON, E., SELHUB, J., and SAKAMI, W. 1969. The substrate specificity of 5-methyltetrahydropteroyl-triglutamate-homocysteine methyltransferase. Biochem. J. *111*, 793–795.

CANTONI, G. L. 1951. Activation of methionine for transmethylation. J. Biol. Chem. *189*, 745–755.

CANTONI, G. L. 1953. S-adenosylmethionine; a new intermediate formed enzymatically from L-methionine and adenosinetriphosphate. J. Biol. Chem. *204*, 403–416.

CASTLE, M., BLONDIN, G., and NES, W. R. 1963. Evidence for the origin of the ethyl group of β-sitosterol. J. Am. Chem. Soc. *85*, 3306–3308.

DUBNOFF, J. W., and BORSOOK, H. 1948. Dimethylthetin and dimethyl-β-propiothetin in methionine synthesis. J. Biol. Chem. *176*, 789–796.

DUERRE, J. A. 1968. *In vivo* and *in vitro* metabolism of S-adenosyl-L-homocysteine by *Saccharomyces cerevisiae*. Arch. Biochem. Biophys. *124*, 422–430.

DUERRE, J., and SCHLENK, F. 1962. Formation and metabolism of S-adenosyl-L-homocysteine in yeast. Arch. Biochem. Biophys. *96*, 575–579.

DU VIGNEAUD, V., CHANDLER, J. P., MOYER, A. W., and KEPPEL, D. M. 1939. The ability of homocysteine plus choline to support growth of the white rat on a methionine-free diet. J. Biol. Chem. *128*, CViii.

DU VIGNEAUD, V. P., and RACHELE, J. R. 1965. The concept of transmethylation in mammalian metabolism and its establishment by isotopic labeling through "*In Vivo*" Experimentation. *In* Transmethylation and Methionine Biosynthesis, S. K. Shapiro and F. Schlenk (Editors). Univ. Chicago Press, Chicago.

DU VIGNEAUD, V., RACHELE, J. R., and WHITE, A. M. 1956. A crucial test of transmethylation labeling. J. Am. Chem. Soc. *78*, 5131–5132. (1952)

FLAVIN, M., and KERR, D. 1969. Inhibition of cystathionine-γ-synthetase by S-adenosylmethionine: A control mechanism for methionine synthesis in *Neurospora*. Biochim. Biophys. Acta *177*, 177–179.

GREENE, R. C. 1969. Kinetic studies of the metabolism of S-adenosylmethionine synthetase from yeast. Biochemistry *8*, 2255–2265.

GUEST, J. R., HELLEINER, C. W., CROSS, M. J., and WOODS, D. D. 1960. Cobalamin and the synthesis of methionine by ultrasonic extracts of *Escherichia coli*. Biochem. J. *76*. 396–405.

HURWITZ, J., GOLD, M., and ANDERS, M. 1964. The enzymatic methylation of RNA and DNA. IV. The properties of the soluble ribonucleic-acid-methylating enzymes. J. Biol. Chem. *239*, 3474–3482.

JAUREGUIBERRY, G., LAW, J. H., MCCLOSKEY, J. A., and LEDERER, E. 1965. Studies on the mechanism of biological carbon alkylation reactions. Biochemistry *4*, 347–353.

KELLER, E. B., RACHELE, J. R., and DU VIGNEAUD, V. 1949. A study of transmethylation with methionine containing deuterium and $C^{14}$ in the methyl group. J. Biol. Chem. *177*, 733–738.

LEDERER, E. 1964. The origin and function of some methyl groups in branched-chain fatty acids, plant sterols and quinones. Biochem. J. *93*, 449–468.

LENFANT, M., ZISSMANN, E., and LEDERER, E. 1967. Biosynthesis of the ethyl side chain of stigmasterol derivatives by the slime mold *Dictyostelium discoideum*. Tetrahedron Letters *1967*, 1049–1052.

MUDD, S. H. 1965. The mechanism of the enzymatic synthesis of S-adenosylmethionine. *In* Transmethylation and Methionine Biosynthesis. S. K. Shapiro and F. Schlenk (Editors). Univ. Chicago Press, Chicago.

PARKS, L. W. 1958A. S-adenosylethionine and ethionine inhibition. J. Biol. Chem. *232*, 169–176.

PARKS, L. W. 1958B. S-adenosylmethionine and ergosterol synthesis. J. Am. Chem. Soc. *80*, 2023–2024.

PARKS, L. W., TURNER, J. R., and LARSON, R. L. 1965. Transmethylation in yeast sterol synthesis. *In* Transmethylation and Methionine Biosynthesis, S. K. Shapiro and F. Schlenk (Editors). Univ. Chicago Press, Chicago.

PIGG, C. J., SPENCE, K. D., and PARKS, L. W. 1962. Methionine biosynthesis in yeast. Arch. Biochem. Biophys. *97*, 491–496.

RUSSELL, P. T., VAN ALLER, R. T., and NES, W. R. 1967. The mechanism of introduction of alkyl groups at C-24 of sterols. II. The necessity of the $\Delta^{24}$-bond. J. Biol. Chem. *242*, 5802–5806.

SCHLENK, F., DAINKO, J. L., and STANFORD, S. M. 1959. Improved procedure for the isolation of S-adenosylmethionine and S-adenosylethionine. Arch. Biochem. Biophys. *83*, 28–34.

SCHLENK, F., and DE PALMA, R. E. 1957A. The formation of S-adenosylmethionine in yeast. J. Biol. Chem. *229*, 1037–1050.

SCHLENK, F., and DE PALMA, R. E. 1957B. The preparation of S-adenosylmethionine. J. Biol. Chem. *229*, 1051–1057.

SCHLENK, F., and ZYDEK, C. R. 1968. The metabolism of sulfur amino acid derivatives in yeast. Arch. Biochem. Biophys. *123*, 438–446.

SHAPIRO, S. K. 1965. The function of S-adenosylmethionine in methionine biosynthesis. *In* Transmethylation and Methionine Biosynthesis, S. K. Shapiro and F. Schlenk (Editors). Univ. Chicago Press, Chicago.

SHAPIRO, S. K., and EHNINGER, D. J. 1969. Biosynthesis of adenosylmethionine from adenosylhomocysteine in *Candida utilis*. Biochim. Biophys. Acta *177*, 67–77.

SHAPIRO, S. K., and SCHLENK, F. 1960. The biochemistry of sulfonium compounds. Advan. Enzymol. *22*, 237–280.

SHAPIRO, S. K., and SCHLENK, F. 1965. Transmethylation and Methionine Biosynthesis. Univ. Chicago Press, Chicago.

SHAPIRO, S. K., and YPHANTIS, D. A. 1959. Assay of S-adenosylmethionine homocysteine transmethylase. Biochim. Biophys. Acta *36*, 241–244.

SHAPIRO, S. K., YPHANTIS, D. A., and ALMENAS, A. 1964. Biosynthesis of methionine in *Saccharomyces cerevisiae:* partial purification and properties of S-adenosylmethionine: homocysteine methyltransferase. J. Biol. Chem. *239*, 1551–1556.

TAYLOR, R. T., and WEISSBACH, H. 1969A. *Escherichia coli* B $N^5$-methyltetrahydrofolate-homocysteine methyl transferase: sequential formation of bound methylcobalamin with S-adenosyl-L-methionine and $N^5$-methyltetrahydrofolate. Arch. Biochim. Biophys. *129*, 728–744.

TAYLOR, R. T., and WEISSBACH, H. 1969B. *Escherichia coli* B $N^5$-methyltetrahydrofolate-homocysteine cobalamin methyltransferase: activation with S-adenosyl-L-methionine and the mechanism for methyl group transfer. Arch. Biochem. Biophys. *129*, 745–766.

TURNER, J. R., and PARKS, L. W. 1965. Transmethylation products as intermediates in ergosterol biosynthesis in yeast. Biochim. Biophys. Acta *98*, 394–401.

WOODS, D. D., FOSTER, M. A., and GUEST, J. R. 1965. Cobalamin-dependent and independent methyl transfer in methionine biosynthesis. *In* Transmethylaation and Methionine Biosynthesis, S. K. Shapiro and F. Schlenk (Editors). Univ. Chicago Press, Chicago.

ZAPPIA, V., SALVATORE, F., ZYDEK, C. R., and SCHLENK, F. 1968. The production of $^{15}N$-labeled S-adenosylmethionine and adenine by yeast biosynthesis. J. Labeled Compds. *4*, 230–239.

ZAPPIA, V., ZYDEK-CWICK, C. R., and SCHLENK, F. 1969. The specificity of S-adenosylmethionine derivatives in methyl transfer reactions. J. Biol. Chem. *244*, 4499–4509.

## CHAPTER 5

James D. Finkelstein

# Control of Sulfur Metabolism in Mammals[1]

## INTRODUCTION

Methionine is an essential amino acid in mammalian nutrition, yet we know little about the factors which regulate its metabolism. Greater knowledge of methionine metabolism would contribute, not only to an understanding of the problems of nutrition, but to a more sophisticated understanding of human diseases which are caused by specific abnormalities of sulfur amino acid metabolism.

In this chapter, I intend to raise some questions and explore some hypotheses about the control of methionine metabolism in mammals. The discussion will be limited to the pathway between methionine and cysteine, neglecting the interesting subjects of the transport of sulfur compounds across cell membranes and the catabolism of cysteine. I will not describe the methodology in detail and would refer those interested to the original publications.

## METHIONINE METABOLISM IN MAMMALS—THE PATHWAY

The major metabolic functions of methionine are: (a) utilization for protein synthesis; (b) conversion to S-adenosylmethionine—the primary biological methyl group donor; (c) conversion, by means of the transsulfuration pathway to cystathionine, cysteine and further derivatives of cysteine (Fig. 5.1). This last function is also the major route for methionine catabolism in normal adult humans (Laster *et al.* 1965).

Methionine-activating enzyme (MAE, ATP:L-methionine-S-adenosyltransferase, EC 2.5.1.6) catalyzes the conversion of methionine to S-adenosylmethionine (Reaction 1, Fig. 5.1). MAE is highly selective for methionine. Methionine derivatives prepared by oxidation of the sulfur atom, by acylation of the amino group, or by variation in the chain length, are poor substrates. MAE activates ethionine and trifluoromethionine. The requirement for ATP is specific and a divalent cation, $Mg^{++}$, is also necessary. The immediate products of methionine activation are S-adenosyl-L-methi-

[1]This research is supported in part by research grant AM-13,048 from the National Institute of Arthritis and Metabolic Diseases.

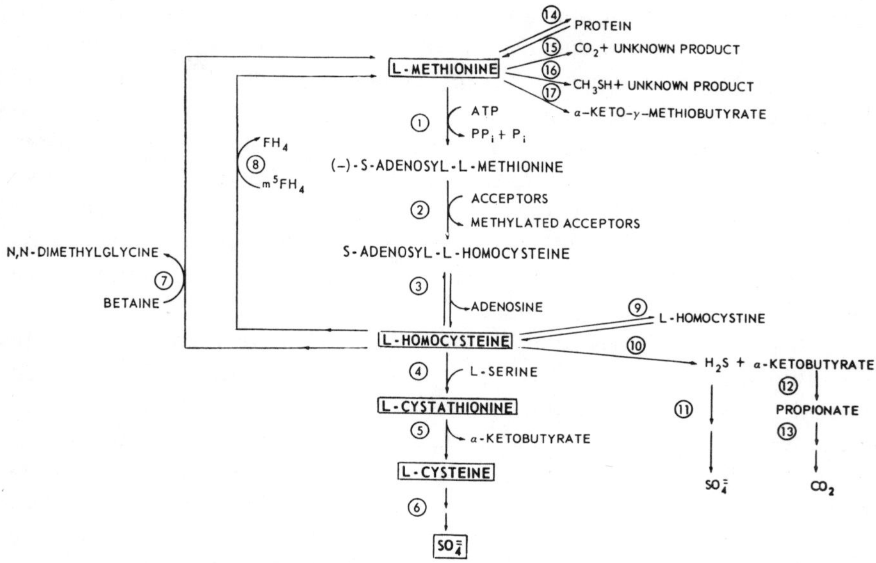


FIG. 5.1. CURRENT CONCEPTS OF METHIONINE METABOLISM IN MAMMALIAN TISSUES

$P_i$ and $PP_i$ are inorganic phosphate and pyrophosphate. $FH_4$ is tetrahydrofolic acid and $m^5FH_4$ is $N^5$-methyltetrahydrofolic acid.

onine (sAMe) and tripolyphosphate. MAE binds both products and in this form is inhibited. Hydrolysis of enzyme-bound polyphosphate to pyrophosphate and inorganic phosphate is catalyzed by MAE itself. Mudd (1965) has reviewed in detail the mechanism of MAE.

S-adenosyl-L-methionine is the primary methyl-group donor in mammalian metabolism. The numerous transmethylation reactions have been reviewed by Mudd and Cantoni (1964). But sAMe can also transfer a propylamine group. In recent studies using rat prostate, Pegg and Williams-Ashman (1969) have shown that in the presence of putrescine sAMe is first decarboxylated. The propylamine group is then transferred to the putrescine-yielding spermidine and methylthioadenosine. Unlike the *E. coli* system, which requires two enzymes to catalyze decarboxylation and transfer, the mammalian system requires only one which may be pyridoxal phosphate dependent. The significance of this reaction to mammalian methionine metabolism is not yet known, but we assume that the bulk of synthesized sAMe is utilized in transmethylation reactions yielding S-adenosyl-L-homocysteine (s-AHSR).

s-AHSR is cleaved to adenosine and homocysteine by a specific hydrolase (EC 3.3.1.1) (de la Haba and Cantoni 1959). The reaction is reversible and the equilibrium favors s-AHSR synthesis *in vitro*. Only if the products homocysteine and adenosine are removed will the equilibrium shift toward hydrolysis. For this reason we could expect to find s-AHSR, but not homocysteine, in mammalian tissues (Salvatore *et al.* 1968). Duerre and his co-workers (1969) suggest that a major portion of s-AHSR may be directly deaminated to S-adenosyl-$\gamma$-thio-$\alpha$-ketobutyrate. Their conclusion is incompatible with the failure to find the deamination product in urine of normal animals and also with the known conversion of dietary methionine to sulfate. It is possible that they have observed the fate of intravenously-administered, extracellular s-AHSR rather than the metabolism of the endogenously synthesized compound.

At least two enzymes catalyze the remethylation of L-homocysteine to L-methionine. We know that there is greater utilization of preformed methyl groups than of newly formed methyl groups in rats that are fed diets adequate in choline, folic acid, vitamin $B_{12}$, serine and glycine (du Vigneaud 1952). Consequently, we have assumed that the dominant remethylation reaction requires the enzyme betaine-homocysteine methyltransferase (BH Enz, EC 2.1.1.5) (Fig. 5.1, Reaction 7). Betaine, the methyl group donor in this reaction, is derived from endogenous or exogenous choline by oxidation of the alcohol group. BH Enz is inhibited by dimethylglycine—the reaction-product—and by homoserine, which is competitive with homocysteine (Ericson 1960).

The second reaction which methylates homocysteine uses $N^5$-methyltetrahydrofolate as the methyl donor ($N^5$-methyltetrahydrofolate-homocysteine methyltransferase, mTHF Enz) (Fig. 5.1, Reaction 8). Vitamin $B_{12}$ is a cofactor for mTHF Enz. sAMe is required in trace amounts. The transferred methyl group is derived from the "one-carbon pool" and represents a "newly-formed" methyl group. mTHF Enz is of considerable interest to biochemists, nutritionists, and clinicians, since it represents an interdigitation of folic acid, vitamin $B_{12}$, methionine, and "one-carbon fragment" metabolism. The failure of the mTHF reaction due to vitamin $B_{12}$ deficiency may lead to signs of folate deficiency as available folate is trapped as the $N^5$-methyltetrahydrofolate derivative.

Mammalian liver contains additional enzymes which are capable of transmethylating homocysteine using thetin and S-adenosylmethionine as methyl donors. The two enzymes are of dubious physiological significance. Thetin does not occur in the diet nor is it found in mammalian tissues. sAMe methylation of homocysteine yields no net gain in available methionine.

L-homocysteine may react with serine to form cystathionine in a unidirectional reaction catalyzed by cystathionine synthase (Cyst. Synth.) (Fig. 5.1, Reaction 4). The physiological significance of cystathionine synthesis may be due, largely, to the irreversibility of the reaction. Homocysteine converted to cystathionine can no longer serve as a methionine precursor. The hypothesis that cystathionine synthase is identical to serine (threonine) dehydratase (EC 4.2.1.21) has been proven wrong (Mudd *et al.* 1965A; Brown *et al.* 1966; Finkelstein 1967; Nakagawa and Kimura 1968). For this reason there is little data about the enzyme. Even the question of a pyridoxal phosphate cofactor requirement is being questioned (Nakagawa and Kimura 1968; Brown *et al.* 1968).

Once it has been synthesized, cystathionine is cleaved by the enzyme cystathionase yielding the products cysteine and α-ketobutyrate. This reaction translocates the original sulfur atom of methionine (and homocysteine) to the 3 carbon moiety of cysteine (serine). Although cystathionase is identical to L-homoserine hydro-lyase (homoserine dehydratase; L-cysteine desulfurase; EC 4.1.1.15) there is no evidence to suggest that the enzyme acts on cystathionine to form free homoserine as an intermediate. Cystathionase can catalyze the synthesis of cystathionine from homoserine and cysteine both *in vitro* (Matsuo and Greenberg 1959) and in the intact animal (Wong *et al.* 1968).

Mammalian metabolism of cysteine proceeds by several pathways by means of which the synthesis of various significant derivatives (including taurine) occurs. Inorganic sulfate is the final product of methionine catabolism. Since the control mechanism which regulates methionine metabolism in mammals seems to be proximal to cysteine metabolism, these pathways will not be described further.

## DISTRIBUTION OF ENZYMES IN MAMMALIAN TISSUES

A pathway for methionine metabolism is present in every tissue studied (Table 5.1) (Mudd *et al.* 1965B). In view of the essential role of trans-

TABLE 5.1

SPECIFIC ACTIVITIES OF ENZYMES OF SULFUR AMINO ACID METABOLISM IN RAT TISSUES[1,2]

| | MAE mμmoles/mg[3] 60 Min | Cyst. Synth. mμmoles/mg 135 Min | BH Enz mμmoles/mg 60 Min | mTHF Enz mμmoles/mg 60 Min | Cystathionase mμmoles/mg 30 Min |
|---|---|---|---|---|---|
| Liver | 34 | 415 | 69.3 | 3.9 | 54 |
| Kidney | 9.9 | 188 | 0.6 | 15.8 | 20 |
| Pancreas | 8 | 373 | 0.1 | 5.0 | 28 |
| Brain | 3.3 | 84 | 0 | 4.6 | 0.1 |
| Lung | 1.7 | 0.4 | 0 | 1.3 | 0.4 |
| Heart | 1.0 | 0.02 | 0 | 4.2 | 0 |
| Spleen | 2.4 | 8.6 | 0.3 | 3.0 | 0.02 |
| Adrenal | 4.2 | 4.7 | 0.6 | 10.0 | 0.6 |
| Testes | 3.3 | 0.2 | 0.1 | 7.4 | 0.13 |
| Adipose | 1.5 | 86 | 0.2 | 3.6 | 1.3 |
| Small Int. | 2.2 | 69 | 0 | 0 | 4.3 |

[1]Results are the mean values derived from studies of 5–15 rats.
[2]Values for MAE, Cyst. Synth., and C'ase from Mudd *et al.* (1965).
[3]Millimicromoles per milligram.

methylation reactions, this is not surprising. However, the relative activities of the enzymes vary markedly from tissue to tissue. We can draw several conclusions from the data in Table 5.1.

(1) Only the liver has significant quantities of all five enzymes. BH Enz is almost unmeasurable in the other tissues. mTHF Enz, on the other hand, has a wide distribution (all tissues but intestinal mucosa) and it is likely to be the major homocysteine remethylating enzyme in extrahepatic tissues.

(2) Every tissue has a mechanism for utilizing homocysteine—a finding which is consistent with the fact that homocysteine is absent in blood from normal animals.

(3) In all tissues but brain and spleen, higher levels of cystathionine synthase are associated with higher levels of cystathionase. Cystathionine accumulates in mammalian brain (Tallan *et al.* 1958). In the spleen, where cystathionase is reduced, cystathionine levels need not increase if mTHF Enz competes effectively with cystathionine synthase for the available homocysteine.

(4) Heart, testes and lung are relatively deficient in cystathionine synthase. Methionine metabolism may proceed largely by remethylation of homocysteine with mTHF Enz.

(5) Intestinal mucosa is without either homocysteine methylase. Methionine metabolism must proceed solely by means of the transsulfuration pathway.

The absence of BH Enz in intestinal mucosa was investigated further (Finkelstein and Kyle 1969). We identified an inhibiting factor in intestinal mucosa which directly altered the liver enzyme. Several properties of the inhibitor suggested that it was trypsin. The inhibition was prevented by the presence of trypsin inhibitors. Nevertheless, when we prepared intestinal extracts in the presence of trypsin inhibitors, we detected no BH Enz activities.

## MATURATION OF ENZYME SYSTEMS IN RAT TISSUES

Table 5.2 contains the results of assays of the specific activities of the 5 enzymes in rat liver of 4 ages. Except for the 6–8 gm (suckling) group, all of the animals were fed a standard laboratory ration. The developmental patterns were variable. Specific activities of MAE, BH Enz and mTHF Enz declined with age, while the activities of Cyst. Synth. and cystathionase increased. Chase *et al.* (1968) who studied MAE levels in fetal and neonatal rats, demonstrated that the specific activity of MAE began to rise *in utero.* In our laboratory we studied the developmental patterns for all the enzymes (except BH Enz) in rat kidney, pancreas, and brain (Finkelstein 1967). The pattern for an enzyme did not invariably parallel the pattern in liver. We found many notable variations. In the pancreas there was a 60%

TABLE 5.2

EFFECT OF AGE ON THE SPECIFIC ACTIVITIES OF ENZYMES OF SULFUR AMINO ACID METABOLISM IN RAT LIVER[1]

| Weight (Gm) | MAE mμmoles/mg 60 Min | Cyst. Synth. mμmoles/mg 135 Min | BH Enz mμmoles/mg 15 Min | mTHF Enz mμmoles/mg 60 Min | Cystathionase mμmoles/mg 30 Min |
|---|---|---|---|---|---|
| 6–8 | 56 | 156 | 33 | 10.5 | 40 |
| 70 | 52 | 315 | 25 | 6.4 | 52 |
| 140 | 42 | 318 | 24 | 4.0 | 73 |
| 200 | 34 | 363 | 22 | 3.7 | 63 |

[1]Experimental methods are described in Finkelstein (1967).

reduction in cystathionine synthase in the oldest animals. In brain, we found the specific activities of MAE and cystathionase were constant. In the pancreas, the specific activity of mTHF Enz increased with age.

Expression of the results as total units/organ/gm body weight rather than as specific activity altered the developmental patterns in some instances, particularly in brain. With all 4 enzymes, brain content/gm body weight decreased with age—a finding which reflected the fall in brain weight from 5% (suckling rat) to 0.6% (280 gm rat) of the total body weight.

The results of these studies suggest that the developmental pattern of an enzyme is tissue-specific as well as enzyme-specific. In general, the liver of the growing rat contains higher concentrations of enzymes which are required to utilize methionine (MAE) and to regenerate methionine (BH Enz, mTHF Enz). The liver of the growing rat contains lower concentrations of the enzymes of transsulfuration (Cyst. Synth., cystathionase).

## RESPONSE OF ENZYMES TO ENVIRONMENTAL FACTORS

### Effect of Diet

Dietary alterations significantly affect the specific activities of the five enzymes in rat tissues. In earlier studies of MAE, cystathionine synthase, and cystathionase, I found that diet-induced changes were rarely parallel in the three enzymes in a single tissue (Finkelstein 1967). Nor were changes in a single enzyme the same in the four tissues studied. We now have extended the study to include BH Enz and mTHF Enz and the data for the effects of diet on the specific activities of the enzymes in rat liver appear in Table 5.3. Although the new information does not alter our earlier conclusions, several findings warrant emphasis.

(a) Protein supplementation of a low protein diet produces an increase in the specific activities of MAE, Cyst. Synth., cystathionase, and BH Enz, but the activity of mTHF Enz declines. In the converse experiment, protein restriction after a period of high protein feeding, markedly increases the

TABLE 5.3

EFFECT OF DIET ON SPECIFIC ACTIVITIES OF ENZYMES OF SULFUR AMINO ACID METABOLISM IN RAT LIVER[1]

| Diet | MAE | Cyst. Synth. | BH Enz | mTHF Enz | Cystathionase |
|---|---|---|---|---|---|
| Low protein (LPD) | 100 | 100 | 100 | 100 | 100 |
| LPD + fast | 324 | 138 | 151 | 71 | 75 |
| LPD + fast + casein | 492 | 270 | 124 | 57 | 106 |
| High protein | 142 | 452 | 182 | 27 | 290 |
| Standard | 67 | 390 | 153 | 56 | 351 |
| Standard + fast | 64 | 304 | 113 | 61 | 253 |

[1]Diets and treatment described in Finkelstein (1967). Results are expressed as the percentage of the specific activity in livers obtained from the LPD group.

specific activity of mTHF Enz within three days. The activities of the other enzymes were decreased. These patterns of change were unique to liver. High protein feeding had the opposite effect on pancreatic enzymes. In the kidney MAE fell (43% of control), Cyst. Synth. rose 166%, and the levels of cystathionase and mTHF Enz were unchanged. Of the brain enzymes only cystathionase, which increased to 171%, was affected by the high protein ration.

(b) Responses to fasting were unpredictable but were affected by the antecedent diet. Hepatic MAE rose markedly in animals previously fed a protein restricted diet, but remained unchanged in animals that were prefed the standard diet.

(c) Protein content alone cannot explain the effect of diet on MAE levels in the liver. The specific activity of the enzyme in animals fed the low-protein diet fell between the levels observed in animals on the high protein and the standard rations.

(d) The responses of hepatic mTHF Enz and BH Enz to the protein content of the diet are opposite. The level of mTHF Enz seems to reflect the need for methionine resynthesis which increases with protein deprivation. BH Enz activity, on the other hand, increases with increased dietary protein thus suggesting that its primary function might be the removal of homocysteine rather than the resynthesis of methionine.

**Effect of Hormones**

The effects of various hormones on the specific activities of MAE, Cyst. Synth., and cystathionase demonstrated the same principles. Rarely did a hormone result in the same change in a given enzyme in liver, kidney, pancreas, or brain. Similarly, there were few examples of parallel changes in the three enzymes in a given tissue (Finkelstein 1967). Nevertheless, we observed many significant changes. Foo Pan and his co-workers (Pan 1967, 1968A, 1968B; Shou *et al.* 1969) have studied in detail the effects of hor-

TABLE 5.4

EFFECTS OF HORMONES ON SPECIFIC ACTIVITIES OF ENZYMES OF SULFUR AMINO ACID METABOLISM IN RAT LIVER[1]

| Treatment | MAE | Cyst. Synth. | BH Enz | mTHF Enz | Cystathionase |
|---|---|---|---|---|---|
| Hydrocortisone | 158[3] | 151[3] | 300[3] | 96 | 84 |
| Thyroxine | 83 | 101 | 35[3] | 78 | 27[3] |
| Growth hormone | 160 | 100 | 77 | 98 | 120 |
| Estradiol | 176[3] | 104 | 91 | 82 | 145[3] |
| Testosterone | 123[3] | 98 | 115 | 78 | 109 |
| Alloxan | 298[3] | 141[3] | 170[3] | 70 | 115 |
| Glucagon[2] | 99 | 195[3] | 110 | 63 | 180[3] |

[1]Results are expressed as the percentage of the specific activity of appropriate controls. Data for MAE, Cyst. Synth., and Cystathionase and the methods of treatment were previously reported in Finkelstein (1967).
[2]Glucagon studies were performed with rats fed a low protein diet for 7 days. Hormone treatment and fasting were begun 24 hr. before sacrifice.
[3]Statistical significance relative to controls ($p < .05$) demonstrated by assays using individual livers.

mones on hepatic MAE. Their findings agree, for the most part, with the data recorded in Table 5.4. Increases in hepatic MAE which result from the administration of hydrocortisone and alloxan, are due to new enzyme synthesis. Insulin replacement prevents the alloxan effect. The Table also illustrates the several effects of endocrine alterations on the specific activities of the homocysteine transmethylases. BH Enz activity increases after glucocorticoid and alloxan treatment. It is significantly depressed by thyroxine administration. No hormone increases mTHF Enz and the significance of the decreased specific activity associated with thyroxine, estradiol, testosterone, alloxan, and glucagon is presently under investigation. We should again note that the responses of the two transmethylases are infrequently parallel.

## THE METHIONINE SPARING EFFECT OF CYSTINE

We have shown that the individual enzymes are responsive to environmental factors. But we do not know the consequences of the changes in the levels of enzyme activity. We required a system in which we knew the effect of the experimental conditions on total methionine metabolism. The methionine-sparing effect of cystine fulfills this criterion. Although cystine-feeding can reduce the nutritional requirement for methionine, it cannot totally replace methionine in the diet (Womack *et al.* 1937; Rose *et al.* 1955; Rose and Rice 1939; Womack and Rose 1941; Rose and Wixom 1955; Rama Rao *et al.* 1961). The methionine-sparing effect of cystine is not due to the biosynthesis of methionine. Indeed, there is no evidence to suggest that mammals can incorporate the sulfur of cyst(e)ine into the methionine molecule. "Methionine-sparing" must be the result of the diversion of methionine from the catabolic pathway (transsulfuration) to the pathways for protein synthesis and transmethylation. Therefore, we studied the effect of altering the dietary sulfur amino acid content on the activities of the enzymes in rat liver (Finkelstein and Mudd 1967).

The results of these studies appear in Table 5.5. The nonsupplemented, low-sulfur amino acid diet contained sufficient methionine to constitute an

TABLE 5.5

EFFECT OF DIETARY SULFUR AMINO ACID CONTENT ON SPECIFIC ACTIVITIES OF ENZYMES IN RAT LIVER[1]

| Supplement Added to Low Sulfur Amino Acid Diet | MAE mμmoles/mg 60 Min | Cyst. Synth. mμmoles/mg 135 Min | BH Enz mμmoles/mg 15 Min | mTHF Enz mμmoles/mg 60 Min | Cystathionase mμmoles/mg 30 Min |
|---|---|---|---|---|---|
| None | 95[2] | 161[2] | ... | 9.1 | 26 |
| Methionine | 83[2] | 168[2] | 15.8 | 5.1 | 31 |
| Cystine | 63 | 89 | 13.5 | 6.6 | 27 |
| Methionine and Cystine | 85[2] | 165[2] | ... | ... | 27 |

[1]Methods and data were presented previously in Finkelstein and Mudd (1967).
[2]$p < .005$ relative to cystine-supplemented animals.

adequate diet, when either methionine or cystine was added. Rats on the three supplemented diets gained weight at the same rate. When we replaced the dietary methionine supplement with cystine, a 50% reduction in cystathionine synthase resulted. The level of MAE also fell significantly. Cystathionase and BH Enz activities were slightly lower in cystine fed animals while the activity of mTHF Enz increased. In rats given supplements of cystine as well as methionine, enzyme levels were identical with the animals which had received methionine alone.

We also studied the time-course of the effects of dietary changes and supplements. In the first experiment, we fed the animals methionine-supplemented diets and then changed to a cystine-supplemented ration. Within 1 day the level of cystathionine synthase declined and by 3 days reached a plateau at 42% of the control. MAE and cystathionase activities declined until the fifth day. In a second series of experiments, we studied the effect of the parenteral administration of methionine on levels of MAE, cystathionine synthase and cystathionase. The animals were fed the cystine-supplemented diet during the studies. The levels of MAE and cystathionine synthase increased within 24 hr after intraperitoneal injection of 120 mg of L-methionine. After a second injection, and an additional 24 hr, the levels had increased further. The level of MAE rose to 206% of the control value and cystathionine synthase levels increased to 310%. Twenty-four hours after the first injection, the specific activity of cystathionase increased from 21 to 25 mμmoles per mg of protein per 30 min. A second injection had no additional effect. We also found that the effect of methionine was dose-dependent and that the effect of homocysteine, like that of methionine, was to produce an increase in the levels of MAE and cystathionase. Homocysteine injections, however, did not lead to an increase in the specific activity of cystathionine synthase in liver.

These data suggest that (1) cystine regulates enzyme levels in the liver of methionine-restricted rats and/or, (2) methionine regulates enzyme levels in the liver of cystine-repleted animals.

We conducted several studies to define the mechanism of these effects (Finkelstein and Mudd 1967). Cystathionine synthase activity was inhibited *in vitro* by L-cystine. However, the high concentration of inhibitor required for 50% inhibition is unlikely to occur in livers of cystine-fed rats. The livers do not contain a dissociable inhibitor of cystathionine synthase. Structural alterations of the enzyme, secondary to a reaction of protein thiol groups with half-cystine residues, could cause this inhibition. But we discounted this possibility when we found no evidence of enzyme-inhibitor binding *in vitro* and after we observed no difference in the response of extracts from methionine-supplemented and cystine-supplemented animals to treatment designed to remove protein-bound half-cystine residues. We do not know the mechanisms which allow methionine and cystine to control the activities of these enzymes. Regulation is probably affected at the level of enzyme synthesis. We are investigating this possibility at present.

## CONTROL OF METHIONINE METABOLISM

### The Relationship Between Individual Enzymes

Clearly, the mechanism of control does not involve the mammalian equivalent of an "operon" system. Even if we restrict our attention to three enzymes which are components of the "transsulfuration pathway" (MAE, Cyst. Synth., cystathionase) there are few examples of coordinate changes in a single organ in response to a given stimulus. Indeed, we have observed frequently that changes in the levels of the enzymes are in opposite directions. For example, the change from a low protein diet to the standard ration results in a decrease in hepatic MAE (67%) but in an increase in cystathionine synthase (390%) and cystathionase (351%). Similarly, thyroxine treatment selectively inhibits hepatic cystathionase. Other examples of divergent changes can be found in Tables 5.3 and 5.4. The differences in the developmental patterns of MAE, Cyst. Synth., and cystathionase provide still further evidence of noncoordinate control of the transsulfuration pathway.

The absence of coordinate control also characterizes the responses of transsulfuration enzymes in extrahepatic tissues. Further, the response of a given enzyme in an organ cannot be predicted from the response of the same enzyme in liver. Tissue differences require explanation. Isoenzymes might provide an answer; but there is little data concerning isoenzymes in methionine metabolism. We found that a patient who lacked hepatic cystathionine synthase had a similar deficiency in cerebral tissue (Mudd *et al.* 1966). In the subsequent discussion I will deal only with liver, on the assumption that this is the major site of methionine metabolism. However, the impossibility of extrapolating from liver to other tissues is implicit.

**Regulation by Tissue Concentrations of Sulfur Compounds**

Although we cannot invoke an "operon" concept, it is possible that tissue concentrations of substrates and/or products regulate the activities of the individual enzymes. Such regulation could occur at the level of enzyme synthesis (induction/repression) or at the level of the enzyme protein (inhibition). MAE for example, could respond to the levels of methionine, sAMe and perhaps ATP. However, we have found that hepatic MAE increases after methionine administration only in animals who were previously cystine repleted. Patients with cystathionine synthase deficiency (homocystinuria) also have increased tissue concentrations of methionine and normal levels of hepatic MAE (Finkelstein *et al.* 1964). Similarly sAMe levels in rat liver rise acutely after the intraperitoneal injection of methionine, yet hepatic MAE is not altered (Baldessarini 1966). Finally, large doses of alcohol cause a decrease in the level of hepatic ATP and an increase in the level MAE (Finkelstein and Kyle 1968). There is a possibility that *in vitro* studies mask substrate effects which could occur *in vivo*. Nevertheless, we have no data which suggest that inhibition of MAE by sAMe and polyphosphate has physiological significance.

There is, in contrast, evidence to indicate that accumulation of homocysteine will reverse the sAHSR hydrolase reaction. Patients who have cystathionine synthase deficiency with an increase in tissue homocysteine (Tada *et al.* 1967) will excrete S-adenosylhomocysteine in the urine (Perry *et al.* 1967). Accumulation of homocysteine might be expected to cause an increase in the activity of cystathionine synthase. Similarly, cystathionine excess might inhibit the enzyme. The injection of homocysteine however, does not affect the level of cystathionine synthase. Furthermore, a patient who lacked cystathionase and retained cystathionine showed only a modest reduction in cystathionine synthase activity (Finkelstein *et al.* 1966).

We have less data about the responses either of the homocysteine transmethylases or of cystathionase to changes in tissue concentrations of substrates and products. BH Enz is not affected in homocystinuria (Brenton *et al.* 1965). The level of mTHF Enz is affected, in all probability, by the tissue content of methionine. The specific activity of mTHF Enz decreases when test animals are fed either excessive protein, excessive methionine with adequate protein, or excessive methionine with inadequate protein (Dickerman *et al* 1964; Kutzbach *et al.* 1967; Finkelstein 1969).

**The Regulatory Mechanism: An Hypothesis**

Although we know neither the molecular mechanism for the control mechanism nor the nature of the compounds which trigger the regulatory process, we assume that a control system must exist. Sulfur amino acid homeostasis must be maintained. Methionine must be utilized optimally for protein synthesis, transmethylation reactions, and the generation of cystathionine, cysteine and further derivatives of cysteine. Simultaneously, toxic metabolites must be removed. In order to validate the presence of a control mechanism, it is necessary to show that the changes in individual enzyme activities explain the changes in total methionine metabolism. The measurement of "total methionine metabolism" is difficult. We are developing methods for this determination in the whole animal and in the isolated liver; but for the moment, we are restricted to one well-controlled phenomenon—the methionine-sparing effect of cystine. In this model we can assume that all metabolic requirements which involve methionine are satisfied when the animals are fed either cystine or methionine supplements.

When we feed cystine rather than methionine supplements, we are eliminating one necessary role for methionine—the synthesis of cysteine. We can then look to our enzyme data in order to find a basis for the diversion of available methionine toward protein synthesis and transmethylation, and away from transsulfuration. A marked reduction in cystathionine synthase and a smaller decrease in methionine-activating enzyme represent the most significant changes. There is little change in the homocysteine methylases.

These observations can explain the methionine-sparing by cysteine if we postulate that methionine metabolism is essentially a cycle, with a single unidirectional outlet—the transsulfuration pathway. There exist 2 regulatory loci both of which are formed by the presence of at least 2 enzymes which compete for a single substrate. The first regulatory point is located at the point where methionine is activated either for protein synthesis or transmethylation (MAE). The second locus is formed by the enzymes which utilize homocysteine. Cystathionine synthesis commits the molecule to the transsulfuration pathway whereas homocysteine transmethylation regenerates methionine. We do not know how frequently the cycle is completed nor do we know how much homocysteine is diverted to cystathionine during each cycle. The conversion of almost all of the dietary methionine to sulfate within 24 hr can be explained by a slow cycling with a large shunt to cystathionine, or by a rapid cycling and a small but frequent shunt.

The hypothesis that methionine metabolism is a cycle with a unidirectional outlet can explain the methionine-sparing effect of cystine. Reduced cystathionine synthesis would result in the return of a greater fraction of homocysteine to the methionine pool. The decreased MAE might facilitate protein synthesis.

From the data presented, we can also postulate some relationship between mTHF Enz and BH Enz. We have assumed that, of the two, betaine methylation is more significant to methionine metabolism. This may be true in the liver under basal conditions. Since liver alone contains BH Enz, mTHF Enz must be more significant in other tissues. Of greater importance is the fact that BH Enz could not conserve methionine when we restrict the amount of methionine and choline in the diet. Betaine synthesis requires the metabolism of three methionine molecules but regenerates only one. Maximum efficiency is less than 33%. Efficiency can be increased only when there is an increase in the utilization of the other carbon atoms of betaine; that is, by utilization of dimethylglycine. This can be accomplished through the "one-carbon pool" with the synthesis of $N^5$-methyltetrahydrofolate. In the diet studies, we noted that the administration of a high-protein ration caused BH Enz activity to rise presumably at a time when methionine conservation was unimportant. The level of BH Enz declines on the other hand with protein restriction. I would suggest that the need for methionine synthesis determines mTHF Enz activity while the requirement for homocysteine and/or betaine removal regulates the activity of BH Enz.

## ACKNOWLEDGEMENT

I wish to thank Dr. S. Harvey Mudd in whose laboratory we initiated many of these studies, and to whom I am indebted for many helpful discussions. Mr. Walter E. Kyle and Mrs. Barbara Marcel have provided excellent technical assistance.

## BIBLIOGRAPHY

BALDESSARINI, R. J. 1966. Alterations in tissue levels of S-adenosylmethionine. Biochem. Pharmacol. *15*, 741–748.

BRENTON, D. P., CUSWORTH, D. C., and GAULL, G. E. 1965. Homocystinuria: Biochemical studies of tissues including a comparison with cystathioninura. Pediatrics *35*, 50–56.

BROWN, C., BRENNAN, C., and GORDON, P. H. 1968. Cystathionine synthase. Federation Proc. *27*, 782.

BROWN, F. C., MALLADY, J., and ROSZELL, J. A. 1966. The separation of rat liver serine dehydratase and cystathionine synthase. J. Biol. Chem. *241*, 5220–5224.

CHASE, H. P., VOLPE, J. J., and LASTER, L. 1968. Transsulfuration in mammals: Fetal and early development of methionine-activating enzyme and its relation to hormonal influences. J. Clin. Invest. *47*, 2099–2108.

DE LA HABA, G., and CANTONI, G. L. 1959. The enzymatic synthesis of S-adenosyl-L-homocysteine from adenosine and homocysteine. J. Biol. Chem. *234*, 603–608.

DICKERMAN, H., REDFIELD, B. G., BIERI, J. G., and WEISSBACH, H. 1964. The role of vitamin $B_{12}$ in methionine biosynthesis in avian liver. J. Biol. Chem. *239*, 2545–2552.

DUERRE, J. A., MILLER, C. H., and REAMS, G. G. 1969. Metabolism of S-adenosyl-L-homocysteine *in vivo* by the rat. J. Biol. Chem. *244*, 107–111.

DUVIGNEAUD, V. 1952. A Trail of Research in Sulfur Chemistry and Metabolism. Cornell Univ. Press, Ithaca, N.Y.

ERICSON, L. E. 1960. Betaine-homocysteine-methyl-transferases. I. Distribution in nature. Acta Chem. Scand. *14*, 2102–2112.

FINKELSTEIN, J. D. 1967. Methionine metabolism in mammals. Effects of age, diet and hormones on three enzymes of the pathway in rat tissues. Arch. Biochem. Biophys. *122*, 583–590.

FINKELSTEIN, J. D. 1969. Unpublished data.

FINKELSTEIN, J. D., and KYLE, W. E. 1968. Ethanol effects on methionine metabolism in rat liver. Proc. Soc. Exptl. Biol. Med. *129*, 497–501.

FINKELSTEIN, J. D., and KYLE, W. E. 1969. Unpublished data.

FINKELSTEIN, J. D., and MUDD, S. H. 1967. Transsulfuration in mammals. The methionine-sparing effect of cystine. J. Biol. Chem. *242*, 873–880.

FINKELSTEIN, J. D., MUDD, S. H., IRREVERRE, F., and LASTER, L. 1964. Homocystinuria due to cystathionine synthase deficiency: The mode of inheritance. Science *146*, 785–787.

FINKELSTEIN, J. D., MUDD, S. H., IRREVERRE, F., and LASTER, L. 1966. Deficiencies of cystathionase and homoserine dehydratase activities in cystathioninuria. Proc. Natl. Acad. Sci. *55*, 865–872.

KUTZBACH, C., GALLOWAY, E., and STOKSTAD, E. L. R. 1967. Influence of vitamin $B_{12}$ and methionine on levels of folic acid compounds and folate enzymes in rat liver. Proc. Soc. Exptl. Biol. Med. *124*, 801–805.

LASTER, L., MUDD, S. H., FINKELSTEIN, J. D., and IRREVERRE, F. 1965. Homocystinuria due to cystathionine synthase deficiency: The metabolism of L-methionine. J. Clin. Invest. *44*, 1708–1719.

MATSUO, Y., and GREENBERG, D. M. 1959. A crystalline enzyme that cleaves homoserine and cystathionine. IV. Mechanism of action, reversibility and substrate specificity. J. Biol. Chem. *234*, 516–519.

MUDD, S. H. 1965. The mechanism of the enzymatic synthesis of S-adenosylmethionine. *In* Transmethylation and Methionine Biosynthesis, S. K. Shapiro and F. Schlenk (Editors). Univ. Chicago Press, Chicago.

MUDD, S. H., and CANTONI, G. L. 1964. Biological transmethylation, methyl group neogenesis and other "one-carbon" metabolic reactions dependent upon tetrahydrofolic acid. *In* Comprehensive Biochemistry, Group Transfer Reactions, Vol. 15, M. Florkin and E. H. Stotz (Editors). Elsevier Publishing Co., Amsterdam.

MUDD, S. H., FINKELSTEIN, J. D., IRREVERRE, F., and LASTER, L. 1965A. Threonine dehydratase activity in humans lacking cystathionine synthase. Biochem. Biophys. Res. Commun. *19*, 665–670.

MUDD, S. H., FINKELSTEIN, J. D., IRREVERRE, F., and LASTER, L. 1965B. Transsulfuration in mammals. Microassays and tissue distributions of three enzymes of the pathway. J. Biol. Chem. *240*, 4382–4392.

MUDD, S. H., LASTER, L., FINKELSTEIN, J. D., and IRREVERRE, F. 1966. Studies on homocystinuria. *In* Amines and Schizophrenia. H. E. Himwich, J. R. Smythies and S. S. Kety (Editors). Pergamon Press, New York.

NAKAGAWA, H., and KIMURA, H. 1968. Purification and properties of cystathionine synthetase from rat liver: Separation of cystathionine synthetase from serine dehydratase. Biochem. Biophys. Res. Commun. *32*, 208–214.

PAN, F., CHANG, G. G., LEE, S. C., and TANG, M. S. 1968A. Induction of methionine adenosyltransferase in rat liver by corticosteroids. Proc. Soc. Exptl. Biol. Med. *128*, 611–616.

PAN, F., LEE, S. C., CHANG, G. G., and CHIN, J. R. 1968B. Gonadal hormones in the regulation of methionine adenosyltransferase levels in rat liver. Proc. Soc. Exptl. Biol. Med. *129*, 161–165.

PAN, F., and TARVER, H. 1967. Effects of diet and other factors on methionine adenosyltransferase levels in rat liver. J. Nutr. *92*, 274–280.

PEGG, A. E., and WILLIAMS-ASHMAN, H. G. 1969. On the role of S-adenosyl-L-methionine in the biosynthesis of spermidine by rat prostate. J. Biol. Chem. *244*, 682–693.

PERRY, T. L., HANSEN, S., MACDOUGALL, L., and WARRINGTON, P. D. 1967. Sulfur containing amino acids in the plasma and urine of homocystinurics. Clin. Chem. Acta *15*, 409–420.

RAMA RAO, P. B., NORTON, H. W., and JOHNSON, B. C. 1961. The amino acid composition and nutritive value of protein(s). J. Nutr. *73*, 38–42.

ROSE, W. C., COON, M. J., LOCKHART, H. B., and LAMBERT, G. F. 1955. Amino acid requirements of man. Threonine and methionine requirements. J. Biol. Chem. *215*, 101–110.

ROSE, W. C., and RICE, E. E. 1939. Utilization of certain sulfur containing compounds for growth purposes. J. Biol. Chem. *130*, 305–323.

ROSE, W. C., and WIXOM, R. L. 1955. Amino acid requirements of man. Sparing effect of cystine on methionine requirement. J. Biol. Chem. *216*, 763–773.

SALVATORE, F., ZAPPIA, V., and SHAPIRO, S. K. 1968. Quantitative analysis of S-adenosylhomocysteine in liver. Biochim. Biophys. Acta *158*, 461–464.

SHOU, L., PAN, F., and CHIN, S. F. 1969. Pancreatic hormones and hepatic methionine adenosyltransferase in the rat. Proc. Soc. Exptl. Biol. Med. *131*, 1012–1018.

TADA, K., YOSHIDA, T., HIRONO, H., and ARAKAWA, T. 1967. Homocystinuria: Amino acid pattern of the liver. Tohoku J. Exptl. Med. *92*, 325–332.

TALLAN, H. H., MOORE, S., and STEIN, W. H. 1958. L-cystathionine in human brain. J. Biol. Chem. *230*, 707–716.

WOMACK, M., KREMMERER, K. S., and ROSE, W. C. 1937. Relation of cystine and methionine to growth. J. Biol. Chem. *121*, 403–410.

WOMACK, M., and ROSE, W. C. 1941. Partial replacement of dietary methionine by cystine for purposes of growth. J. Biol. Chem. *141*, 375–379.

WONG, P. W. K., SCHWARZ, V., and KOMROWER, G. M. 1968. The biosynthesis of cystathionine in patients with homocystinuria. Pediat. Res. *2*, 149–160.

CHAPTER 6

H. D. Peck, Jr.

# Sulfur Requirements and Metabolism of Microorganisms

## INTRODUCTION

Inorganic sulfur generally enters into biosynthetic pathways at the oxidation levels of sulfate ($SO_4^=$) and sulfide ($S^=$). Other naturally occurring forms of sulfur such as thiosulfate, polythionates, polysulfides, and elemental sulfur must be either oxidized to sulfate or reduced to sulfide before they are available for biosynthetic reactions. Biologically, the interconversion of sulfate and sulfide is a reversible process and can be considered as at least two broad phases: *sulfate reduction*, the reduction of sulfate to sulfide; and *sulfide oxidation*, the oxidation of sulfide to sulfate. The overall process is termed the "sulfur cycle" and each phase involves different organisms and different enzymatic pathways. It can largely be considered as a microbial process although both plants and animals certainly make significant quantitative contributions to the "sulfur cycle." Because several groups of bacteria utilize sulfate reduction or sulfide oxidation in their energy metabolism, the involvement of inorganic sulfur compounds in the nutrition of microorganisms is more extensive than in the nutrition of either plants or animals.

Mammals are unable to reduce sulfate to sulfide for the biosynthesis of sulfur-containing amino acids and the cofactors such as biotin, thiamine and coenzyme A (Huovinen and Gustafsson 1967). However, mammals are generally capable of incorporating sulfate as such into various lipids, carbohydrates and phenols and metabolizing and incorporating to a limited extent reduced sulfur compounds (Huovinen and Gustafsson 1967; Koj 1968). Animals are considered to be unable to reduce sulfate but this generalization has not been examined in detail. It would be extremely interesting to verify this statement or determine where in the animal kingdom this capability is lost. These experiments can be technically difficult because of the problem of obtaining animals free of bacterial contaminants and the widespread occurrence of bacterial symbionts in certain invertebrates (Henry and Block 1969; Lanham 1968).

In contrast to animals, plants and most bacteria have the ability to reduce sulfate to the level of sulfide as indicated by their growth with sulfate as the sole source of sulfur. The potential for incorporating sulfate directly into various sulfate derivatives is also present in these two groups of organisms. Sulfate reduction that is physiologically characterized by the formation of only enough sulfide to meet nutritional requirements has been termed assimilatory sulfate reduction. This is the only type of sulfate reduction occurring in plants and the most commonly encountered type in bacteria. A small group of anaerobic bacteria, known as the sulfate reducing bacteria, produce massive amounts of sulfide during growth in the presence of sulfate and an electron donor. This process has been identified as dissimilatory or respiratory sulfate reduction in that sulfate serves as the terminal electron acceptor in respiration in much the same manner as oxygen in aerobic respiration. The physiological processes in which sulfate is involved can be summarized as follows: first, direct incorporation in organic materials; second, reduction to sulfide for nutritional purposes (assimilatory sulfate reduction); third, reduction to sulfide during sulfate respiration (dissimilatory sulfate reduction).

Sulfide is the form in which sulfur is placed in direct bonding with carbon atoms for the formation of the usual sulfur-containing metabolites. Although sulfide incorporation has not been extensively or quantitatively studied, animals depend upon external sources of sulfur-containing metabolites, and sulfate is their nutritionally important form of inorganic sulfur. In plants and many bacteria, it is probable that both sulfate and sulfide are directly incorporated into biological materials and both can be considered to be nutritionally important. In other bacteria, enzymatic data suggest that sulfur in the form of sulfate may be of no direct nutritional significance (Nagy *et al.* 1969).

Sulfides and elemental sulfur occur in nature as components of the earth's crust and many of these deposits have been formed by biological action. Organic materials can also serve as a source of sulfide by the anaerobic degradative processes, collectively termed *putrefaction*, which occur under many conditions. In this connection it should be noted that the sulfur in reduced sulfur compounds such as cysteine can be oxidized in this form to yield sulfite ($SO_3^=$) which is readily oxidized to sulfate either nonenzymatically or enzymatically by sulfite oxidase. Sulfide can also be oxidized to the level of sulfate in many tissues but it is not clear whether this is an enzymatic or nonenzymatic process or involves both types of reactions. These processes might well be termed *incidental sulfur oxidation* as they are probably of minor importance and occur in connection with other more important reactions.

In microorganisms reduced inorganic sulfur compounds such as sulfide, elemental sulfur, and thiosulfate can serve as electron donors for aerobic or anaerobic respiration and photosynthesis with the formation of large

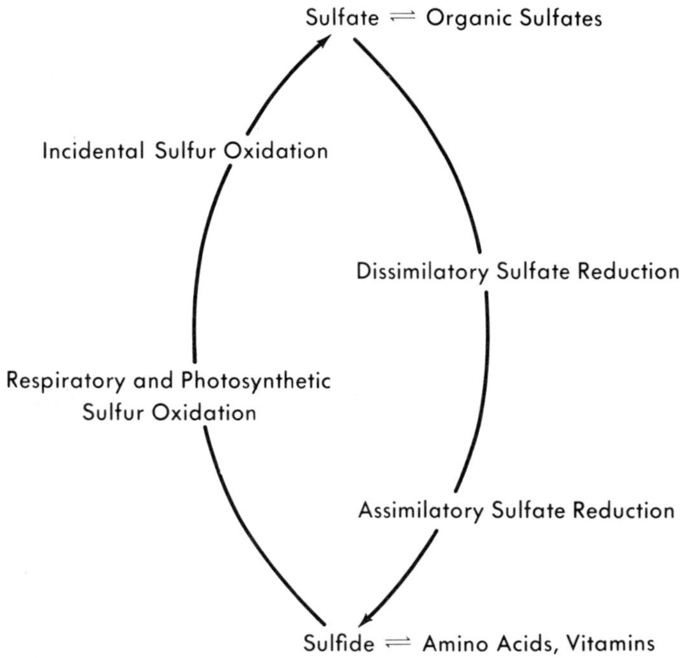


FIG. 6.1. PHYSIOLOGICAL PROCESSES INVOLVED IN THE SULFUR CYCLE

amounts of sulfate and frequently, with formation of extremely acidic conditions. As with dissimilatory sulfate reduction, these processes are limited to bacteria but represent an extremely important phase of the sulfur cycle, namely, oxidation of reduced sulfur back to the level of sulfate. These relationships have been briefly outlined in Fig. 6.1. More detailed information concerning various aspects of inorganic sulfur metabolism can be obtained in selected reviews by Peck (1962), Postgate (1965), Postgate (1969), Thompson (1967), and Wilson (1962).

## SULFATE ACTIVATION AND REDUCTION

### The Activation of Sulfate

The chemical reduction of sulfate is a relatively difficult reaction and little progress was made on the mechanism of biological sulfate reduction until Lipmann and Robbins simultaneously with Wilson and Bandurski elucidated the mechanism of sulfate activation in a number of organisms. Their observations along with the initial observations of DeMeio and Wizerkaniok are summarized in reviews by Gregory and Robbins (1960) and Wilson (1962). It was shown that in its biosynthetically active form,

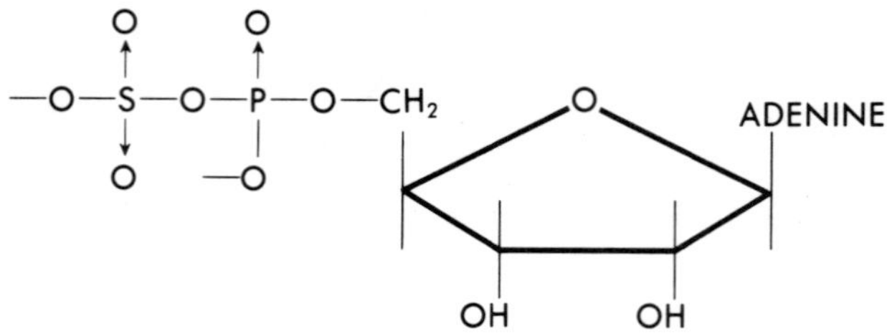


ADENOSINE-5′-PHOSPHOSULFATE (APS)

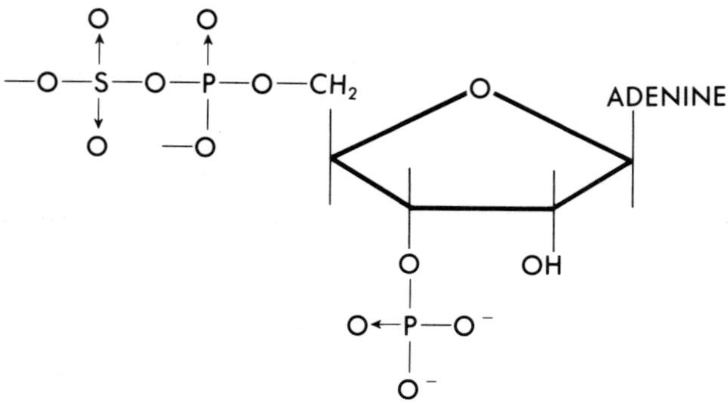


3′-PHOSPHOADENOSINE-5′-PHOSPHOSULFATE (PAPS)

FIG. 6.2. THE STRUCTURES OF ADENOSINE-5′-PHOSPHOSULFATE AND 3′-PHOSPHOADENOSINE-5′-PHOSPHOSULFATE

sulfate is esterified with a nucleotide to form 3′-phosphoadenosine-5′-phosphosulfate (PAPS) and its structure is shown in Fig. 6.2. The nucleotide sulfate has a group potential of 18–19 kcal which is about twice that of the comparable phosphate bond and suggests its possible involvement in other energy-yielding reactions. The synthesis of active sulfate is a two-step reaction involving the intermediate formation of another sulfur-containing nucleotide, adenosine 5′-phosphosulfate (APS), which can be considered to be a sulfate analogue of ADP (Fig. 6.2). In the first step [reaction (1)] ATP reacts with sulfate to yield APS and inorganic pyrophosphate ($PP_i$).

$$ATP + SO_4^= \rightleftharpoons APS + PP_i \qquad (1)$$

The enzyme mediating this reaction is ATP sulfurylase or ATP:sulfate adenylyl transferase and is dependent on the presence of $Mg^{++}$. The equilibrum of the reaction lies in the direction of ATP and sulfate ($K = 10^{-8}$) and it is probably necessary to pull the reaction in the direction of APS formation by hydrolyzing $PP_i$ with pyrophosphatase [reaction (2)].

$$PP_i + H_2O \rightarrow 2\,P_i \tag{2}$$

The second step in the activation of sulfate is the phosphorylation of APS in the 3′- position to yield PAPS and ADP [reaction (3)].

$$APS + ATP \rightleftharpoons PAPS + ADP \tag{3}$$

The enzyme responsible for this phosphorylation has been named APS kinase or ATP: adenyl sulfate 3′-phosphotransferase and again is a $Mg^{++}$ dependent reaction. The overall reaction for the synthesis of PAPS proceeds readily and shows no calculated change in free energy. This pathway is of particular importance as it is the only known mechanism by which sulfate can enter into biosynthetic or energy-yielding reactions.

**Formation of Sulfate Esters**

Sulfate itself is one of the important nutritional forms of sulfur as sulfate esters occur in a number of compounds of extreme biological importance such as chondroitin sulfates, heparin, dermatan sulfate, steroid sulfates and sulfolipids. Our chemical and biochemical knowledge of these esters is scanty and in microorganisms, except for choline sulfate (Itahashi 1961), sulfolipids (Mayers *et al.* 1969), and a new sulfonic acid in spores (Bonsen *et al.* 1969), little is directly known about the importance of these esters of sulfate. In fact, studies on the biosynthesis of cysteine involving both mutants and repression suggest that sulfate may not be a nutritionally important form of sulfur for certain bacteria or that there may be alternative pathways for the formation of sulfate esters. At present, the only known mechanism for the synthesis of sulfate esters involves the transfer of the sulfate group from PAPS to an acceptor molecule in the presence of specific enzymes, termed sulfokinases, as shown in reaction (4) where X-OH is any of a number of acceptor molecules. It remains ques-

$$X\text{-}OH + PAPS \rightleftharpoons PAP + XOSO_3^- \tag{4}$$

tionable as to whether these reactions can be considered to be reversible. If there exist storage forms of sulfate, such as choline sulfate, one might expect the reaction to be reversible although Spencer *et al.* (1968) were unable to detect the formation of PAPS from this compound. Although Hilz *et al.* (1959) were able to observe the reduction of sulfate from *p*-nitrophenylsulfate, Gregory and Lipmann did not directly demonstrate the formation of PAPS in their initial studies with phenol sulfokinase. This problem plus the finding of new sulfate-containing compounds such as ascorbic acid sulfate (Mead and Finamore 1969) suggest that this may well be an exciting area for future research efforts.

**Assimilatory Sulfate Reduction**

Results from the laboratories of Hilz and Bandurski first identified PAPS as the form in which sulfate was reduced by cell-free extracts from yeast. These workers also identified the electron donor as NADPH and established the overall reaction for the enzyme(s) commonly called PAPS reductase as shown in reaction (5).

$$\text{PAPS} + \text{NADPH} + \text{H}^+ \rightarrow \text{PAP} + \text{SO}_3^= + \text{NADP}^+ \quad (5)$$

Wilson and Bandurski have expanded their original observations and demonstrated that three proteins are required for the reduction of the sulfate moiety of PAPS and the intermediary role of FAD in the transfer of electrons from reduced NADP (Asahi *et al.* 1961). One of these is a low molecular weight electron carrier, perhaps similar to thioredoxin (Thelander 1967), which accepts electrons from reduced FAD; a second protein (enzyme A) which catalyzes the transfer of electrons from FAD to the electron transfer protein and PAPS reductase (enzyme B). The product of the reduction does not appear to be free sulfite but rather a bound derivative of sulfite (Torii and Bandurski 1967). The reductase is not specific for PAPS but shows considerable activity for APS (Wilson *et al.* 1961). In addition, the reversibility of the reaction, that is the formation of PAPS from PAP and sulfite, has not as yet been demonstrated. As indicated previously (Peck 1961), PAPS appears to be the form of sulfate that is reduced by most organisms and is characteristic of assimilatory sulfate reduction. More recently PAPS reductase has been studied in a number of different organisms (Schiff and Levinthal 1968; Dreyfuss and Monty 1963; Jones-Mortimer 1968) and found to be under close but complicated metabolic controls. Cysteine generally causes the repression of the entire pathway in both *Salmonella typhimurium* and *Escherichia coli*. Cysteine-less mutants have been extensively studied and the growth of mutants lacking the ability to form PAPS and repression of these enzymes by cysteine suggest that sulfate per se is not important in the nutrition of these organisms, or that there exists another pathway for its direct utilization.

The further metabolism of sulfite to sulfide requires a single enzyme or enzyme complex, termed sulfite reductase, that catalyzes this six electron reduction [reaction (6)] of sulfite to sulfide in a single step. Yeast and

$$\text{SO}_3^= + 3\,\text{NADPH} + 3\,\text{H}^+ (6\text{MVH}) \rightarrow \text{S}^= + 3\,\text{H}_2\text{O} + 3\,\text{NADP}^+ (6\text{MV}) \quad (6)$$

certain bacterial reductases can utilize NADPH as an electron donor while the reductases of certain mutants and higher plants utilize only the reduced form of the dye, methyl viologen. The latter are also lacking in flavin prosthetic groups. Although the NADP-enzyme has been prepared as a high molecular weight homogenous protein, it appears to consist of two or more subunits, one of which contains a heme moiety and is responsible for the reduction of sulfite and another which contains flavins and is responsible for the transfer of electrons from the reduced pyridine nucleotide to sulfite reductase. It has been reported that the sulfite reductase component can be further fractionated (Asada *et al.* 1966). The enzyme also catalyzes the reduction of nitrite and hydroxylamine but repression of the reductase by cysteine strongly indicates that sulfite reduction is the physiological function of the enzyme (Siegel and Kamin 1968; Yoshimoto and Sato 1968; Tamura 1965; Asada 1968; Prabhakararao and Nicholas 1969). The wide distribution of this enzyme in species capable of reducing sulfate and its coincident repression by cysteine along with PAPS reductase indicate that it is involved in the assimilatory pathway of sulfate reduction. No free intermediates have been detected between sulfite and sulfide, and this has led to the concepts that either intermediates are enzyme-bound or that the reduction occurs in a single step. Many organisms have the ability to reduce thiosulfate, presumably by thiosulfate reductase (Kaji and McElroy 1959) and a number of observations suggest that thiosulfate or a bound derivative may be involved in sulfite reduction (Dreyfuss and Monty 1963; Levinthal and Schiff 1968; Findley and Akagi 1969). Although sulfite reductase is one of the most extensively studied enzymes in the pathway of assimilatory sulfate reduction, considerably more work will be required before its mechanism of action is understood.

## Sulfide Utilization

Sulfide can be incorporated into amino acids by a large number of organisms, including mammals (Huovinen and Gustafusson 1967), but it does not accumulate in appreciable quantities during assimilatory sulfate reduction. For a number of years it was believed that the pyridoxal dependent enzyme, cysteine synthase [reaction (7)] first purified by Schlossmann and Lynen (1957), was responsible for the incorporation of sulfide by the synthesis of cysteine from serine and sulfide.

$$\text{L-serine} + H_2S \rightleftharpoons \text{L-cysteine} + H_2O \qquad (7)$$

Although sulfide can be incorporated by means of this reaction, observations on the Km for sulfide and genetic analysis for this enzyme presumably involved in the biosynthesis of cysteine suggested that this was not the sole or most important pathway for the incorporation of sulfide by microorganisms. Subsequently, Kredich and Tomkins (1966) described a two-step pathway involving *o*-acetyl-L-serine from *E. coli* and *S. typhimurium* [reactions (8) and (9)]. The first enzyme has been given the trivial name of serine transacetylase, and the second, *o*-acetyl serine sulfhydrylase.

$$\text{L-serine} + \text{acetyl CoA} \rightarrow o\text{-acetyl-L-serine} + \text{CoA} \quad (8)$$

$$o\text{-acetyl-L-serine} + H_2S \rightarrow \text{L-cysteine} + H_2O + \text{acetate} \quad (9)$$

The properties and occurrence of these enzymes can be reconciled with the genetic and physiological behavior of whole cells and thus these reactions appear to constitute the major pathway for cysteine biosynthesis. These enzymes have been further characterized and have been reported to occur as a bifunction protein complex in *S. typhimurium* (Kredich *et al.* 1969; Becker *et al.* 1969). The finding of these two enzymes in plants indicates that this pathway may be the principal mechanism for the assimilation of sulfide in biological material (Giovanelli and Mudd 1967; Smith and Thompson 1969). It also should be noted that *o*-acetyl serine has been reported to have a regulatory function in the synthesis of enzymes responsible for assimilatory sulfate reduction in *E. coli* (Jones-Mortimer 1968).

Although the activity of the assimilatory pathway is usually low, in a given organism, the widespread occurrence of this pathway of sulfate reduction in plants and bacteria probably makes it quantitatively, the most important source of reduced sulfur in nature. It is extremely interesting from the aspect of comparative biochemistry and further studies on the regulation and enzymology of this pathway should shortly elevate our understanding of assimilatory sulfate reduction to a level comparable to that of other biosynthetic pathways.

**Dissimilatory Sulfate Reduction**

The ability to utilize sulfate as a terminal electron acceptor in anaerobic respiration is limited to two genera of bacteria, *Desulfovibrio* and *Desulfotomaculum*. These bacteria can reduce sulfate, sulfite, thiosulfate and certain polythionates with molecular hydrogen or a limited number of organic electron donors. The reaction for the reduction of sulfate is shown in reaction (10).

$$4H_2 + SO_4^{=} \rightarrow S^{=} + 4H_2O \quad (10)$$

The sulfate-reducing bacteria are widely distributed in nature and largely responsible for the massive formations of hydrogen sulfide that occur in a number of ecological situations. This formation of sulfide is also responsible for a number of economically important activities that are harmful to man and other aerobic organisms. An excellent discussion of these relationships can be found in a review by Postgate (1960).

The pathway for dissimilatory sulfate reduction in these organisms requires the initial activation of the sulfate ion as in assimilatory sulfate reduction; however, APS (Fig. 6.1) rather than PAPS is the form in which the sulfate is reduced (Ishimoto 1959; Peck 1959). The enzymes involved have been separated and the pathway established as shown in reactions (11), (12), and (13).

$$ATP + SO_4^{=} \rightleftharpoons APS + PP_i \tag{11}$$

$$PP_i + H_2O \rightarrow 2P_i \tag{12}$$

$$APS + 2e \rightarrow AMP + SO_3^{=} \tag{13}$$

The first enzyme in the pathway, ATP sulfurylase, has been purified by Akagi and Campbell (1962) and appears to be comparable to similar enzymes found in other organisms. Although the exact nature of the electron donor is unknown, cytochrome $c_3$, ferredoxin, and flavodoxin stimulate electron transfer between hydrogen and sulfate in various cell-free extracts of these microorganisms. Under experimental conditions, reduced methyl viologen is usually utilized as the electron donor. The enzyme catalyzing the reductive step has been named APS reductase and extensively purified by Peck *et al.* (1965). It was shown to be an iron containing flavoprotein and, in contrast to PAPS reductase, the reduction has been shown to be reversible in the presence of artificial electron acceptors. Extracts lack APS kinase, and neither the extracts nor purified enzyme show any activity toward PAPS reduction. The first steps in dissimilatory sulfate reduction are certainly analogous to those in assimilatory sulfate reduction but the enzyme catalyzing the reductive step, APS reductase, seems, at the present state of our knowledge, to be quite different from PAPS reductase.

Realization of the ATP-requirement for sulfate reduction and examination of reaction (10) for the reduction of sulfate with hydrogen, requires that biologically utilizable energy in the form of ATP must be generated during the oxidation of hydrogen with sulfate. Phosphorylation coupled to electron transfer occurring during the oxidation of hydrogen has been demonstrated in cell-free extracts of one of these organisms, *Desulfovibrio gigas*. It is uncoupled by low concentrations of established uncoupling agents and appears to be quite similar to cell-free systems from aerobic organisms capable of carrying out oxidative phosphorylation. This was anticipated and is completely consistent with the respiratory nature of dissimilatory sulfate reduction (Peck 1966).

Whole cells and extracts of the sulfate-reducing bacteria reduce sulfite with hydrogen, as shown in reaction (14), and it has been assumed by many

$$3H_2 + SO_3^{=} \rightarrow S^{=} + 3H_2O \tag{14}$$

authors that this pathway is identical to the pathway of sulfite reduction in assimilatory sulfate reduction. Although this may eventually prove to be correct, the evidence available at the present time does not justify this assumption. The work of Woolfolk (1962) suggested an inorganic pathway of sulfite reduction which would explain the constant ability of whole cells and extracts of these organisms to reduce thiosulfate. As shown in Fig. 6.3,

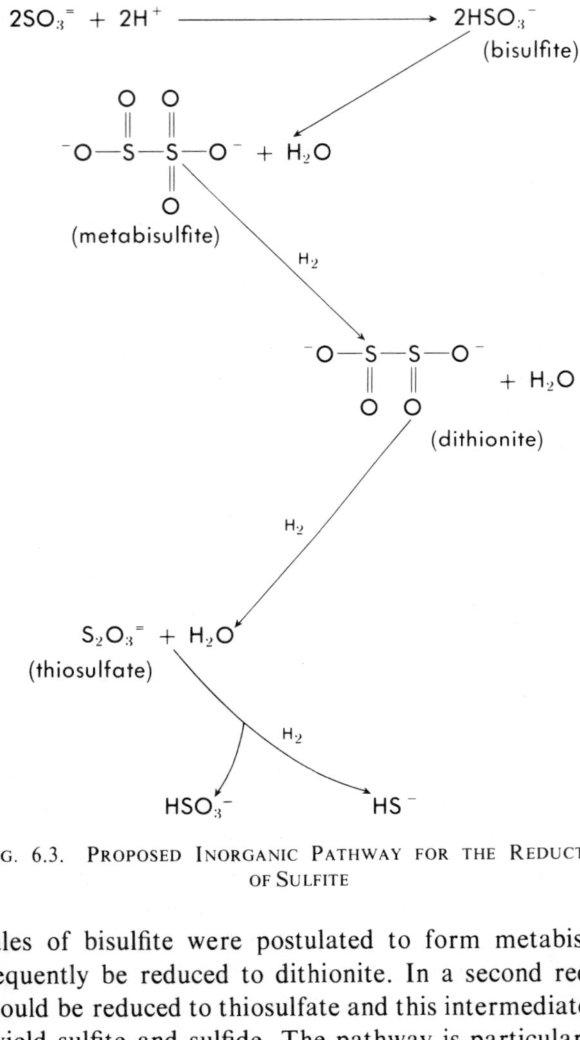


FIG. 6.3. PROPOSED INORGANIC PATHWAY FOR THE REDUCTION OF SULFITE

two molecules of bisulfite were postulated to form metabisulfite which would subsequently be reduced to dithionite. In a second reductive step, dithionite would be reduced to thiosulfate and this intermediate reductively cleaved to yield sulfite and sulfide. The pathway is particularly attractive as it obviates the necessity of assuming a six electron reduction or entirely enzyme-bound intermediates and provides a rationale for thiosulfate reduction.

The sulfite reductase from the sulfate reducing bacteria can utilize reduced methyl viologen as electron donor but exhibits no activity with reduced pyridine nucleotide and is thus comparable to the sulfite reductases found in plants (Ishimoto and Yagi 1961). Cytochrome $c_3$ stimulates the oxidation of hydrogen with sulfite as electron acceptor and, more recently, both ferredoxin (Akagi 1965; LeGall and Dragoni 1966) and flavodoxin (LeGall and Hatchikian 1967) have been shown to stimulate this reaction in extracts. Nakatsukasa and Akagi (1969) have partially purified and separated thiosulfate reductase from sulfite reducing activity. Kobayashi *et al.* (1969) and Findley and Akagi (1969) have obtained preparations from these bacteria which consume hydrogen in the presence of sulfite but do not produce sulfide and appear to be accumulating one or more intermediates. The latter investigators have identified the intermediate accumulating as trithionate while the former have reported the accumulation of thiosulfate. These observations are not irreconcilable and suggest the actual operation of the inorganic pathway of sulfite reduction shown in Fig. 6.4. The

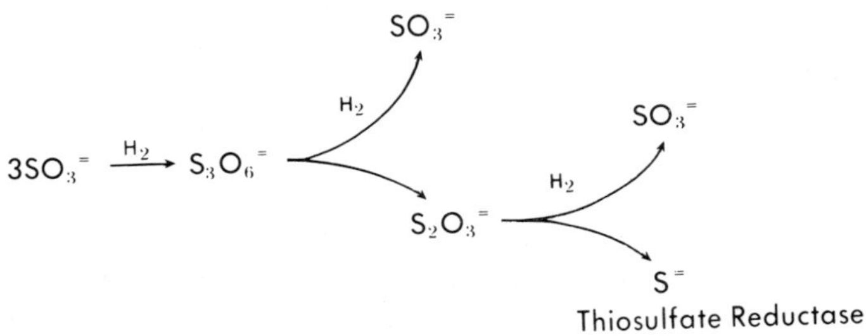


FIG. 6.4. TRITHIONATE PATHWAY OF SULFITE REDUCTION

only enzyme that has been studied to any extent is the thiosulfate reductase. These recent reports suggest that the entire pathway of dissimilatory sulfate may be different from the pathway of assimilatory sulfate reduction.

Several investigators have speculated on the primitive nature and evolutionary significance of dissimilatory sulfate reduction and its possible role in the evolution of aerobic respiration (Peck 1967; Postgate 1968). These ideas lead to the assumption that the assimilatory pathway has evolved from the dissimilatory pathway and to the possibility that further biochemical investigations of these pathways will discover more similarity than is evident at the present time. If these speculations are proven to have some reality, the study of the enzymes involved in sulfate reduction will be extremely exciting from the standpoint of biochemical evolution.

## SULFUR AND SULFIDE OXIDATION

### General Considerations

In addition to being continually produced by biological sulfate reduction and putrefaction, sulfide in the form of pyrites and elemental sulfur occur as components of the earth's crust. In contrast to the reduction of sulfate, the oxidation of sulfide is chemically a relatively easy reaction and in alkaline solutions elemental sulfur and more oxidized forms of sulfur are produced from sulfide in the presence of oxygen. This instability has complicated the biochemical study of sulfur oxidation and when the oxidation of sulfide is reported to occur in a tissue or extract, it is not always clear whether this is the result of a specific or nonspecific reaction. Our most detailed and reliable information concerning the biological pathway of sulfide and sulfur oxidation has come from the study of microorganisms that specifically utilize reduced sulfur compounds as electron donors in their respiratory metabolism. The primary physiological groups involved are the thiobacilli which utilize oxygen, but in one case nitrate, as terminal electron acceptor [reaction (15)] and

$$S_2O_3^{=} + 2O_2 + H_2O \rightarrow 2SO_4^{=} + 2H^+ \qquad (15)$$

the green and purple sulfur photosynthetic bacteria [reaction (16)].

$$2CO_2 + S_2O_3^{=} + 3H_2O \xrightarrow{h\nu} 2(CH_2O) + 2SO_4^{=} + 2H^+ \qquad (16)$$

Although with the latter organisms, the oxidation of reduced sulfur compounds is usually tightly coupled with carbon dioxide fixation or results in the production of molecular hydrogen, the end product of these oxidations is normally sulfate. The enzyme systems found in the thiobacilli have been most intensively studied and described as the general pathway for sulfide and sulfur oxidation. The pathway of sulfur oxidation in the photosynthetic bacteria appears to be identical with that in the thiobacilli (Thiele 1968) and an excellent consideration of these bacteria can be found in a recent review by Schlegel and Jannasch (1967).

The thiobacilli are a relatively small group of bacteria that very early attracted the attention of microbiologists because of their simple nutritional requirements: inorganic salts, reduced sulfur compounds as electron donors, oxygen as their terminal electron acceptor and carbon dioxide as their sole source of carbon. This growth in a completely inorganic media and toleration of the very acid environment produced during the oxidation of reduced sulfur compounds led early microbiologists to speculate that these organisms were very special forms of life. Today they are generally regarded as being among the most biosynthetically competent of bacteria plus possessing specific enzyme systems for the utilization of reduced sulfur compounds as electron donors. However, it should be expected that there has been some modification of regulatory mechanisms and pathways, such as the TCA cycle, as a requisite for the autotrophic or chemosynthetic mode of existence (Smith *et al.* 1967).

The thiobacilli are capable of oxidizing a number of different reduced compounds of sulfur. Most important of these are sulfide, thiosulfite, sulfite, polythionates ($S_{n+2}O_6^{=}$), and elemental sulfur.

### Oxidation of Elemental Sulfur

Study of specific oxidations in different species of the thiobacilli has led to the postulation of a number of different pathways, i.e., the polythionate pathway, the thiosulfate pathway. The thiobacilli do not appear to be a homogenous group of bacteria and there are probably some differences in the pathways among different species. Suzuki and Silver (1966) have recently isolated and studied an enzyme from *Thiobacillus thioparus* that catalyzes the oxidation of elemental sulfur to sulfite [reaction (17)].

$$S + O_2 + H_2O \rightarrow SO_3^{=} + 2\,H^+ \quad (17)$$

This appears to be the central reaction in all pathways for the oxidation of reduced inorganic sulfur compounds and emphasizes the key role of sulfite in these oxidations. This enzyme has been purified about 25-fold and shown to require reduced glutathione for activity. It is probably a mixed function oxygenase as oxygen-18 from molecular oxygen is incorporated into its products (Suzuki 1965) and it contains nonheme iron. Furthermore, the oxidation can occur under anaerobic conditions. The formation of hydrogen peroxide and absence of cytochrome in the purified enzyme indicates that this oxidation is not coupled to the formation of ATP by oxidative phosphorylation. This reaction has been reported to occur in other thiobacilli (Adair 1966; Silver and Lundgren 1968) and other sulfur oxidizing systems described by Adair (1966) and Taylor (1968) can be resolved with the pathway of Suzuki and Silver by assuming the presence of a bound naturally occurring thiol which can be replaced by reduced glutathione in accepting $S_8$ to form, probably, a polysulfide.

### Reactions of Sulfite

Sulfite can be oxidized directly to the level of sulfate by two pathways, one of these, the APS pathway, involving a substrate phosphorylation.

A cytochrome-linked sulfite oxidase [reaction (18)] has been studied in extracts of *T. novellus* by Charles and Suzuki (1965), and *T. neapolitanus* by Hempfling and Trudinger (1967). This is a direct oxidation of sul-

$$SO_3^{=} + \tfrac{1}{2}O_2 \rightarrow SO_4^{=} \quad (18)$$

fite and does not involve a substrate phosphorylation. However, the reaction is cytochrome-linked and electrons derived from the oxidation are probably involved in oxidative phosphorylation.

The second pathway for the direct oxidation of sulfite, the APS pathway, results in the formation of high energy phosphate from a substrate phosphorylation in addition to being cytochrome-linked and thus probably participating in oxidative phosphorylation [reactions (19), (20) and (21)].

$$2SO_3^{=} + 2AMP + 4\ \text{cytochrome ox} \rightleftharpoons 2APS + 4\ \text{cytochrome red} \quad (19)$$

$$2APS + 2P_i \rightleftharpoons 2SO_4^{=} + 2ADP \quad (20)$$

$$2ADP \rightleftharpoons ATP + AMP \quad (21)$$

In the first step, sulfite is oxidized to APS in the presence of AMP by APS reductase, the same enzyme required in the first step of dissimilatory sulfate reduction (Bowen *et al.* 1966). Cytochrome has been shown to function as electron acceptor for this oxidation and this indicates that electrons from the oxidation probably participate in oxidative phosphorylation. APS then undergoes phosphorolysis catalyzed by the enzyme, ADP sulfurylase [reaction (20)] to yield ADP and sulfate. ATP can be produced for use in biosynthetic reactions by the action of adenylic kinase on ADP [reaction (21)]. The APS pathway occurs simultaneously with the sulfite oxidase pathway and their relationship remains obscure. Although the APS pathway does not occur in all thiobacilli, in certain species it appears to be the major or sole mechanism operating in the oxidation of sulfite [see Peck (1962) and (1968) for a more complete discussion]. This is the only instance where a substrate phosphorylation occurs in connection with the oxidation of an inorganic compound and probably permits the thiobacilli to obtain more energy than would be available solely from oxidative phosphorylation.

A third reaction of sulfite which appears to be physiological is its nonenzymatic reaction with elemental sulfur to yield thiosulfate. The function of this reaction is unknown but it does represent a third mechanism for the utilization of sulfite and may be important in maintaining sulfite at nontoxic levels [reaction (22)]. This reaction proceeds so readily that for a

$$SO_3^{=} + S^0 \rightarrow SSO_3^{=} \quad (22)$$

number of years thiosulfate was considered to be the first product in the oxidation of elemental sulfur.

### Metabolism of Thiosulfate

Thiosulfate can either be reductively cleaved to yield sulfite and sulfide (or a bound derivative) or oxidized via the polythionate pathway. The cleavage of thiosulfate is probably catalyzed by the enzyme, rhodanase [reaction (23)] and its occurrence has been reported in several species of thiobacilli.

$$SSO_3^{=} + CN^{-} \rightarrow SO_3^{=} + SCN^{-} \quad (23)$$

The naturally occurring acceptor has not been identified but may be identical to the bound thiol involved in the oxidation of elemental sulfur. This enzyme is assumed to be utilized in oxidations of thiosulfate that do not involve the formation of polythionates.

Alternatively, thiosulfate can be oxidized to tetrathionate and other polythionates via the so called polythionate pathway (London and Rittenberg 1964) shown in Fig. 6.5. The step from thiosulfate to tetrathionate is

$$4S^{=} \longrightarrow 2S_2O_3^{=} \longrightarrow S_4O_6^{=} \longrightarrow SO_3^{=} + S_3O_5^{=} \longrightarrow 4SO_3^{=} \longrightarrow 4SO_4^{=}$$

FIG. 6.5. POLYTHIONATE PATHWAY

cytochrome linked and it may serve as an alternative pathway for electrons to enter into oxidative phosphorylation. The operation of the polythionate pathway, indicated by the accumulation of polythionates, is extremely dependent on growth conditions and its significance has been widely questioned because of a general lack of enzymological studies.

**Oxidation of Sulfide**

The pathway for the oxidation of sulfide is somewhat in doubt due to its instability towards molecular oxygen. It certainly can be oxidized to elemental sulfur and can be oxidized to thiosulfate by the thiobacilli and other tissues. However, this may involve the sulfur-oxidizing system of Suzuki and the intermediate formation of sulfite. The relationship of various compounds between sulfide and sulfite are shown in Fig. 6.6. The major assumption is the existence at the level of elemental sulfur of a thiol-carrier which is the actual substrate for the sulfur-oxidizing enzyme and the product of both sulfide oxidation and the metabolism of thiosulfate by rhodanase.

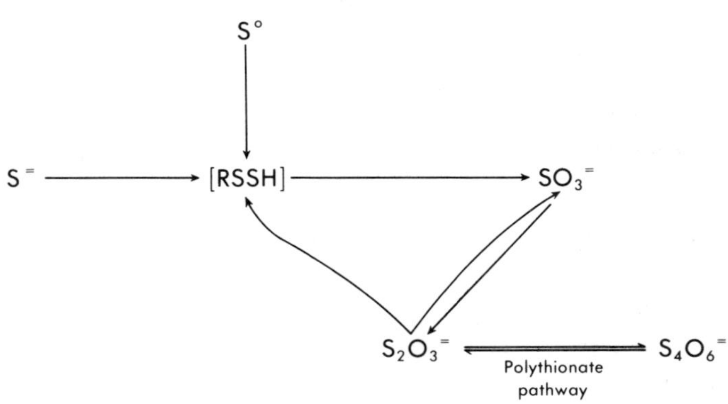


FIG. 6.6. METABOLISM OF REDUCED SULFUR COMPOUNDS

## CONCLUSION

The reactions that have been discussed constitute the sulfur cycle and account for the metabolism of sulfur compounds in nature. The reactions in the cycle are in part catalyzed by microorganisms whose nutrition and physiology are extremely interesting in their own right, i.e., the sulfate-reducing bacteria, the chemosynthetic thiobacilli and the photosynthetic sulfur bacteria. It is also curious that the reactions involved in the reduction of sulfate seem to be so similar to those involved in the oxidation of sulfide. Aside from various biochemical and enzymological questions, the most important nutritional question involving the sulfur cycle concerns the possibility that the cycle can serve in the presence of light as a primary system for carbon dioxide fixation during the growth of sulfate-reducing bacteria and photosynthetic sulfur bacteria in a closed system.

## BIBLIOGRAPHY

ADAIR, F. W. 1966. Membrane-associated sulfur oxidation by the autotroph *Thiobacillus thiooxidans*. J. Bacteriol. *92*, 899–904.

AKAGI, J. M. 1965. The participation of a ferredoxin of *Clostridium nigrificans* in sulfite reduction. Biochem. Biophys. Res. Commun. *21*, 72–77.

AKAGI, J. M., and CAMPBELL, L. L. 1962. Studies on thermophilic sulfate-reducing bateria. III. Adenosine triphosphate-sulphurylase of *Clostridium nigrificans* and *Desulfovibrio desulfuricans*. J. Bacteriol. *84*, 1194–1201.

ASADA, K., TAMURA, G., and BANDURSKI, R. S. 1966. Cleavage of sulfite reductase into two non-dialyzable components and stimulation of reductase activity by nucleotides. Biochem. Biophys. Res. Commun. *25*, 529–534.

ASADA, K., TAMURA, G., and BANDURSKI, R. S. 1968. Methyl-viologen linked sulfite reductase from spinach leaves: A hemoprotein. Biochem. Biophys. Res. Commun. *30*, 554–559.

ASAHI, T., BANDURSKI, R. S., and WILSON, L. G. 1961. Yeast sulfate-reducing system. II. Enzymatic reduction of protein disulfide. J. Biol. Chem. *236*, 1830–1835.

BECKER, M. A., KREDICH, N. M., and TOMPKINS, G. M. 1969. The purification and characterization of *O*-acetylserine sulfhydrylase A from *Samonella typhimurium*. J. Biol. Chem. *244*, 2418–2427.

BONSEN, P. P. M., SPUDICH, J. A., NELSON, D. L., and KORNBERG, A. 1969. Biochemical studies of bacterial sporulation and germination. XII. A sulfonic acid as a major sulfur compound of *Bacillus subtilis* spores. J. Bacteriol. *98*, 62–68.

BOWEN, T. J., HAPPOLD, F. C., and TAYLOR, B. F. 1966. Studies on adenosine-5′-phosphosulphate reductase from *Thiobacillus denitrificans* Biochim. Biophys. Acta *118*, 566–576.

CHARLES, A. M., and SUZUKI, I. 1965. Sulfite oxidase of a facultative autotroph, *Thiobacillus novellus*. Biochem. Biophys. Res. Commun. *19*, 686–690.

DREYFUSS, J., and MONTY, K. J. 1963. Coincident repression of the reduction of 3′-phosphoadenosine 5′-phosphosulfate, sulfite and thiosulfate in the cysteine pathway of *Salmonella typhimurium*. J. Biol. Chem. *238*, 1019–1024.

FINDLEY, J. E., and AKAGI, J. M. 1969. Evidence for thiosulfate formation during sulfite reduction by *Desulfovibrio vulgaris*. Biochem. Biophys. Res. Commun. *36*, 266–271.

GIOVANELLI, J., and MUDD, S. H. 1967. Synthesis of homocysteine and cysteine by enzyme extracts of spinach. Biochem. Biophys. Res. Commun. *27*, 150–156.

GREGORY, J. D., and LIPMANN, F. 1957. The transfer of sulfate among phenolic compounds with 3′,5′-diphosphoadenosine as coenzyme. J. Biol. Chem. *229*, 1081–1090.

GREGORY, J. D., and ROBBINS, P. W. 1960. Metabolism of sulfur compounds. Ann. Rev. Biochem. *29*, 347–364.

HEMPFLING, W. P., and TRUDINGER, P. A. 1967. Purification and some properties of sulfite oxidase from *Thiobacillus neapolitanus*. Arch. Mikrobiol. *59*, 149–157.

HENRY, S. M., and BLOCK, R. J. 1969. The sulfur metabolism of insects. IV. The conversion of inorganic sulfate to organic sulfur compounds in cockroaches. The role of intracellular symbionts. Contrib. Boyce Thompson Inst. *20*, 317–330.

HILZ, H., KITTLER, M., and KNAPE, G. 1959. The reduction of sulfate in yeast. Biochem. Z. *332*, 151–166. (German)

HOUVINEN, J. A., and GUSTAFSSON, B. F. 1967. Inorganic sulphate, sulphite and sulphide as sulphur donors in the biosynthesis of sulphur amino acids in germ-free and conventional rats. Biochim. Biophys. Acta *136*, 441–447.

ISHIMOTO, M. 1959. Sulfate reduction in cell-free extracts of *Desulfovibrio*. J. Biochem. (Tokyo) *49*, 103–109.

ISHIMOTO, M., and YAGI, T. 1961. Biochemical studies on sulfate-reducing bacteria. IX. Sulfite reductase. J. Biochem. (Tokyo) *49*, 103–109.

ITAHASHI, M. 1961. Comparative biochemistry of choline sulfate metabolism. J. Biochem. (Tokyo) *50*, 52–61.

JONES-MORTIMER, M. C. 1968. Positive control of sulphate reduction in *Escherichia coli*. Isolation, characterization and mapping of cysteineless mutants of *E. coli* K 12. Biochem. J. *110*, 589–595.

JONES-MORTIMER, M. C., WHELDRAKE, J. F., and PASTERNAK, C. A. 1968. The control of sulphate reduction in *Escherichia coli* by *O*-acetyl-*L*-serine. Biochem. J. *107*, 51–53.

KAJI, A., and MCELROY, W. D. 1959. Mechanism of hydrogen sulfide formation from thiosulfate. J. Bacteriol. *77*, 630–637.

KELLY, D. P. 1968. Biochemistry of oxidation of inorganic sulphur compounds by microorganisms. Australian J. Sci. *31*, 165–173.

KOBAYASHI, K., TACHIBANA, S., and ISHIMOTO, M. 1969. Intermediary formation of trithionate in sulfite reduction by a sulfate-reducing bacterium. J. Biochem. (Tokyo) *65*, 155–157.

KOJ, A. 1968. Enzymic reduction of thiosulphate in preparations from beef liver. Acta Biochim. Polon. *15*, 161–169.

KREDICH, N. M., BECKER, N. A., and TOMKINS, G. M. 1969. Purification and characterization of cysteine synthetase, a bifunctional protein complex, from *Salmonella typhimurium*. J. Biol. Chem. *244*, 2428–2439.

KREDICH, N. M., and TOMKINS, G. M. 1966. The enzymic synthesis of *L*-cysteine in *Escherichia coli* and *Salmonella typhimurium*. J. Biol. Chem. *241*, 4955–4965.

LANHAM, U. M. 1968. The Blockmann Bodies: Hereditary intracellular symbionts of insects. Biol. Rev. *43*, 269–286.

LEGALL, J., and DRAGONI, N. 1966. Dependence of sulfite reduction on a crystallized ferredoxin from *Desulfovibrio gigas*. Biochem. Biophys. Res. Commun. *23*, 145–149.

LEGALL, J., and HATCHIKIAN, E. C. 1967. Purification and properties of a flavoprotein functioning in the reduction of sulfite by the *Desulfovibrio gigas*. Comptes Rendus *264*, 2580–2583. (French)

LEVINTHAL, M., and SCHIFF, J. A. 1968. Studies of sulfate utilization by algae. V. Identification of thiosulfate as a major acid-volatile product formed by a cell-free sulfate reducing system from *Chlorella*. Plant Physiol. *43*, 555–562.

LONDON, J., and RITTENBERG, S. C. 1964. Path of sulfur in sulfide and thiosulfate oxidation by thiobacilli. Proc. Natl. Acad. Sci. *52*, 1183–1190.

MAYERS, G. L., POUSADA, M., and HAINES, T. H. 1969. Microbial sulfolipids. III. The disulfate of ( + ) -1,14-docosanediol in *Orchromonas danica*. Biochemistry *8*, 2981–2986.

MEAD, C. A., and FINAMORE, F. J. 1969. The occurrence of ascorbic acid sulfate in the brine shrimp *Artemia salina*. Biochemistry *8*, 2652–2655.

NAGY, Z., KARI, C., and HERNADI, F. 1969. Growth of *Escherichia coli* in the presence of cysteine on sulphate-deficient media. Arch. Mikrobiol. *65*, 391–400.

NAKATSUKASA, W., and AKAGI, J. M. 1969. Thiosulfate reductase isolated from *Desulfotomaculum nigrificans*. J. Bacteriol. *98*, 429–433.

PECK, H. D., JR., 1959. The ATP-dependent reduction of sulfate with hydrogen in extracts of *Desulfovibrio desulfuricans*. Proc. Natl. Acad. Sci. *45*, 701–708.

PECK, H. D., JR., 1961. Enzymatic basis for assimilatory and dissimilatory sulfate reduction. J. Bacteriol. *82*, 933–939.

PECK, H. D., JR. 1962. Comparative metabolism of inorganic sulfur compounds. Bacteriol. Rev. *26*, 67–94.

PECK, H. D., JR. 1966. Phosphorylation coupled with electron transfer in extracts of the sulfate-reducing bacterium, *Desulfovibrio gigas*. Biochem. Biophys. Res. Commun. *22*, 112–118.

PECK, H. D., JR. 1967. Some evolutionary aspects of inorganic sulfur metabolism. Lectures on theoretical and applied aspects of modern microbiology. Univ. Maryland, College Park, Md.

PECK, H. D., JR. 1968. Energy-coupling mechanisms in chemolithotrophic bacteria. Ann. Rev. Microbiol. *22*, 489–518.

PECK, H. D., JR., DEACON, T. E., and DAVIDSON, J. T. 1965. Studies on adenosine 5′-phosphosulfate reductase from *Desulfovibrio desulfuricans* and *Thiobacillus thioparus*. I. Assay and purification. Biochim. Biophys. Acta *96*, 429–446.

POSTGATE, J. R. 1960. The economic activities of sulfate reducing bacteria. Progr. Ind. Microbiol. *2*, 49–69.

POSTGATE, J. R. 1965. Recent advances in the study of sulphate-reducing bacteria. Bacteriol. Rev. *29*, 425–441.

POSTGATE, J. R. 1968. Fringe biochemistry among microbes. Proc. Roy. Soc. (London) Series B *171*, 67–76.

POSTGATE, J. R. 1969. The sulfur cycle. *In* Inorganic Sulphur Chemistry, G. Nickless (Editor). Elsevier Publishing Co., New York.

PRABHAKARARAO, K., and NICHOLAS, D. J. D. 1969. Sulphite reductase from baker's yeast: a haemoflavoprotein. Biochim. Biophys. Acta *180*, 253–263.

SCHIFF, J. A., and LEVINTHAL, M. 1968. Studies of sulfate utilization of algae. IV. Properties of a cell-free sulfate reducing system from *Chlorella*. Plant Physiol. *43*, 547.

SCHLEGEL, H. G., and JANNASCH, H. W. 1967. Enrichment cultures. Ann. Rev. Microbiol. *21*, 49–70.

SCHLOSSMANN, K., and LYNEN, F. 1957. Biosynthesis of cysteine from serine and hydrogen sulfide. Biochem. Z. *328*, 591–594. (German)

SIEGEL, L. M., and KAMIN, H. 1968. *E. coli* TPNH-sulfite reductase. *In* Flavins and Flavoproteins, K. Yagi (Editor). Univ. Tokyo Press.

SILVER, M., and LUNDGREN, D. G. 1968. Sulfur-oxidizing enzyme of *Ferrobacillus ferrooxidans* (*Thiobacillus ferrooxidans*). Can. J. Biochem. *46*, 457–461.

SMITH, A. J., LONDON, J., and STANIER, R. Y. 1967. Biochemical basis of obligate autotrophy in blue-green algae and thiobacilli. J. Bacteriol. *94*, 972–983.

SMITH, I. K., and THOMPSON, J. F. 1969. The synthesis of *O*-acetylserine by extracts prepared from higher plants. Biochem. Biophys. Res. Commun. *35*, 935–945.

SPENCER, B., HUSSEY, E. C., ORSI, B. A., and SCOTT, J. M. 1968. Mechanism choline *O*-sulphate utilization in fungi. Biochem. J. *106*, 461–470.

SUZUKI, I. 1965. Incorporation of atmospheric oxygen-18 into thiosulfate by the sulfur-oxidizing enzyme of *Thiobacillus thiooxidans*. Biochim. Biophys. Acta *110*, 97–101.

SUZUKI, I., and SILVER, M. 1966. The initial product and properties of the sulfur-oxidizing enzyme of thiobacilli. Biochim. Biophys. Acta *122*, 22–33.

TAMURA, G. 1965. Studies on the sulfite reducing system of higher plants. II. Purification and properties of sulfite reductase from *Allium ordorum*. J. Biochem. (Tokyo) *57*, 207–214.

TAYLOR, B. F. 1968. Oxidation of elemental sulfur by an enzyme system from *Thiobacillus neapolitanus*. Biochim. Biophys. Acta *170*, 112–122.

THELANDER, L. 1967. Thioredoxin reductase. Characterization of a homogeneous preparation from *Escherichia coli* B. J. Biol. Chem. *242*, 852–859.

THIELE, H. H. 1968. The sulfur metabolism in Thiorhodaceae. V. Enzymes of sulfur metabolism in *Thiocapsa floridana* and *Chromatium* species. Antonie Van Leeuwenhoek J. Microbiol. Ser. *34*, 350–356.

THOMPSON, J. F. 1967. Sulfur metabolism in plants. Ann. Rev. Plant Physiol. *18*, 59–84.

TORII, K., and BANDURSKI, R. S. 1967. Yeast sulfate-reducing system. III. An intermediate in the reduction of 3′-phosphoryl-5′-adenosine phosphosulfate to sulfite. Biochim. Biophys. Acta *136*, 286–295.

TRUDINGER, P. A. 1967. The metabolism of inorganic sulphur compounds by thiobacilli. Rev. Pure Appl. Chem. *17*, 1–24.

WILSON, L. G. 1962. Metabolism of sulfate: sulfate reduction. Ann. Rev. Plant Physiol. *13*, 201–224.

WILSON, L. G., TADASHI, A. T., and BANDURSKI, R. S. 1961. Yeast sulfate-reducing system. I. Reduction of sulfate to sulfite. J. Biol. Chem. *236*, 1822–1829.

WOOLFOLK, C. A. 1962. Reduction of inorganic compounds with molecular hydrogen by *Micrococcus lactilyticus*. II. Stoichiometry with inorganic sulfur compounds. J. Bacteriol. *84*, 659–668.

YOSHIMOTO, A., and SATO, R. 1968. Studies on yeast sulfite reductase. II. Partial purification and properties of genetically incomplete sulfite reductases. Biochim. Biophys. Acta *153*, 576–588.

SECTION III

# Sulfur in Plant Nutrition

CHAPTER 7

**John F. Thompson**
**Ivan K. Smith**
**David P. Moore**

## Sulfur Requirement and Metabolism in Plants[1,2]

### INTRODUCTION

Plants are noted for the diversity of sulfur compounds contained therein (Young and Maw 1958). Of the considerable number of compounds that have been found in plants, only a few have been recognized as required for normal cell function. These vital sulfur compounds include the sulfur-containing protein amino acids (cysteine-cystine and methionine), glutathione, S-adenosylmethionine, and the cofactors, thiamine-pyrophosphate, biotin, lipoic acid and coenzyme A. Other sulfur compounds play crucial roles in certain tissues, for example, the sulfolipids of the chloroplasts. Nonessential sulfur compounds found in plants include the mustard oil glycosides, choline sulfate, thetins, S-methylmethionine, penicillin, and the S-alkylated cysteines, their sulfoxides and compounds derived therefrom. Many of the nonessential compounds are important as flavor constituents, odoriferous compounds, transport forms of sulfur, and antibiotics. The nonessential compounds can represent a significant part (up to 3%) of the dry weight of the plant; even so, they apparently are not responsible for the major portion of the sulfur requirement.

The plant normally synthesizes all of its organic sulfur compounds from sulfate. On the other hand, nonruminant animals must acquire certain reduced sulfur compounds (or their precursors) from plants. Since cysteine and methionine are responsible for the bulk (>90%) of the organic sulfur

[1]This work was supported in part by the Sulphur Institute, Washington, D.C., and London, England.

[2]Abbreviations used which are not defined in the text: ATP (adenosine triphosphate), FAD (flavin adenine dinucleotide), NADP (nicotinamide adenine dinucleotide phosphate), PAP (3′-phospho-adenosine-5′-phosphate).

in plants, the sulfur requirements of plants are determined primarily by the sulfur content of their proteins which is, in turn, largely genetically controlled. Because of the overwhelming quantitative importance of cysteine and methionine, the discussion of sulfur metabolism in plants will be concerned primarily with the metabolism of these two compounds.

## SULFUR REQUIREMENT

The sulfur requirement of a plant has been defined as the "minimum uptake of this nutrient associated with maximum yield of dry matter" (Stanford and Jordan 1966). The concentration of the nutrient present in the plant when this condition is met is usually referred to as the critical percentage (Ulrich 1952). The critical percentage varies widely and is usually determined empirically for each plant species. For example, the critical sulfur content of sugar cane is probably 0.04% or less (Stanford and Jordan 1966; Catani *et al.* 1959). On the other hand, the minimum sulfur requirement for white clover is about 0.26% (McNaught and Chrisstoffels 1961), and for alfalfa it is about 0.22% (Bear and Wallace 1950; Harward *et al.* 1962; Pumphrey and Moore 1965B). The bulk of the other species for which information is available lie between these extremes (Martin and Walker 1966; Stanford and Jordan 1966; Beaton 1966; Ensminger and Freney 1966). The wide variability in sulfur content of plants is not surprising when it is recognized that the requirement depends almost entirely on the amount of sulfur in protein (Thompson *et al.* 1960; Jordan 1964; Allaway and Thompson 1966). The protein content, in turn, varies widely with species, age, environment, and nutrition.

Evidence is accumulating which suggests that the ratio of sulfur to nitrogen in the plant is a much more reliable measure of the sulfur requirement than is the absolute level of sulfur. The sequence and number of amino acids in the polypeptide chains of a specific protein are determined by genetic information. While nutritional and environmental factors may determine the amounts and relative proportions of proteins formed, genetic factors control the amino acid composition of the specific proteins. Therefore, the amount of sulfur amino acids in plant proteins can vary *only* when there are changes in the relative amounts of the various proteins formed (e.g., cytoplasmic versus chloroplastic proteins). Generally speaking, these changes are not great enough to significantly affect the ratio of sulfur to nitrogen in the overall protein fraction (Fujiwara and Torii 1961; Dijkshoorn and van Wijk 1967; Stewart and Porter 1969).

The ratio of nitrogen to sulfur in plant proteins from vegetative tissues has been reported to be approximately 15 to 1 for a wide range of species and conditions (Dijkshoorn *et al.* 1960; Stewart and Whitfield 1965; Dijkshoorn and van Wijk 1967; Stewart and Porter 1969). Although there is some variation with species, the range in values is considerably less than for the absolute sulfur content. Legumes had a ratio of about 17 to 1 and the grasses about 14 to 1 (Dijkshoorn and van Wijk 1967). A ratio of about 15 to 1 was reported for protein from wheat and bean tops, and corn tops had a ratio of about 11 (Stewart and Whitfield 1965; Stewart and Porter 1969). While these values may reflect actual species differences in the protein fraction, the data reviewed by Dijkshoorn and van Wijk (1967) are based on organic N and organic S rather than on the protein fraction, and this could account for some of the variation. Additional data on the N to S ratio in the protein fraction for a wider range of plant species might clarify this variation.

The relative constancy of N to S in plant proteins is further evidence that the absolute requirement for sulfur is determined primarily by those factors affecting the amount of protein formed. Turnip plants grown with adequate sulfur nutrition contained more sulfur amino acids than the deficient plants, but this was almost entirely due to the fact that the sulfur-adequate plants synthesized more total protein (Thompson *et al.* 1960). The methionine content of the protein in the plants was about the same for the two levels of sulfur nutrition. Similar results were reported for sulfur-adequate and sulfur-deficient oat plants (Jordan 1964). While the addition of sulfur resulted in better growth and, therefore, more total protein production, the amino acid ratios for the protein were almost identical in the two sets of plants. The proportion of sulfur amino acids to other amino acids was found to remain about the same between sulfur-deficient and normal alfalfa plants (Mertz *et al.* 1952). Large differences in the total amount of protein produced by wheat plants resulted when the nitrogen and sulfur fertility levels were varied, yet the N to S ratio of the protein fraction remained essentially constant at 15 (Stewart and Porter 1969). In another study, the growth and the nitrogen and sulfur uptake were markedly affected by soil temperature and fertility while the N-S composition of the protein remained largely unaffected (Stewart and Whitfield 1965). These workers concluded that the primary effect of the treatments was on the amount of protein formed and not on its composition. The absolute sulfur and nitrogen content of alfalfa was found to be dependent on both the stage of maturity and on sulfur fertilization (Pumphrey and Moore 1965A). However, the sulfur and nitrogen contents declined steadily throughout the growth of the plant, and the decrease was largely independent of sulfur fertilization. As a result, the N to S ratio remained nearly constant during the growth of the plant. This finding is consistent with the results of other investigators which show that the protein content is highest early in the growth cycle and declines as the plant grows to maturity

(Wedin *et al.* 1956; van Ripper and Smith 1959; Kust and Smith 1961). The N to S ratio in canary grass and rye grass was not greatly affected by sulfur nutrition and age even though there were wide differences in sulfur content due to treatment (Wood and Barrien 1939A, 1939B). There is some evidence that the N to S ratio may decrease with age although this effect is not large (Dijkshoorn and van Wijk 1967; Barrien and Wood 1939).

The close relationship between nitrogen and sulfur is also evident when viewed from the standpoint of nitrogen requirement. The N content of sulfur-deficient alfalfa was increased by sulfur fertilization presumably due to a greater N fixation and a higher protein content (Harward *et al.* 1962; Pumphrey and Moore 1965B). Nitrogen utilization of wheat was found to be largely dependent upon the sulfur status of the plant (Stewart and Porter 1969).

The relative constancy of N to S in plant protein has prompted a number of workers to use the total N to total S ratio to assess the sulfur status of plants (Dijkshoorn *et al.* 1960; Walker *et al.* 1956; Jordan and Bardsley 1958; Pumphrey and Moore 1965B; Stewart and Whitfield 1965; Stewart and Porter 1969). This approach is based on the idea that, when sulfur is limiting protein formation, nonprotein nitrogen will accumulate in the plant resulting in a total N to total S ratio higher than that found in protein. Sulfur deficient plants have been shown to accumulate nitrate, certain amino acids, and amides (Adams and Sheard 1966; Mertz *et al.* 1952; Mertz and Matsumoto 1956; Rendig and McComb 1961; Stewart and Porter 1969). Practically all of the sulfur in sulfur deficient plants is found in the protein fraction (Dijkshoorn and van Wijk 1967; Stewart and Porter 1969). Conversely, when S is adequate (or when nitrogen is limiting) non-protein sulfur (largely sulfate) accumulates and the total N to total S ratio is below that found in plant protein (Dijkshoorn and van Wijk 1967; Stewart and Porter 1969). The extent to which this concept is valid when both N and S are in abundant supply is not known. In this case, a ratio based on total N and total S may depart significantly from the ratio found in plant protein and, in the absence of other criteria, could be misleading. It is unlikely that high N to S ratios could be attained under these circumstances, thereby giving a false indication of sulfur deficiency, since this would require a buildup of large quantities of nonprotein nitrogen in order to maintain a ratio above that found in protein. Indeed, total N to total S ratios substantially below 15 were obtained where both nitrogen and sulfur were in abundant supply (Stewart and Porter 1969). Whether the concept of a critical ratio based on total N to total S is applicable to those species which accumulate significant quantities of nonprotein organic sulfur is not clear (Dijkshoorn and van Wijk 1967), since it is not generally known if these organic sulfur compounds accumulate to any large degree in sulfur-deficient plants. In one report, turnip leaves were shown to contain substantial methylcysteine sulfoxide when the plants were adequately supplied with sulfur but not when the plants were sulfur deficient (Thompson *et al.* 1960).

Current evidence would suggest that sulfate sulfur does not build up in the plant to any significant degree until the sulfur requirement of the plant for protein formation has been met. The sulfate sulfur content of the tissue, therefore, may be a sensitive indicator of the sulfur status of the plant (Ulrich 1961). The critical level of sulfate sulfur in alfalfa at early bloom stage has been set at 500 to 700 ppm (Rendig, unpublished). Sulfate sulfur did not begin to accumulate in any significant quantity until the total sulfur in the plant reached 1500 ppm; above this point, the increase in total sulfur was almost directly proportional to the increase in sulfate sulfur (Rendig and Martin, unpublished). Therefore, the total sulfur content corresponding to the critical sulfate sulfur level was 2000 to 2200 ppm which agrees quite well with the 0.22% S reported by others to be the critical level of sulfur in alfalfa (Harward *et al.* 1962; Pumphrey and Moore 1965B). In subterranean clover, the critical sulfate sulfur content was found to be 170 ppm with little difference noted between leaves, stems or petioles (Jones 1962). Although the sulfate sulfur level appears to show promise as an indicator of the sulfur status in the plant, this information alone is not sufficient to assess the sulfur requirement. The N to S ratio of the plant's protein may be employed to estimate the sulfur requirement when the nitrogen requirement for maximum yield is known (Stanford and Jordan 1966). The buildup of sulfate sulfur in the plant is entirely consistent, however, with the use of an N to S ratio to evaluate the sulfur needs of the plant.

## SULFUR METABOLISM

Inasmuch as the function of this article is to provide background material for the examination of the role of sulfur in animal nutrition, the discussion will be limited to the formation of cysteine and methionine because these are quantitatively the most important sulfur compounds. For the same reason, the differences between plants and nonruminant animals with regard to sulfur metabolism will be discussed.

An outstanding metabolic difference between plants and nonruminant animals is the inability of the latter to reduce sulfate (in which sulfur has a valence of +6) to sulfide (valence state of −2). The vital compounds listed previously contain sulfur in the reduced state. The nonruminant animal must get these compounds or their precursors from plants. Hence, the subsequent discussion emphasizes the metabolic pathway of sulfur from sulfate to methionine via cysteine.

As far as possible, studies of sulfur metabolism in higher plants will be stressed in the ensuing discussion. Unfortunately, there are important areas in which information from higher plants is inadequate or lacking. In these areas, it will be necessary to utilize information from yeast, fungi, and bacteria. Fortunately, where there is comparable data, higher plants metabolize sulfur compounds in a similar fashion to microorganisms.

**Formation of Active Sulfate**

Reactions involved in the reduction of sulfate have been demonstrated *in vitro*; however, the pathway of sulfate reduction *in vivo* is in doubt.

Some of the early work on sulfate reduction (Wilson and Bandurski 1958) demonstrated a requirement for ATP in yeast extracts. The ATP requirement suggested that "active" sulfate, 3′-phosphoadenosine-5′-phosphosulfate (PAPS in Fig. 7.1), is the real substrate as was subsequently demonstrated (Wilson *et al.* 1961). PAPS had been identified (Robbins and Lipmann 1956) in animal systems as the compound responsible for the biosynthesis of sulfate esters. Sulfate esters are important in animals as components of connective tissue, in brain lipids, as steroid sulfates, and as detoxification products.

PAPS is formed in both plants and animals from sulfate and ATP in two steps through the intermediate, adenosine-5′-phosphosulfate (APS in Fig. 7.1) (Bandurski *et al.* 1956). In the first step, APS is formed by the sul-

ADENINE—RIBOSE—PHOSPHATE—SULFATE

APS

ADENINE—RIBOSE—PHOSPHATE—SULFATE
|
PHOSPHATE

PAPS

FIG. 7.1. DIAGRAMMATIC STRUCTURES OF APS AND PAPS

furylation of ATP [equation (1)] (Asahi 1964). The ribose portion of APS is then phosphorylated to yield PAPS [equation (2)] (Bandurski *et al.* 1956). As indicated by arrow lengths, the reaction mediated by ATP

$$ATP + SO_4^{=} \underset{\text{ATP sulfurylase}}{\rightleftharpoons} APS + PPi \qquad (1)$$

$$APS + ATP \underset{\text{APS kinase}}{\rightleftharpoons} PAPS + ADP \qquad (2)$$

sulfurylase has an equilibrium away from APS formation. This reaction [equation (1)] goes to completion in nature either by the action of APS kinase (where the equilibrium is towards PAPS formation) or by the action of pyrophosphorylase, which removes the pyrophosphate (PPi) product.

The only known function of APS in higher plants is as an intermediate in PAPS formation. In some anaerobic bacteria, APS acts as an electron acceptor in dissimilatory sulfate reduction (i.e., where the main function is to accept electrons rather than to form sulfide) (Peck 1962).

**Reduction of PAPS to Sulfite**

In yeast, PAPS rather than sulfate, is the substrate for sulfate reduction (Wilson *et al.* 1961). The reductant is an unusual one, a protein dithiol [$Pr(SH)_2$] and the products are apparently a nondialyzable bound sulfite (Torii and Bandurski 1964A) and a protein disulfide ($PrS_2$) [equation (3)] (Wilson *et al.* 1961). The protein dithiol is regenerated by $FADH_2$ [equation (4)]. Neither of the reactions represented in equations

$$PAPS + Pr(SH)_2 \rightarrow PAP + [SO_3^{=}] + PrS_2 \quad (3)$$

$$PrS_2 + FADH_2 \rightarrow Pr(SH)_2 + FAD \quad (4)$$

(3) and (4) has been demonstrated in higher plant extracts.

In *Chlorella*, thiosulfate is a product of *in vitro* sulfate reduction (Levinthal and Schiff 1968) and PAPS is probably an intermediate (Hodson *et al.* 1968). NADPH or various thiols can act as reductants (Hodson and Schiff 1969). The reduction by thiols is presumably comparable to that by $Pr(SH)_2$ in yeast.

The significance of either of these two sulfate reduction systems to the higher plant is unknown since sulfate reduction in higher plant extracts has not been accomplished.

**Reduction of Sulfite to Sulfide**

There is evidence that bound sulfite is a product of PAPS reduction and also a substrate for sulfite reduction (Torii and Bandurski 1964B). However, free sulfite is an active substrate for sulfite reduction *in vitro* (Tamura 1965; Asada 1967). The product of sulfite reduction is sulfide. Although most biological redox reactions take place with a 1 or 2 electron change, the reduction of sulfite apparently involves a 6 electron change with no detectable free intermediate sulfur compounds. Highly purified spinach sulfite reductase can be resolved into two components, a flavoprotein and a hemoprotein (Asada *et al.* 1968). The hemoprotein can reduce sulfite with methylviologen but not with pyridine nucleotide. NADPH can reduce sulfite when the flavoprotein is also present (Fig. 7.2). Although ferredoxin can also be an active reductant (Tamura *et al.* 1967) and is active in bacterial extracts (Akagi 1965), it is likely that NADPH is the natural reduc-

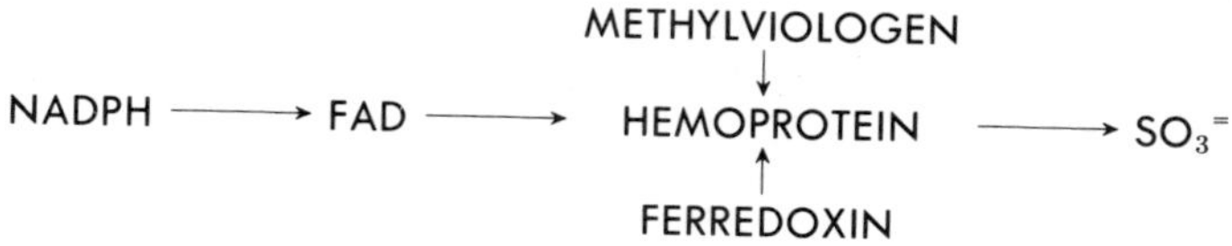


FIG. 7.2. POSSIBLE PATHWAY OF ELECTRON TRANSPORT TO SULFITE

tant [equation (5)] since sulfite reduction takes place in tissues in which ferredoxin has not been demonstrated.

$$3\,NADPH + [SO_3^{=}] + 3H^+ \rightarrow 3NADP^+ + S^{=} + 3H_2O \qquad (5)$$

**Incorporation of Sulfide into Cysteine**

Sulfide is the most reduced form of inorganic sulfur found in plants. Sulfide is incorporated into an organic compound with the formation of cysteine. Until 1966, it was considered that serine was the natural sulfide acceptor (Schlossman and Lynen 1957), with pyridoxal phosphate (PALP) acting as cofactor [equation (6)]. In 1966, it was shown that *o*-acetyl-

$$S^{=} + \text{serine} \xrightarrow[\text{serine sulfhydrase}]{\text{PALP}} \text{cysteine} + H_2O \qquad (6)$$

serine was the probable normal sulfide acceptor in bacteria [equation (7)] (Kredich and Tomkins 1966). Recent evidence indicates that *o*-acetylserine may also be the natural precursor of cysteine in higher plants because:

$$S^{=} + o\text{-acetylserine} \xrightarrow[o\text{-acetylserine sulfhydrase}]{\text{PALP}} \text{cysteine} + \text{acetate} \qquad (7)$$

(1) *o*-acetylserine is more active than serine in cysteine formation (Giovanelli and Mudd 1968; Thompson and Moore 1968), and (2) plants contain a serine acetylase (Smith and Thompson 1969A). More direct evidence is not yet available. However, a marked inhibition of serine acetylase by L-cysteine (Smith *et al.* 1969) strengthens other evidence that *o*-acetylserine is the natural sulfide acceptor. This inhibition would provide appropriate feedback control of cysteine biosynthesis and may account for the low level of unbound cysteine in plants (Nightingale *et al.* 1932).

**Formation of Homocysteine**

Quantitatively, methionine is the most important compound derived from cysteine and it is formed by the methylation of homocysteine. The transfer of the sulfur atom from cysteine to homocysteine and vice versa is termed transsulfuration. Higher plants can form homocysteine from cysteine. Animals can only do the reverse which explains why methionine

$$\text{CYSTEINE} \underset{\substack{\text{ANIMALS}\\ \text{FUNGI}}}{\overset{\substack{\text{FUNGI}\\ \text{HIGHER PLANTS}}}{\rightleftarrows}} \text{HOMOCYSTEINE}$$

FIG. 7.3. TRANSSULFURATION IN VARIOUS ORGANISMS

(as a precursor of homocysteine) but not cysteine is an essential amino acid. Fungi can carry out transsulfuration in both directions (see Fig. 7.3). The key intermediate in transsulfuration is the thioether, cystathionine. By the use of mutants, it has been established that bacteria and fungi utilize an ester of homoserine to combine with cysteine to form cystathionine. In bacteria, the ester is *o*-succinylhomoserine (Rowbury and Woods 1964), whereas in fungi it is *o*-acetylhomoserine (Nagai and Flavin 1966). In higher plants, both *o*-acetylhomoserine and *o*-succinylhomoserine are active in cystathionine synthesis (Giovanelli and Mudd 1966A) though no enzyme for homoserine esterification has been reported [equation (8)].

$$o\text{-acetylhomoserine} + \text{cysteine} \xrightarrow[\text{cystathionine-}\gamma\text{-synthase}]{\text{PALP}} \text{cystathionine} + \text{acetate} \quad (8)$$

Higher plants (Giovanelli and Mudd 1966B) and fungi (Flavin and Slaughter 1964) split cystathionine under the influence of $\beta$-cystathionase to yield homocysteine [equation (9)].

$$\text{cystathionine} + H_2O \xrightarrow[\beta\text{-cystathionase}]{\text{PALP}} \text{homocysteine} + \text{pyruvate} + NH_3 \quad (9)$$

Animals (Binkley 1944) and fungi (Flavin and Slaughter 1964) carry out a $\gamma$-cleavage of cystathionine [equation (10)].

$$\text{cystathionine} + H_2O \xrightarrow[\gamma\text{-cystathionase}]{\text{PALP}} \text{cysteine} + \alpha\text{-ketobutyrate} + NH_3 \quad (10)$$

It has also been suggested that cystathionine may not be a normal intermediate in homocysteine biosynthesis (Giovanelli and Mudd 1967; Wiebers and Garner 1967). It has been shown that spinach and *Neurospora* extracts can catalyze the formation of homocysteine from *o*-acetylhomoserine and sulfide [equation (11)] thus bypassing cystathionine and cysteine. How-

$$o\text{-acetylhomoserine} + S^{=} \underset{o\text{-acetylhomoserine sulfhydrase}}{\rightarrow} \text{homocysteine} + \text{acetate} \quad (11)$$

ever, it appears likely that cystathionine is an intermediate in the normal pathway of homocysteine (and methionine) biosynthesis, because one mutant that requires homocysteine or methionine for growth accumulates cystathionine (Horowitz 1947) and lacks the β-cystathionase that cleaves cystathionine to homocysteine (Flavin and Slaughter 1964). Furthermore, two other mutants that require cystathionine, homocysteine or methionine for growth are deficient in the enzyme for synthesizing cystathionine from cysteine (Kerr and Flavin 1968).

**Methylation of Homocysteine to Methionine**

Knowledge of methylation of homocysteine initially was derived from investigations of bacterial mutants (Davis and Mingioli 1950). The discovery that an *E. coli* auxotroph could grow when given either methionine *or* vitamin $B_{12}$ was the first clue that $B_{12}$ might be involved in methylation of homocysteine. Since then, it has been established that in the above mutant and in animals, homocysteine can be methylated by 5-methyltetrahydrofolic acid through the mediation of cobalamine coenzyme (Woods *et al.* 1965; Elford *et al.* 1965). Further work has revealed that the methyl donor can be 5-methyltetrahydrofolate monoglutamate (methyl-THFA-GLU) or the triglutamate (Fig. 7.4). However, extracts of wild type *E. coli* (which

$H_4$-PTERIN-p-AMINO BENZOIC-GLU-GLU-GLU
|
METHYL

5-METHYL-TETRAHYDROFOLATE-TRIGLUTAMATE

FIG. 7.4. DIAGRAMMATIC STRUCTURE FOR 5-METHYL-TETRAHYDROFOLATE-TRIGLUTAMATE (OR 5-METHYL-TETRAHYDROPTEROYL-TRIGLUTAMATE)

do not require vitamin $B_{12}$) can methylate homocysteine with the di- and triglutamates but *not* with the monoglutamate (Burton *et al.* 1969). Higher plants can utilize 5-methyltetrahydrofolate monoglutamate for homocysteine methylation but only about 10% as well as with the triglutamate (Burton and Sakami 1969). There is no evidence that higher plants can make vitamin $B_{12}$ or need cobalamine coenzyme for homocysteine methylation. Hence, the requirement for cobalamine coenzyme for homocysteine methylation is a characteristic of animals but not plants.

Enzymatic and tracer studies have revealed that the methyl group is derived from the beta carbon of serine. This carbon is transferred to a folic acid derivative to form methylene-THFA-$GLU_3$ which is reduced to 5-methyl-THFA-$GLU_3$ [equations (13) and (14)] (Woods *et al.* 1965).

$$\text{THFA-GLU} + 2\,\text{ATP} + 2\,\text{GLU} \rightarrow \text{THFA-GLU}_3 + 2\,\text{ADP} + 2\,\text{Pi} \quad (12)$$

tetrahydrofolate monoglutamate-glutamate ligase

$$\text{serine} + \text{THFA-GLU}_3 \xrightarrow[\text{serine hydroxymethylase}]{} \text{glycine} + \text{methylene-THFA-GLU}_3 \quad (13)$$

$$\text{methylene-THFA-GLU}_3 + \text{FADH}_2 \xrightarrow[\text{methylene-THFA-GLU}_3\text{ reductase}]{} \text{5-methyl-THFA-GLU}_3 + \text{FAD} \quad (14)$$

$$\text{5-methyl-THFA-GLU}_3 + \text{homocysteine} \xrightarrow[\text{methyl-THFA-GLU}_3\text{ homocysteine transmethylase}]{} \text{THFA-GLU}_3 + \text{methionine} \quad (15)$$

The accuracy of the above formulation is supported by the report that *Neurospora* methionineless mutants lacking enzymes to catalyze reactions (12), (14), and (15) have been identified (Selhub *et al.* 1969).

An alternative pathway for methionine biosynthesis has been demonstrated in *Neurospora* where methylmercaptan reacts with *o*-acetylhomoserine to form methionine [equation (17)] (Moore and Thompson 1967; Smith and Thompson 1969B). Methylmercaptan can be derived from S-methylcysteine in *Neurospora* extracts (Moore and Thompson 1967) by a methylcysteine lyase (Smith and Thompson 1969B).

$$\text{methylcysteine} + H_2O \xrightarrow[\text{methylcysteine lyase}]{} CH_3SH + \text{pyruvate} + NH_3 \quad (16)$$

$$CH_3SH + o\text{-acetylhomoserine} \xrightarrow[\text{methionine synthetase}]{} \text{methionine} + \text{acetate} \quad (17)$$

The significance of this pathway in providing a net synthesis of methionine cannot be ascertained at present.

**Formation of S-adenosylmethionine**

Methionine is most important as a constituent of protein and as a precursor of the principal methyl donor, S-adenosylmethionine (SAM). SAM is formed from ATP and methionine in a reaction [equation (18)] that removes all of the phosphates from ATP (Cantoni 1953).

$$\text{ATP} + \text{methionine} \xrightarrow[\text{SAM synthetase}]{} \text{SAM} + \text{PPi} + \text{Pi} \quad (18)$$

SAM is the most versatile methyl donor in nature (Mudd and Cantoni 1964). The methylation of homocysteine is one of the few methylations not carried out by SAM. Equation (19) gives the general reaction for methylation (where A is a methyl acceptor) and equation (20) shows how homocysteine is regenerated from S-adenosylhomocysteine (SAHC).

$$\text{S-adenosylmethionine} + \text{A} \xrightarrow[\text{transmethylase}]{} \text{S-adenosylhomocysteine} + CH_3A \quad (19)$$

$$\text{S-adenosylhomocysteine} + H_2O \xrightarrow[\text{S-adenosylhomocysteine lyase}]{} \text{adenosine} + \text{homocysteine} \quad (20)$$

### Control of Sulfur Metabolism

It is a common observation that higher plants have low levels of cysteine-cystine and methionine as compared to other amino acids (Thompson *et al.* 1960). This observation implies that the levels of cysteine and methionine are closely controlled. There is little data that give clues as to the mechanisms that operate in higher plants, but there is more information from *Neurospora*, yeast, and bacteria that may well be found relevant to the higher plant.

It has been shown that L-cysteine is a potent inhibitor of serine acetylase from higher plants (Smith *et al.* 1969) and bacteria (Kredich and Tomkins 1966). This inhibition appears to be specific, indicating that feedback inhibition may be one way in which cysteine levels are regulated in the higher plant. In *E. coli*, cysteine and sulfide inhibit the synthesis of all the enzymes responsible for the transformation of sulfate to cysteine (Jones-Mortimer 1968). *o*-Acetyl-L-serine induces these same enzymes and apparently does not act by counteracting the cysteine effect. The actual effective compound is in doubt, since some results indicate that sulfide derived from cysteine may be the true repressor (Dreyfuss and Monty 1963).

In an analogous fashion, it has been observed that methionine inhibits the formation of cystathionine-γ-synthase in *Neurospora* (Kerr and Flavin 1968). However, in bacteria, control is effected at an earlier step (i.e., in the synthesis of *o*-succinylhomoserine where methionine and SAM exert a synergistic inhibition of the enzyme) (Lee *et al.* 1966).

## CONCLUSION

The sulfur requirement of plants is largely dependent on the amount of sulfur found in protein and is, therefore, influenced by those factors which affect protein production. The incorporation of the sulfur amino acids, cysteine and methionine, into protein is under genetic control and as a result, the nitrogen to sulfur ratio of plant protein is relatively constant. For this reason, the discussion of sulfur metabolism in plants has been limited to the transformation of sulfate sulfur to cysteine and methionine sulfur.

The differences in sulfur metabolism between plants and nonruminant animals have been discussed and are shown diagramatically in Fig. 7.5. Both plants and animals can synthesize PAPS and utilize it as a sulfate donor. Since only plants reduce PAPS to sulfite and sulfide, plants must be the source of reduced sulfur compounds for nonruminant animals. Higher plants carry out transsulfuration from cysteine to homocysteine while animals do the transsulfuration in the opposite direction. Hence, plants must

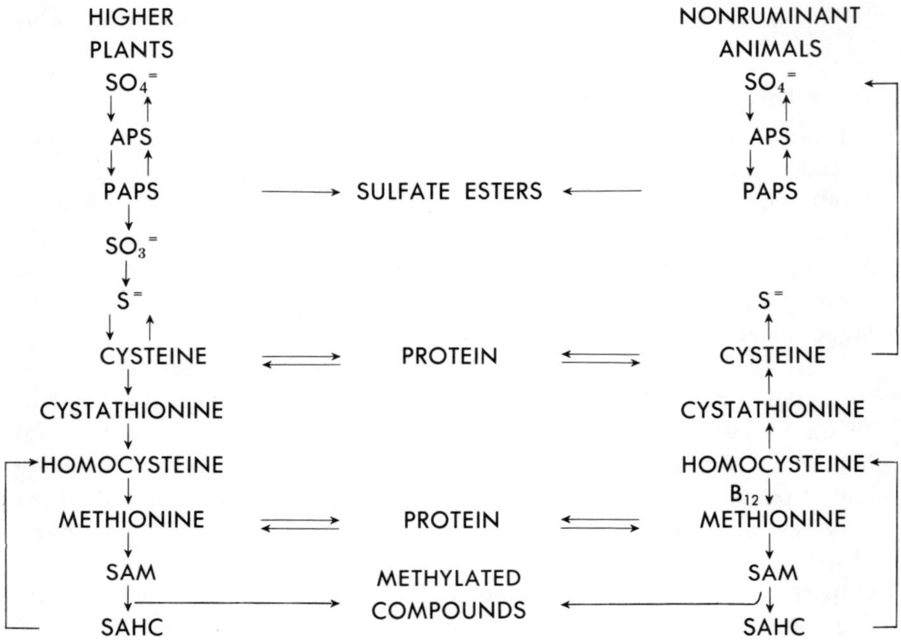


FIG. 7.5. DIAGRAM OF METABOLIC PATHWAYS OF SULFUR IN HIGHER PLANTS AND NONRUMINANT ANIMALS

be the source of homocysteine (or methionine) for animals but animals can fulfill their cysteine requirement from methionine in the diet. Both animals and plants can methylate homocysteine but the mechanisms are different.

## BIBLIOGRAPHY

ADAMS, C. A., and SHEARD, R. W. 1966. Alterations in the nitrogen metabolism of *Medicago sativa* and *Dactylis glomerata* as influenced by potassium and sulfur nutrition. Can. J. Plant Sci. *46*, 671–680.

AKAGI, J. M. 1965. The participation of a ferredoxin of *Clostridium nigrificans* in sulfite reduction. Biochem. Biophys. Res. Commun. *21*, 72–77.

ALLAWAY, W. H., and THOMPSON, J. F. 1966. Sulfur in the nutrition of plants and animals. Soil Sci. *101*, 240–247.

ASADA, K. 1967. Purification and properties of a sulfite reductase from leaf tissue. J. Biol. Chem. *242*, 3546–3654.

ASADA, K., TAMURA, G., and BANDURSKI, R. S. 1968. Methylviologen-linked sulfite reductase from spinach leaves: a hemoprotein. Biochem. Biophys. Res. Commun. *30*, 554–559.

ASAHI, T. 1964. Sulfur metabolism in higher plants. IV. Mechanism of sulfate reduction in chloroplasts. Biochim. Biophys. Acta *82*, 58–66.

BANDURSKI, R. S., WILSON, L. G., and SQUIRES, C. L. 1956. The mechanism of "active sulfate" formation. J. Am. Chem. Soc. *78*, 6408–6409.

BARRIEN, B. S., and WOOD, J. G. 1939. Studies on the sulphur metabolism of plants. II. The effect of nitrogen supply on the amounts of protein sulphur, sulphate sulphur and on the value of the ratio of protein nitrogen to protein sulphur at different stages during the life cycle of the plant. New Phytologist *38*, 257–264.

BEAR, F. E., and WALLACE, A. 1950. Alfalfa—its mineral requirements and chemical composition. New Jersey Agr. Expt. Sta. Bull. *748*.

BEATON, J. D. 1966. Sulfur requirements of cereals, tree fruits, vegetables, and other crops. Soil Sci. *101*, 267–282.

BINKLEY, F. 1944. Cleavage of cystathionine by an enzyme system from rat liver. J. Biol. Chem. *155*, 39–43.

BURTON, E. G., and SAKAMI, W. 1969. The formation of methionine from the monoglutamate form of methyltetrahydrofolate by higher plants. Biochem. Biophys. Res. Commun. *36*, 228–234.

BURTON, E. G., SELHUB, J., and SAKAMI, W. 1969. The substrate specificity of 5-methyltetrahydropteroyltriglutamate-homocysteine methyl transferase. Biochem. J. *111*, 793–795.

CANTONI, G. L. 1953. S-adenosylmethionine; a new intermediate formed enzymatically from L-methionine and adenosinetriphosphate. J. Biol. Chem. *204*, 403–416.

CATANI, R. A., ARRUDA, H. C., and PELEGRINO, D. 1959. *Cited by* STANFORD, G., and JORDAN, H. V. 1966. Sulfur requirements of sugar, fiber, and oil crops. Soil Sci. *101*, 258–266.

DAVIS, B. D., and MINGIOLI, E. S. 1950. Mutants of *Escherichia coli* requiring methionine or vitamin $B_{12}$. J. Bacteriol. *60*, 17–28.

DIJKSHOORN, W., LAMPE, J. E. M., and VAN BURG, P. F. J. 1960. A method of diagnosing the sulphur nutrition status of herbage. Plant Soil *13*, 227–241.

DIJKSHOORN, W., and VAN WIJK, A. L. 1967. The sulphur requirements of plants as evidenced by the sulphur-nitrogen ratio in the organic matter. A review of published data. Plant Soil *26*, 129–157.

DREYFUSS, J., and MONTY, K. J. 1963. Coincident repression of the reduction of 3′-phosphoadenosine-5′-phosphosulfate, sulfite and thiosulfate in the cysteine pathway of *Salmonella typhimurium*. J. Biol. Chem. *238*, 3781–3783.

ELFORD, H. L. *et al.* 1965. Enzymatic synthesis of the methyl group of methionine. *In* Transmethylation and Methionine Biosynthesis, S. K. Shapiro and F. Schlenk (Editors). Univ. Chicago Press, Chicago.

ENSMINGER, L. E., and FRENEY, J. R. 1966. Diagnostic techniques for determining sulfur deficiencies in crops and soils. Soil Sci. *101*, 283–290.

FLAVIN, M., and SLAUGHTER, C. 1964. Cystathionine cleavage enzymes of *Neurospora*. J. Biol. Chem. *239*, 2212–2219.

FUJIWARA, A., and TORII, K. 1961. Physiology of sulfate on higher plants. I. Effect of sulfur deficiency on metabolism of higher plants. Tohoku J. Agr. Res. *12*, 277–290.

GIOVANELLI, J., and MUDD, S. H. 1966A. Enzymatic synthesis of cystathionine by extracts of spinach, requiring *O*-acetylhomoserine or *O*-succinylhomoserine. Biochem. Biophys. Res. Commun. *25*, 366–371.

GIOVANELLI, J., and MUDD, S. H. 1966B. Transsulfuration in higher plants. Plant Physiol. Suppl. *41*, 13–14.

GIOVANELLI, J., and MUDD, S. H. 1967. Synthesis of homocysteine and cysteine by enzyme extracts of spinach. Biochem. Biophys. Res. Commun. *27*, 150–156.

GIOVANELLI, J., and MUDD, S. H. 1968. Sulfuration of *O*-acetylhomoserine and *O*-acetylserine by two enzyme fractions from spinach. Biochem. Biophys. Res. Commun. *31*, 275–280.

HARWARD, M. E., CHAO, T. T., and FANG, S. C. 1962. The sulfur status and sulfur supplying power of Oregon soils. Agron. J. *54*, 101–106.

HODSON, R. C., and SCHIFF, J. A. 1969. Reduction of sulfate to thiosulfate *in vitro*. XI Intern. Botan. Congr. Abstr. *92*.

HODSON, R. C., SCHIFF, J. A., SCARSELLA, A. J., and LEVINTHAL, M. 1968. Studies of sulfate utilization by algae. 6. Adenosine-3′-phosphate-5′-phosphosulfate (PAPS) as an intermediate in thiosulfate formation from sulfate by cell-free extracts of *Chlorella*. Plant Physiol. *43*, 563–569.

HOROWITZ, N. H. 1947. Methionine synthesis in *Neurospora*. The isolation of cystathionine. J. Biol. Chem. *171*, 255–264.

JONES, M. B. 1962. Total sulfur and sulfate sulfur content in subterranean clover as related to sulfur responses. Soil Sci. Soc. Am. Proc. *26*, 482–484.

JONES-MORTIMER, M. C. 1968. Positive control of sulphate reduction in *Escherichia coli*. The nature of the pleiotropic cysteineless mutants of *E. coli* K 12. Biochem. J. *110*, 597–602.

JORDAN, H. V. 1964. Sulfur as a plant nutrient in the Southern United States. U.S. Dept. Agr. Tech. Bull. *1297*.

JORDAN, H. V., and BARDSLEY, C. E. 1958. Response of crops to sulfur on southeastern soils. Soil Sci. Soc. Am. Proc. *22*, 254–256.

KERR, D. S., and FLAVIN, M. 1968. Synthesis of cystathionine from *O*-acetylhomoserine in *Neurospora:* A step in methionine biosynthesis. Biochem. Biophys. Res. Commun. *31*, 124–130.

KREDICH, N. M., and TOMKINS, G. M. 1966. The enzymic synthesis of L-cysteine in *Escherichia coli* and *Salmonella typhimurium*. J. Biol. Chem. *241*, 4955–4965.

KUST, C. A., and SMITH, D. 1961. Influence of harvest management on level of carbohydrate reserves, longevity of stands and yields of hay and protein from Vernal alfalfa. Crop Sci. *1*, 267–269.

LEE, L., RAVEL, J., and SHIVE, W. 1966. Multimetabolite control of a biosynthetic pathway by sequential metabolites. J. Biol. Chem. *241*, 5479–5480.

LEVINTHAL, M., and SCHIFF, J. A. 1968. Studies of sulfate utilization by algae. 5. Identification of thiosulfate as a major acid-volatile product formed by a cell-free sulfate-reducing system from *Chlorella*. Plant Physiol. *43*, 555–562.

MARTIN, W. E., and WALKER, T. W. 1966. Sulfur requirements and fertilization of pasture and forage crops. Soil Sci. *101*, 248–257.

MCNAUGHT, K. J., and CHRISSTOFFELS, P. J. E. 1961. Effect of sulphur deficiency on sulphur and nitrogen levels in pastures and lucerne. New Zealand J. Agr. Res. *4*, 177–196.

MERTZ, E. T., and MATSUMOTO, H. 1956. Further studies on the amino acids and proteins of sulfur-deficient alfalfa. Arch. Biochem. Biophys. *63*, 50–63.

MERTZ, E. T., SINGLETON, V. L., and GAREY, C. L. 1952. The effect of sulfur deficiency on the amino acids of alfalfa. Arch. Biochem. Biophys. *38*, 139–145.

MOORE, D. P., and THOMPSON, J. F. 1967. Methionine biosynthesis from S-methylcysteine by thiomethyl transfer in *Neurospora*. Biochem., Biophys. Res. Commun. *28*, 474–479.

MUDD, S. H., and CANTONI, G. L. 1964. Biological transmethylation, methyl-group neogenesis and other "one-carbon" metabolic reactions dependent upon tetrahydrofolic acid. *In* Comprehensive Biochemistry, M. Florkin, and E. Stotz (Editors). Elsevier Press, Amsterdam.

NAGAI, S., and FLAVIN, M. 1966. Acetylhomoserine and methionine biosynthesis in *Neurospora*. J. Biol. Chem. *241*, 3861–3863.

NIGHTINGALE, G. T., SCHERMERHORN, L. G., and ROBBINS, W. R. 1932. Effects of sulphur deficiency on metabolism in tomatoes. Plant Physiol. *7*, 565–595.

PECK, H. D., Jr. 1962. The role of adenosine-5′-phosphosulfate in the reduction of sulfate to sulfite by *Desulfovibrio desulfuricans*. J. Biol. Chem. *237*, 198–203.

PUMPHREY, F. V., and MOORE, D. P. 1965A. Sulfur and nitrogen content of alfalfa herbage during growth. Agron. J. *57*, 237–239.

PUMPHREY, F. V., and MOORE, D. P. 1965B. Diagnosing sulfur deficiency of alfalfa (*Medicago sativa* L.) from plant analysis. Agron. J. *57*, 364–366.

RENDIG, V. V. unpublished. *Cited by* MARTIN, W. E., and WALKER, T. W. 1966. Sulfur requirements and fertilization of pasture and forage crops. Soil Sci. *101*, 248–257.

RENDIG, V. V., and MARTIN, W. E. unpublished. *Cited by* MARTIN, W. E., and WALKER, T. W. 1966. Sulfur requirements and fertilization of pasture and forage crops. Soil Sci. *101*, 248–257.

RENDIG, V. V., and MCCOMB, E. A. 1961. Effect of nutritional stress on plant composition. II. Changes in sugar and amide nitrogen content of normal and sulfur deficient alfalfa during growth. Plant Soil *14*, 176–186.

ROBBINS, P. W., and LIPMANN, F. 1956. Identification of enzymatically active sulfate as adenosine-3′-phosphate-5′-phosphosulfate. J. Am. Chem. Soc. *78*, 2652–2653.

ROWBURY, R. J., and WOODS, D. D. 1964. *O*-Succinylhomoserine as an intermediate in the synthesis of cystathionine by *Escherichia coli*. J. Gen. Microbiol. *36*, 341–358.

SCHLOSSMAN, K., and LYNEN, F. 1957. Biosynthesis of cysteine from serine and hydrogen sulfide. Biochem. Z. *328*, 591–594.

SELHUB, J., BURTON, E., and SAKAMI, W. 1969. Identification of three enzymes specifically involved in the *de novo* methionine methyl biosynthesis of *N. Crassa*. Federation Proc. *28*, 352.

SMITH, I. K., and THOMPSON, J. F. 1969A. The synthesis of *O*-acetylserine by extracts prepared from higher plants. Biochem. Biophys. Res. Commun. *35*, 939–945.

SMITH, I. K., and THOMPSON, J. F. 1969B. Utilization of S-methylcysteine and methylmercaptan by methionineless mutants of *Neurospora* and the pathway of their conversion to methionine. II Enzyme studies. Biochim. Biophys. Acta *184*, 130–138.

SMITH, I. K., THOMPSON, J. F., and MORRIS, C. J. 1969. Synthesis of *O*-acetylserine in higher plants and its inhibition by cysteine. XI. Intern. Botan. Congr. Abstr. *204*.

STANFORD, G., and JORDAN, H. V. 1966. Sulfur requirements of sugar, fiber and oil crops. Soil Sci. *101*, 258–266.

STEWART, B. A., and PORTER, L. K. 1969. Nitrogen-sulfur relationships in wheat (*Triticum aestivum* L.), corn (*Zea mays*), and beans (*Phaseolus vulgaris*). Agron. J. *61*, 267–271.

STEWART, B. A., and WHITFIELD, C. J. 1965. Effects of crop residue, soil temperature and sulfur on the growth of winter wheat. Soil Sci. Soc. Am. Proc. *29*, 752–755.

TAMURA, G. 1965. Studies on the sulfite reducing system of higher plants. II. Purification and properties of sulfite reductase from *Allium odorum*. J. Biochem. *57*, 207–214.

TAMURA, G., ASADA, K., and BANDURSKI, R. S. 1967. Linkage of sulfite reduction to pyridine nucleotide. Plant Physiol. Suppl. *42*, S-36.

THOMPSON, J. F., and MOORE, D. P. 1968. Enzymatic synthesis of cysteine and S-methylcysteine in plant extracts. Biochem. Biophys. Res. Commun. *31*, 281–286.

THOMPSON, J. F., MORRIS, C. J., and GERING, R. K. 1960. The effect of mineral supply upon the amino acid composition of plants. Qualitas Plant. Mater. Vegetabiles *6*, 261–275.

TORII, K., and BANDURSKI, R. S. 1964A. A possible intermediate in the reduction of 3′-phosphoryl-5′-adenosine phosphosulfate to sulfite. Biochem. Biophys. Res. Commun. *14*, 537–542.

TORII, K., and BANDURSKI, R. S. 1964B. Yeast sulfate-reducing system. III. An intermediate in the reduction of 3′-phospho-5′-adenosine phosphosulfate to sulfite. Biochim. Biophys. Acta *136*, 286–295.

ULRICH, A. 1952. Physiological bases for assessing the nutritional requirements of plants. Ann. Rev. Plant Physiol. *3*, 207–228.

ULRICH, A. 1961. Plant analysis in sugar beet nutrition. *In* Plant Analysis and Fertilizer Problems, Walter Reuther (Editor), Publ. *8*, Am. Inst. Biol. Sci., Washington, D.C.

VAN RIPPER, G. E., and SMITH, D. 1959. Changes in the chemical composition of the herbage of alfalfa, medium red clover, ladino clover and bromegrass with advanced maturity. Wisconsin Agr. Expt. Sta. Res. Rept. *4*.

WALKER, T. W., ADAMS, A. F. R., and ORCHISTON, H. D. 1956. Effect of level of calcium sulphate on yield and composition of a grass and clover pasture. Plant Soil *7*, 290–300.

WEDIN, W. F., BURGER, A. W., and AHLGREN, H. L. 1956. Effect of soil type, fertilization and stage of growth on yield, chemical composition, and biological value of ladino clover (*Trifolium repens* L.) and alfalfa (*Medicago sativa*). Agron. J. *48*, 147–152.

WIEBERS, J. L., and GARNER, H. R. 1967. Acyl derivatives of homoserine as substrates for homocysteine synthesis in *Neurospora crassa*, yeast and *Escherichia coli*. J. Biol. Chem. *242*, 5644–5649.

WILSON, L. G., ASAHI, T., and BANDURSKI, R. S. 1961. Yeast sulfate-reducing system. I. Reduction of sulfate to sulfite. J. Biol. Chem. *236*, 1822–1829.

WILSON, L. G., and BANDURSKI, R. S. 1958. Enzymic reduction of sulfate. J. Am. Chem. Soc. *80*, 5576–5577.

WOOD, J. G., and BARRIEN, B. S. 1939A. Studies on the sulphur metabolism of plants. I. Preliminary investigations on the effects of different external concentrations of sulphate, ammonia and cystine on the amounts of sulphur-containing compounds in leaves. New Phytologist *38*, 125–149.

WOOD, J. G., and BARRIEN, B. S. 1939B. Studies on the sulphur metabolism of plants. III. On changes in amounts of protein sulphur and sulphate sulphur during starvation. New Phytologist *38*, 265–272.

WOODS, D. D., FOSTER, M. A., and GUEST, J. R. 1965. Cobalamine-dependent and -independent methyl transfer in methionine biosynthesis. *In* Transmethylation and Methionine Biosynthesis, S. K. Shapiro and F. Schlenk (Editors). Univ. Chicago Press, Chicago.

YOUNG, L., and MAW, G. A. 1958. The Metabolism of Sulphur Compounds. Methuen and Co., London.

SECTION IV

# Sulfur in Animal Nutrition

CHAPTER 8

Dominic D. Dziewiatkowski

## Metabolism of Sulfate Esters

### INTRODUCTION

In 1829 Henry noted in his textbook, *The Elements of Experimental Chemistry*, that there were two forms of sulfate in urine. One was immediately precipitable by the addition of $BaCl_2$ to acidified urine and was thought to be inorganic sulfate; the other could only be precipitated after acid hydrolysis. It was almost 50 yr later that Bauman (1876A, B) demonstrated that this second fraction consisted primarily of sulfate esters. Since then, several types of sulfate esters have been found in urines and in plant and in animal tissues, namely of phenols, of phenolic steroids, and of carbohydrates. Moreover, it has been shown that exogenous sulfate can be utilized in the formation of such esters (Hele 1924, 1931; Laidlaw and Young 1948, 1953; Dziewiatkowski 1949B). After intraperitoneal injection of $^{35}S$-sulfate to rats, Laidlaw and Young recovered 5 to 6% of the radioactivity in the ester sulfate fraction of the urine. A much larger fraction of the radioactivity, nearly ½ of the dose, was recovered in the ester sulfate fraction when 2-naphthylamine or 2-naphthol was administered subcutaneously at the same time.

In the period 1899 to 1941, many workers (Herter and Wakeman 1899; Embden and Glassner 1902; Embden 1902; Marenzi 1931; Arnoldt and De Meio 1941) described experiments on the conjugation of phenols with sulfate. From these experiments it was apparent that sulfate esters of phenols could be synthesized in liver and in intestinal tissue. With the advent of $^{35}S$-sulfate as an indicator it became clear that a variety of sulfate esters could be synthesized by animals and that some of these could be retained for considerable periods of time in nearly every tissue of an animal (Dziewiatkowski 1958). Some sulfate esters, such as those of carbohydrates (chondroitin sulfates and keratin sulfates, for example) are retained within organisms as structural components in the connective tissues for considerable periods of time. On the other hand, sulfate esters such as those of phenols and steroids have a transient existence in the animal body and are quickly excreted, primarily in the urine. Although transient, such esters, particularly those of steroids, may play important but as yet incompletely delineated roles in the regulation of enzymatic reactions, transport of amino acids and other solutes, and as anti-inflammatory agents (Mason 1970).

The esterification process *in vivo* very probably involves the biochemical mechanisms described *in vitro* (Bandurski *et al.* 1956; Segal 1956; Wilson and Bandurski 1956; Robbins and Lipmann 1956A, B). In the first place, it is dependent on the activation of sulfate in two steps to adenosine-3′-phosphate-5′-phosphosulfate:

$$(1)\ ATP + SO_4{}^{=} \xrightleftharpoons{\text{ATP-sulfurylase}} \text{adenosine-5}'\text{-phosphosulfate (APS)} + P_2O_7H_4$$

$$(2)\ ATP + APS \xrightleftharpoons{\text{APS-kinase}} \text{adenosine-3}'\text{-phosphate-5}'\text{-phosphosulfate (PAPS)} + ADP$$

De Meio *et al.* (1955) suggested that there are a number of distinct enzymes which can transfer sulfate from a metabolically "active" form of sulfate (PAPS) to a variety of acceptors. This suggestion is supported by the observations of Bernstein and McGilvery (1952) and those of Gregory and Lipmann (1957), among others. Once PAPS is formed its sulfate may be transferred under the influence of transferases to many kinds of phenols (Baldwin *et al.* 1959; Sato *et al.* 1956), arylamines (Boyland *et al.* 1957), steroids (Schneider and Lewbart 1956; De Meio *et al.* 1956), mucopolysaccharides (Pasternak and Kent 1958; Suzuki and Strominger 1960; Eiber and Danishefsky 1957), cerebrosides (Holmgard 1955), bilirubin (Isselbacher and McCarthy 1958), and a variety of alcohols (Vestermark and Boström 1959; Boström 1961).

When a higher organism ingests sulfate in its diet it must first absorb it from its intestinal tract; second, transport it to those sites within its body where it is to be used; and finally, metabolize it at these sites. This chapter briefly considers each of these three aspects: (1) absorption from the intestine, (2) association of sulfate with certain blood components, and (3) utilization in some tissues.

## ABSORPTION OF SULFATE IN THE INTESTINAL TRACT

It has been noted repeatedly that inorganic sulfate is absorbed from the intestinal tracts of animals. The rat excretes in the urine 41 to 64% of an oral dose of inorganic $^{35}$S-sulfate within 8 hr (Morrow *et al.* 1952) and about 75% within 24 hr (Dziewiatkowski 1949A). Humans also rapidly excrete ingested sulfate in the urine (Borsook *et al.* 1937; Greengard and Wooley 1940). Of particular interest was the observation that, when the $^{35}$S-sulfate excreted in the urine by rats was fractionated into inorganic and ester sulfates, each of these fractions had essentially the same specific activity. Such an agreement in the specific activities suggested that both fractions were derived from the same metabolic pool and further that, if sulfur from other sources (such as cystine or methionine) was used in the synthesis of sulfate esters *in vivo*, the sulfur was first converted to sulfate before it was used in the formation of sulfate esters. Boström (1961) subsequently demonstrated that the number of sulfate esters excreted in urine is large. Moreover, the pattern of excretion may vary, depending on the physiological state of the animal.

The unequivocal demonstration that sulfate was absorbed from the intestinal tract raised the question as to whether it could be absorbed equally well at all levels of the intestinal tract. The elegant studies of Crane (1960) on the transport of sugars and of Schachter *et al.* (1959, 1960) and of Dowdle *et al.* (1960) on the transport of calcium in the intestinal tract by the technique of Wilson and Wiseman (1954) suggested the applicability of this technique to an evaluation of where sulfate is optimally absorbed. Thus, a series of experiments was initiated in which adult rats were used. The food intake of the rats, male (280 to 330 gm) and female (200 to 250 gm) was not restricted. It consisted of Purina Lab Chow and water. At about 10:00 A.M. of the day an experiment was run, the rats were decapitated and the intestinal tract from the pylorus to the anus was immediately removed. Segments approximately 4 cm long were cut out. The distance between segments was about 10 cm; the caecum was not included. After a brief immersion of the segments in ice-cold isotonic saline, they were everted, and again washed in ice-cold isotonic saline. Sacs were prepared from the everted segments, as described by Wilson and Wiseman (1954), such that the mucosal layer was on the outside. Into each sac 0.5 ml of a buffered solution of salts was delivered with a syringe and a blunted hypodermic needle. Each sac was then immersed in 5 ml of the same solution[1] of buffered salts contained in a 25 ml Erlenmeyer flask. The air in the flasks was replaced with oxygen and the flasks were then stoppered and shaken gently in a water-bath at 37° C. After incubation for certain periods of time, the sacs were removed from the flasks and they were briefly im-

[1]The composition of the solution of salts per liter was: NaCl, 135 mM; KCl, 11 mM; $CaCl_2$, 1 mM; $Na_2SO_4$, 1 mM; D-Glucose, 20 mM; and $Na_2HPO_4/NaH_2PO_4$, pH 7.4, 10 mM. $^{35}$S-sulfate was present initially in either the solution bathing the mucosal surface or in a few experiments inside the sacs on the serosal surface.

mersed in each of two 25 ml portions of ice-cold, unlabeled solution of the buffered salts. The sacs were then gently blotted on filter paper and their contents were drained into small tubes. The radioactivities of the contents of the sacs and of the bathing fluids were determined. Representative samples of the bathing solutions and of the solutions in the sacs were also analyzed by electrophoresis on paper. Autoradiograms of the electrophoretograms were prepared using Kodak No-Screen X-Ray film.

For comparison with the results with $^{35}S$-sulfate, $^{32}P$-phosphate and $^{45}Ca$ were used in experiments on 12 rats and 4 rats respectively. These isotopes, $^{32}P$ and $^{45}Ca$, were present initially only in the fluid bathing the mucosa.

Of some concern was a question as to whether the transfer of sulfate from the mucosal to the serosal side depended upon the viability of the tissue *in vitro*. To answer this question, sacs of the terminal ileum were incubated in the $^{35}S$-medium for up to 4 hr. As can be seen in Fig. 8.1, sulfate was transferred from the mucosal to the serosal side for about 2 hr as a linear function of time. Within 1 hr the concentration of $^{35}S$-sulfate on the serosal side was thereby increased to values of the ratio cpm per 100 $\mu$l inside/cpm per 100 $\mu$l per 100 $\mu$l originally outside to greater than one. These data imply that the transfer was an active one for the values are greater than would be predicted at a diffusion equilibrium. Moreover, since the values at 3 and 4 hr are less than at 2 hr and tend to decrease (are less at 4 hr than at 3 hr), the data imply that an integrated metabolism of the intestinal tissue is a requirement for the active transport of sulfate from the mucosal to the serosal side.

A suggestion that an integrated metabolism of the intestinal tissue was a requirement for the active transport of sulfate from the mucosal to the

TABLE 8.1

EFFECT OF INHIBITORS ON THE TRANSFER OF $^{35}S$-SULFATE FROM THE MUCOSAL TO THE SEROSAL SIDE OF THE ILEUM[1]

| Inhibitor | Concentration *M* | Inhibition % of Activity Shown by Controls |
|---|---|---|
| NaF (12) | $10^{-1}$ | 28–35 |
| NaF (12) | $10^{-2}$ | 50–63 |
| NaF (12) | $10^{-3}$ | 94–100 |
| $NaN_3$ (12) | $10^{-3}$ | 44–56 |
| $NaN_3$ (12) | $10^{-4}$ | 45–55 |
| $NaN_3$ (12) | $10^{-5}$ | 79–95 |
| $N_2$ (12) | . . . | 39–54 |

[1]Everted sacs prepared from adjacent segments of the ileum were filled and incubated in the buffered solution of salts with and without added inhibitor. The $^{35}S$-sulfate was initially presented only to the mucosal side. Incubation was for 2 hr at 37°C, with shaking. The number of pairs of sacs used at each concentration is given in parenthesis.

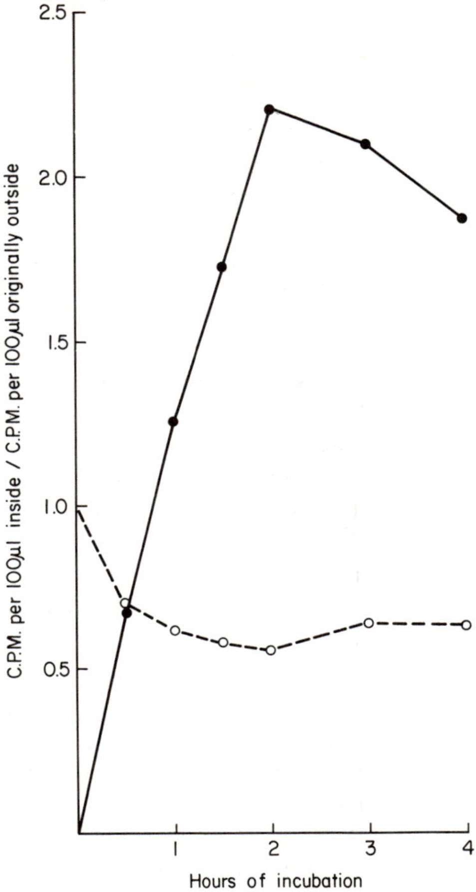


FIG. 8.1. RELATION OF DURATION OF INCUBATION ON TRANSPORT OF $^{35}$S-SULFATE IN THE ILEUM OF THE ADULT RAT

Everted sacs prepared from segments of the distal ileum were incubated for varying periods of time in 5 ml of an oxygenated solution of salts containing, in addition, $^{35}$S-sulfate. The sacs were filled with 0.5 ml of the same solution, except that it did not contain the isotope. Each point is the mean value from four experiments. The concentration of $^{35}$S on the serosal side is shown as •----•, on the mucosal side as o----o. In each case it is the concentration relative to the concentration originally present in the solution on the outside of the sac.

serosal side also derives from experiments in which metabolic inhibitors were used, Table 8.1. In the presence of sodium azide or of sodium fluoride the transfer of sulfate from the mucosal side to the serosal side was markedly decreased. As an inhibitor, sodium azide was more effective than sodium fluoride; approximately the same effect, 50% inhibition, was obtained with sodium azide at a concentration of $10^{-4}$ molar as with sodium fluoride at a concentration of $10^{-2}$ molar. In an atmosphere of nitrogen, too, the transfer of sulfate from the mucosal to the serosal side was inhibited to values about 50% of the values obtained when the gas phase was oxygen. These observations suggest that oxidative metabolism is essential for the active transport of the sulfate from the mucosal to the serosal side.

Based on the above, sacs of the intestinal tract were incubated in an atmosphere of oxygen for no more than 2 hr at 37° C. It was found that inorganic sulfate could pass from the mucosal to the serosal side, and vice-versa, at all levels of the intestinal tract. However, the rates of these transfers were of different magnitudes from one region to another of the intestinal tract. Transfer of $^{35}$S-sulfate from the mucosal side to the serosal side after 1 hr was greatest in the ileum, segments 6, 7, and 8 of Fig. 8.2. In the case of the transfer of $^{35}$S-sulfate from the serosal side to the muco-

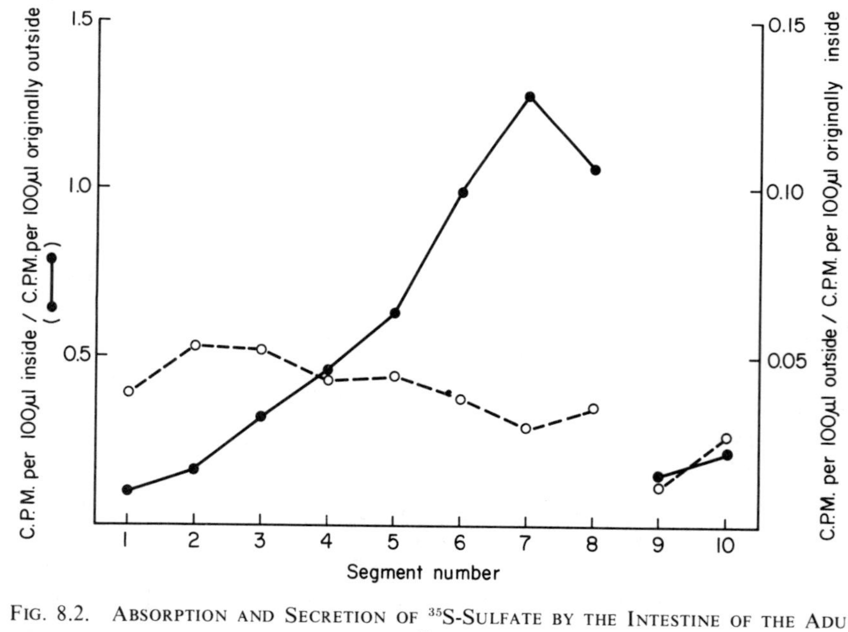


FIG. 8.2. ABSORPTION AND SECRETION OF $^{35}$S-SULFATE BY THE INTESTINE OF THE ADULT RAT

Everted sacs of the intestine prepared from segments approximately 10 cm apart were incubated for 1 hr at 37° C in an oxygenated solution of salts buffered to pH 7.4. $^{35}$S-sulfate was initially present only on the mucosal side, ●----●, or the serosal side, o----o. Each point represents the mean value from at least six experiments. Segment 1 is the proximal duodenum and segment 8 is the terminal ileum. Segments 9 and 10 are the proximal and distal colon, respectively.

sal side, it was most pronounced in the jejunum, segments 2, 3 and 4 of Fig. 8.2. Whether the $^{35}$S-sulfate was presented to the mucosal side or the serosal side, electrophoretic-autoradiographic analysis of the fluids indicated that only inorganic $^{35}$S-sulfate was present in them at the end of 1 hr. Furthermore, by this analytical procedure it was also qualitatively demonstrated that more $^{35}$S-sulfate was transferred in one hour from the mucosal side to the serosal side of the ileum than elsewhere in the intestinal tract.

Under similar experimental conditions $^{45}$Ca was transported from the mucosal side to the serosal side to a greater degree in the duodenum and in the proximal colon than elsewhere in the intestinal tract and phosphate passed most rapidly from the mucosal side to the serosal side in the jejunum, Fig. 8.3.

At the end of an incubation, the sum of the activities of $^{35}$S on the muco-

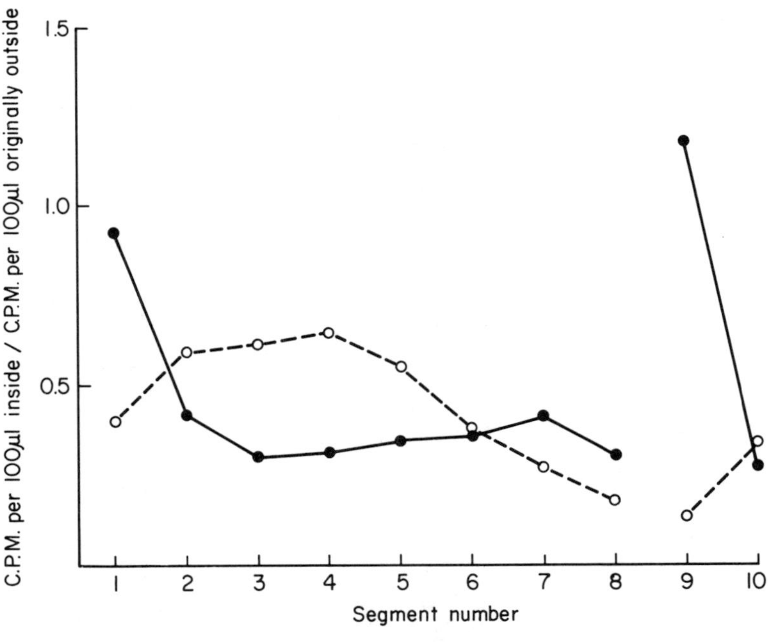


FIG. 8.3. ABSORPTION OF $^{45}$CA AND $^{32}$P-PHOSPHATE BY THE INTESTINE OF THE ADULT RAT

Everted sacs of the intestine were prepared from segments approximately 10 cm apart. Incubation was for 1 hr at 37° C in an oxygenated solution of salts buffered to pH 7.4. The labeled ions, $^{45}$Ca, ●----●, and $^{32}$P-phosphate, ○----○, were initially present only on the mucosal side. Each point represents the mean value from at least four experiments. Segment 1 is the proximal duodenum and segment 8 is the distal ileum. The proximal and distal colon are shown as segments 9 and 10, respectively.

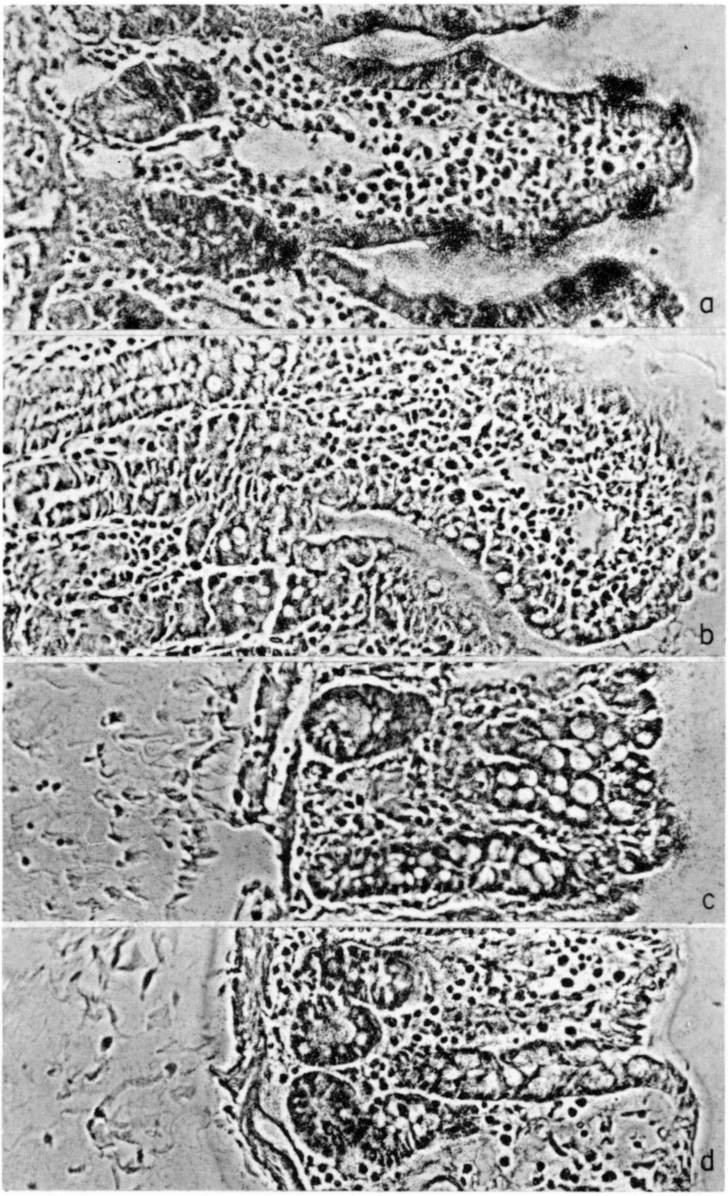


FIG. 8.4. COATED AUTORADIOGRAMS OF THE RAT'S INTESTINE INCUBATED IN A SOLUTION OF SALTS CONTAINING $^{35}$S-SULFATE

Everted sacs of intestinal segments were incubated for 2 hr at 37° C in a buffered solution of salts containing $^{35}$S-sulfate. After washing in an unlabeled solution of the buffered salts, portions of the segments were fixed in formalin. Deparaffinized sections mounted on glass slides were dipped in a 1%

sal and serosal sides was invariably less than the total radioactivity originally present in the medium. To account for the differences, the accumulation of $^{35}S$ in the tissues was ascertained. After the solutions were drained from sacs prepared from segments of the ileum, the sacs were slit lengthwise, rinsed in ice-cold isotonic saline, blotted and weighed. Fixation in 10% formalin for 24 hr at room temperature followed. Such treatment was expected to remove inorganic sulfate. The tissues were then hydrolyzed with 6 *N* hydrochloric acid for 16 hr at 110°C and the $^{35}S$-sulfate was isolated from the hydrolysates as $BaSO_4$ for counting (Dziewiatkowski 1953). The results, given in Table 8.2, suggest that about 5 to 10% of

TABLE 8.2

RETENTION OF $^{35}S$-SULFATE BY INTESTINAL TISSUE OF THE RAT[1]

| Region of Intestine | Number of Samples | "Fixed" $^{35}S$-Sulfate %[2]/100 Mg Tissue |
|---|---|---|
| Adjacent to ICJ[3] | 8 | 3.3 (2.5 to 4.6) |
| 10 cm above ICJ | 8 | 2.5 (1.2 to 4.2) |
| 20 cm above ICJ | 8 | 1.4 (0.6 to 2.5) |

[1]Segments of the ileum, everted and tied at both ends after filling with 0.5 ml of an unlabeled solution of salts, were incubated for 2 hr at 37°C in 5 ml of an oxygenated solution of salts containing 0.5 microcuries of $^{35}S$-sulfate. The tissue samples weighed 325 to 510 mg. After washing and fixation in 10% formalin, the samples of tissue were hydrolyzed with HCl. $^{35}S$-sulfate was isolated as $BaSO_4$ for counting.

[2]The values are given in percentage of the $^{35}S$-sulfate initially present in the incubation medium. The range of values is given in parenthesis.

[3]ICJ stands for ileocaecal junction.

the total radioactvity was "bound" by the tissues in 2 hr, presumably as sulfated protein-polysaccharides in view of the treatment of the tissues.

The question of the localization of the "bound" $^{35}S$-sulfate in the intestinal tissue was thereafter considered. Sacs from adjacent segments along the entire length of the intestinal tracts from two rats were prepared. Pairs of the sacs were incubated with $^{35}S$-sulfate initially on the mucosal side or the serosal side. Incubation was for 1 hr at 37°C in an atmosphere of oxygen. Rings of the sacs were then fixed for 24 hr in 10% formalin or in an ethanolic solution of phosphomolybdic acid (Heatley *et al.* 1956). Following dehydration in ethanol and clearing in xylol, the tissues were em-

---

solution of colloidin and coated with $NTB_3$ emulsion. (× 360) Fig. a and b are autoradiograms of the jejunum; a, when the $^{35}S$-sulfate was presented initially to the mucosal side and b, when it was presented initially to the serosal side. The concentration, microcuries per milliliter was the same in each instance. Note that the reactivity in a is many times greater than in b. Figures c and d are autoradiograms of the distal colon. Here the overall intensities are comparable but the distribution of $^{35}S$ is different. In c, $^{35}S$ is seen rather uniformly distributed in all the goblet cells when the isotope was initially on the mucosal side. In d, note that more $^{35}S$ is seen deep in the crypts than at the tips of the villi when the $^{35}S$ was initially presented to the serosal side.

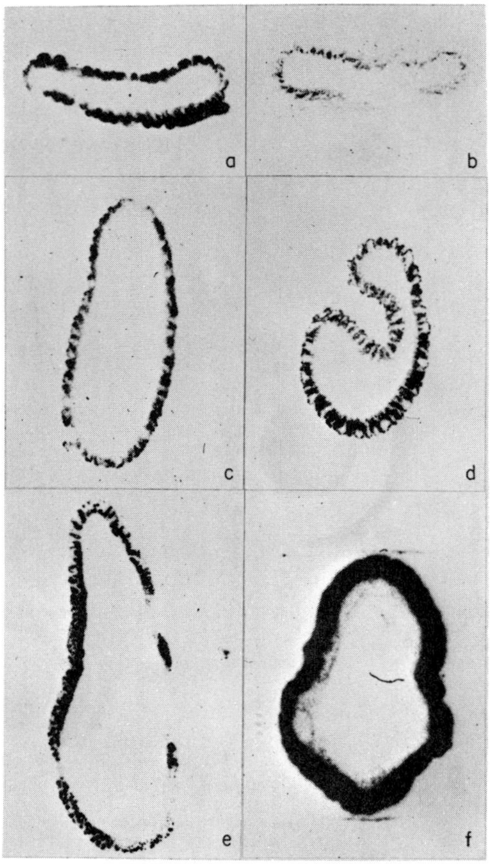


FIG. 8.5. CONTACT AUTORADIOGRAMS OF THE RAT'S INTESTINE AFTER ICUBATION IN A SOLUTION OF SALTS CONTAINING $^{35}$S-SULFATE

Everted sacs of intestinal segments were incubated for 2 hr at 37° C. After washing in an unlabeled solution of buffered salts, portions of the segments were fixed in formalin. Deparaffinized sections, 7 $\mu$ thick, were apposed against strips of Kodak Contrast Process Ortho film ( × 10). Figures a and b are typical of the autoradiograms produced by the duodenum when the $^{35}$S-sulfate was presented initially to the mucosal side, a, or the serosal side, b. The concentrations of $^{35}$S per ml were the same in both instances, but whereas the sac was incubated in 5 ml of the solution it contained only 0.5 ml of the solution. Note that under these conditions much less $^{35}$S was fixed in the tissue when the isotope was presented initially to the serosal side. This was seen to be the case throughout the small intestine. Figures c and d are typical of the autoradiograms produced by the terminal ileum when the amount of $^{35}$S-sulfate, in microcuries, presented initially to the mucosal side, c, was the same as that presented initially to the serosal side, d; the specific activity, micro-

bedded in paraffin and sectioned at 7 $\mu$. Contact autoradiograms (Dziewiatkowski 1951) and coated autoradiograms (Gross *et al.* 1951) of the deparaffinized sections were prepared. The experiment was repeated with 2 more rats; this time, however, the specific activity of the $^{35}$S-sulfate in the solution which was placed in the sac, serosal side, was 10 times that in which the adjacent sac was incubated.

Autoradiograms of the tissues revealed that the $^{35}$S-sulfate was present predominantly in the mucosa in all cases (Fig. a through d of Fig. 8.4). Moreover, the disposition of the $^{35}$S in the tissues also varied according to whether the $^{35}$S-sulfate was presented initially to the mucosal or to the serosal side. At all levels of the intestine more $^{35}$S was demonstrable autoradiographically in the mucosa when the $^{35}$S-sulfate was presented to the mucosal side (compare Fig. a with Fig. b, and Fig. c with Fig. d of Fig. 8.4). Although less $^{35}$S-sulfate was visualized in the mucosa when it was presented to the serosal side more $^{35}$S was seen in the goblet cells proximal to the serosa than in those at a greater distance from the serosa (Fig. d of Fig. 8.4).

It is of interest but the reasons are not apparent for the following. When the specific activity of the $^{35}$S-sulfate presented initially to the serosal side was ten times as great as that presented initially to the mucosal side, autoradiograms of comparable density were obtained from sections of the small intestine (Fig. c and d of Fig. 8.5). Autoradiograms of the colon, on the other hand, were now remarkably more intense (compare Fig. e with Fig. f of Fig. 8.5).

After the injection of $^{35}$S-sulfate into animals, it has been found that $^{35}$S was localized in "bound" form to a greater extent in the mucosa than in the underlying layers of the intestinal tracts. Odeblad and Boström (1952) suggested that sulfate is utilized in the mucosa for the synthesis of sulfate polysaccharides. Bélanger (1953, 1954) described the ingress and exit of $^{35}$S-sulfate, as seen autoradiographically, in different regions of rats' and hamsters' stomachs, and confirmed the observations of Davies and Young (1954) that mucus-secreting cells of the salivary glands rapidly accumulate a large amount of $^{35}$S-sulfate but only transiently. The rapidity with which the intestinal mucosa handles labeled sulfate was empha-

---

curies per millimole of sulfate was thereby ten times as great on the serosal side as on the mucosal side. Similar results were obtained throughout the small intestine. Figures e and f are autoradiograms of the distal colon. The concentrations of $^{35}$S-sulfate presented to the mucosal side, e, and serosal side, f, were as described above in Fig. c and d. The sac of the colon, prepared so that the $^{35}$S-sulfate was presented initially to the serosal side, fixed much more sulfate than the sac prepared so that the $^{35}$S-sulfate was presented initially to the mucosal side.

sized by Boström and Odeblad (1954), but according to Bélanger (1954) not all of the mucosal cells are involved, the goblet cells but not the Brunner glands were active.

That the turnover of sulfate in the intestinal mucosa is rapid has been confirmed by studies in which sulfated polysaccharides in crude form were isolated from rats at intervals of time following the intraperitoneal injection of $^{35}$S-sulfate. The specific activity of the sulfate of the isolated materials increased and then decreased rapidly (Dziewiatkowski 1954). In line with these results are autoradiographic observations on the turnover of $^{35}$S-sulfate in the intestine (Bélanger 1954; Boström and Odeblad 1954; Dziewiatkowski 1956). Additionally, it was observed in rats that the goblet cells farthest from the intestinal lumen accumulate $^{35}$S-sulfate given intraperitoneally sooner than those closer to the lumen. However, this may not be the case in all animal species (Jennings and Florey 1956).

*In vitro*, in the small intestine of the adult rat, particularly in the region of the ileum, the transfer of sulfate from the mucosal side to the serosal side occurs even when the concentration of sulfate on the serosal side is initially greater than that on the mucosal side. As noted in Table 8.3, when the concentration of sulfate on the serosal side was initially ten times that on the mucosal side, transfer of the $^{35}$S-sulfate to the serosal side was only slightly decreased, to 85% of the value found when no concen-

TABLE 8.3

EFFECT OF THE CONCENTRATION OF SULFATE IN THE MEDIUM ON THE TRANSFER OF SULFATE BY THE ILEUM OF THE RAT[1]

| Sulfate Conc. mM/Liter | Relative Transfer to Serosal Side |
|---|---|
| On serosal side—hypertonic[2] | |
| 1 | 1.00 (720 $\mu\mu$M) |
| 10 | 0.85 |
| 100 | 0.44 |
| On mucosal side—hypertonic[2] | |
| 1 | 1.0 (7.30 $\mu\mu$M) |
| 10 | 2.1 |
| 50 | 7.5 |
| 75 | 15.5 |
| On mucosal side—isotonic[2] | |
| 1 | 1.0 (630 $\mu\mu$M) |
| 10 | 5.5 |
| 50 | 16.4 |
| 75 | 31.9 |

[1]In all experiments the $^{35}$S-sulfate was initially in the medium bathing the mucosal side of the everted sacs. Incubation was for 2 hr at 37°C with gentle shaking. The relative amounts of sulfate transferred to the serosal side have been calculated in reference to the amount transferred when the sulfate concentrations on the mucosal and serosal side were initially 1 mM. The $\mu\mu$M of sulfate transferred under the latter conditions are given in parenthesis. At least four animals were used at each concentration of sulfate to obtain the mean values listed.

[2]In one series of experiments additional sodium sulfate was added to the basic solution of buffered salts to obtain the desired concentration of sulfate (hypertonic). In a second series of experiments the increase in sulfate concentration was compensated for by removal of an equivalent amount of sodium chloride; for each mM of sodium sulfate added, 1.5 mM of sodium chloride was withheld (isotonic).

tration gradient was imposed. Even when the concentration of the sulfate on the serosal side was 100 times as great as on the mucosal side, the amount of $^{35}$S-sulfate transferred from the mucosal to the serosal side was decreased only to 44% of the control value.

The effects of increasing the concentration of sulfate on the mucosal side in two ways are also given in Table 8.3. In one series of experiments additional sodium sulfate was added to the solution of buffered salts and in a second series of experiments the increase in sulfate concentration was compensated for by the removal of an equivalent amount of sodium chloride. In either case, as the concentration of sulfate was increased the amount of $^{35}$S-sulfate transferred from the mucosal to the serosal side increased. Within the limits tested, the capacity of the ileum to transfer sulfate from the mucosal to the serosal side was not exceeded; at the highest concentration of sulfate used, 75 mM., more sulfate was transferred than at a concentration of 50 mM. This may simply be a reflection of greater diffusion with increased concentration.

## ASSOCIATION OF SULFATE WITH BLOOD COMPONENTS

The fact that sulfate esters are excreted in the urine strongly suggests that they are present in plasma. Plasma is the medium in which they would be transported from the tissues in which they are synthesized to the kidneys for excretion. Indeed sulfate esters of low molecular weight as well as inorganic sulfate have been shown to be present in plasma. Additionally, some sulfate is strongly bound to plasma proteins. For, after intraperitoneal injection of $^{35}$S-sulfate into rats, it was shown by electrophoresis in starch blocks that some of the $^{35}$S was associated with serum globulins and albumins (Dziewiatkowski 1954). Smith *et al.* (1955) extended these observations. They found that at 24 to 48 hr after intravenous injection of $^{35}$S-sulfate, about 40% of the activity in the serum was associated with a constituent which on filter paper had the electrophoretic mobility of an $\alpha_1$-globulin and which, moreover, stained intensely with fuchsin after treatment with periodic acid. On the basis of these observations they suggested that the $^{35}$S-labeled component in serum was a glycoprotein with sulfate groups.

After the oral administration of $^{35}$S-sulfate, Block *et al.* (1951) isolated $^{35}$S-cystine and $^{35}$S-methionine from the serum albumins of goats. It seemed possible that some of the $^{35}$S present in the serum proteins of rats was also in the form of cystine and/or methionine. To check on this possibility, carrier-free $^{35}$S-sulfate was injected intraperitoneally into adult rats (Dziewiatkowski and Di Ferrante 1957). It was found that as the $^{35}$S in the sera decreased, an increasing fraction of the remaining $^{35}$S was associated with serum proteins. Electrophoretic analysis of the sera, using starch blocks and filter paper, showed that most of the $^{35}S$ was associated with the $\alpha_1$-globulins and the albumins. When plasma was used, it was found that $^{35}S$ was also associated with fibrinogen. The latter is probably due, in part at least, to the presence in rat fibrinogen, as in bovine fibrinogen, of tyrosinyl-*O*-sulfate (Bettelheim 1954). Indeed, this suggestion was subsequently confirmed by Blömback *et al.* (1960). When they isolated peptide B chromatographically from fibrinogen from rabbits given $^{35}S$-sulfate, it contained $^{35}S$. On acid hydrolysis of this peptide, sulfate was liberated simultaneously with the appearance of the tyrosine absorption.

After establishing that $^{35}S$ was strongly bound to plasma proteins, attention was directed to the possible presence of $^{35}S$ as cystine and/or methionine in the proteins. Chromatography of a hydrolysate of serum proteins on Dowex-50 showed that about 5% of the total $^{35}S$ associated with the proteins 24 hr after injection of $^{35}S$-sulfate could be accounted for as cystine and methionine sulfur. This is in agreement with reports that the amount of a dose of sulfate-sulfur which can be recovered from rat tissues as cystine and methionine sulfur is exceedingly small.

Carbohydrate-protein complexes have been isolated from human plasma and it has been suggested that they contain sulfate groups (Winzler *et al.* 1948; Weimer *et al.* 1950; Schmid, 1953). Indeed, Schiller and Dewey (1956) isolated chondroitin sulfate from human plasma. It was no surpirse, therefore, that about 30% of the $^{35}S$ associated with the serum proteins was recovered as sulfate after the following treatment. The serum proteins were extensively digested with trypsin, residual proteins were precipitated with trichloracetic acid, and from the filtrate fragments of small molecular weight were removed by dialysis. The $^{35}S$-sulfate was in the retentate, which also contained glucuronic acid and hexosamines. Apparently it was part of a large molecule.

Dziewiatkowski and Di Ferrante (1957) also described experiments which suggested the presence of $^{35}S$-sulfate in blood cells. An estimate of the concentrations of $^{35}S$ in the cells showed that on a volume basis the mixture of red cells and white cells had less $^{35}S$ than the plasma and that the white cells had significantly more $^{35}S$ than the red cells. Odell *et al.* (1955) have also reported that the uptake of $^{35}S$-sulfate by white cells is greater than by red cells of rats.

To recapitulate, sulfate in plasma is present as inorganic sulfate and in the form of esters of low molecular weight as well as part of high molecular weight polysaccharides. Undoubtedly all of these are in a dynamic state of flux with like molecules in the tissues. It should be apparent, as well, that too little is known of the kinds of sulfate esters in plasma. What role they may play is unknown as well.

## SULFATE ESTERS OF CONNECTIVE TISSUES

The connective tissues contain by far the highest concentration of sulfate esters of all the tissues in an animal body. These are predominantly in the form of chondroitin sulfates and keratin sulfates. In these polysaccharides, which are covalently bound to proteins, the sulfate is in ester linkage with hydroxyl groups of glucosamine and galactosamine. The anionic nature of keratin sulfates derives from them and, in the case of the chondroitin sulfates is accentuated by them.

It is beyond the scope of this presentation to detail the chemistry of the protein polysaccharides of connective tissues. The recent *A Primer on Connective Tissues* by Schubert and Hamerman (1968) is recommended for this purpose. But, even this provocative review is already outdated; the progress in this area is so rapid and multifaceted that it is an herculean task to keep abreast of all aspects. Instead, in line with the topic, "Metabolism of Sulfate Esters," it is pertinent to consider only the metabolism of the sulfate esters in the connective tissues of a limited number of loci.

### Cartilages, Bones, and Teeth

Autoradiographically, one can determine precisely the localization of $^{35}S$ in a tissue and even in the cells of a tissue. If the tissue is fixed in solutions in which inorganic sulfate or other sulfated compounds are soluble, one can reasonably begin to suggest the nature of the remaining $^{35}S$-labeled components of the tissue. A discriminating treatment of tissue sections with enzymes may be of further help in this regard as well.

In cartilages, bones, and teeth the accumulation of $^{35}S$-sulfate in bound forms is dependent on the metabolic activity of the cells. In all three of these tissues, most of the sulfate is linked to carbohydrates which are part of macromolecules, the protein polysaccharides. In cartilages, within a few minutes after administration, $^{35}S$-sulfate is already present in bound form in the cells; almost immediately thereafter it is found in the matrix as well (Bélanger 1954; Godman and Lane 1964). It has been repeatedly determined by chemical analysis that most of the $^{35}S$-sulfate which thus accumulates in the cartilage *in vivo* and *in vitro* is part of the chondroitin sulfate in the tissue (Dziewiatkowski 1951; Boström 1952; Boström and Månsson 1952; Dziewiatkowski 1962). Since chondroitin-4-sulfate and chondroitin-6-sulfate are hydrolyzable by hyaluronidase, the removal of $^{35}S$ from sections of cartilages by incubation with hyaluronidase also implies that the $^{35}S$ is in chondroitin sulfate. It is noteworthy that in the case of epiphyseal cartilages of rats, Bélanger (1954) and Greulich (1956) found that after such enzymatic treatment $^{35}S$ was no longer demonstrable in the matrix or in the cells. On the other hand, although in loci where tendons inserted there was $^{35}S$, its concentration was diminished but it was not completely eliminated. This suggests that in the latter loci there is some type of sulfated polysaccharide in addition to chondroitin-4-sulfate and/or chondroitin-6-sulfate.

$^{35}S$-sulfate in bound form accumulates rapidly in the epiphyses of very young rats. Here the highest concentration is found in the epiphyseal plate. Up to about 24 hr the concentration in the epiphyses continues to increase in the same pattern; more $^{35}S$-sulfate is found in the epiphyseal plate than in the rest of the epiphyseal cartilage. Beyond the 24th hr, the $^{35}S$-sulfate disappears more rapidly from the epiphyseal plate than from the rest of the epiphysis. By the end of 96 hr a nearly uniform distribution of the label is seen in the whole cartilaginous epiphysis. In the region of an epiphysis where an ossification center develops the content of $^{35}S$ decreases remarkably with time.

After equivalent dosages of $^{35}S$-sulfate on a body weight basis, the epiphyses of the long bones from very young rats produce autoradiograms which are more dense than those produced by comparable sections from old rats. Indeed, the epiphyses of a bone from the same rat do not necessarily produce autoradiograms of the same density. For example, the proximal epiphysis of a humerus from a 30-day-old rat is more reactive autoradiographically than the distal epiphysis. In general, such decreased metabolic activity accompanies epiphyseal closure.

The disappearance of $^{35}S$-sulfate from the epiphyseal plate and the concurrent appearance of $^{35}S$-sulfate at a decreased concentration in the adjacent metaphyseal bone suggests that sulfated protein polysaccharides may be involved in the regulation of the development of bone. Chemical analysis of $^{35}S$-labeled metaphysis 24 hr after administration of $^{35}S$-sulfate show that about 20% of the $^{35}S$ is in the form of inorganic sulfate; the remainder is in polysaccharides (Engfeldt *et al.* 1954; Kent *et al.* 1956; Dziewiatkowski *et al.* 1957). Endochondral ossification appears as a process in which the catabolism of sulfated protein polysaccharides is related with subsequent calcification.

In membranous ossification, synthesis of sulfated protein-polysaccharides may antecede calcification; autoradiograms of long bones show $^{35}S$-sulfate as deposited in a thin line in proximity to the periosteum shortly after injection. The tissue in this region of the bone also stains metachromatically with toluidine blue. Because of appositional growth this line of $^{35}S$ moves inward, across the shaft toward the marrow cavity. Once embedded in the calcified shaft the line of $^{35}S$ is no longer stained metachromatically. This is also the case with metaphyseal bone, unless the sections are decalcified prior to staining.

The Haversian systems which are actively developing are outlined with $^{35}$S-sulfate in autoradiograms (Engfeldt *et al.* 1954). From a careful evaluation of autoradiograms of compact bones from adult dogs, Vincent (1955) and Lacroix (1956) concluded that the $^{35}$S-sulfate accumulated in regions of the Haversian canals which were as yet uncalcified and which stained metachromatically. Contiguous with such $^{35}$S-labeled regions were regions where calcification was taking place, as indicated by the deposition of $^{45}$Ca. That in bone as in cartilage $^{35}$S-sulfate may be indicative of sulfated polysaccharides is in line with the fact that polysaccharides labeled with $^{35}$S-sulfate have been extracted from compact bone (Kent *et al.* 1956; Dziewiatkowski *et al.* 1957).

As early as 1952, Verne *et al.* (1952) demonstrated that $^{35}$S-sulfate accumulated in an organically bound form in predentin and in dentin but very little has been done subsequently to ascertain the nature of the labeled components. A more detailed autoradiographic study of these tissues in rats by Bélanger (1954) showed that, 1 to 2 hr after administration, $^{35}$S-sulfate was present throughout the predentine and dentine. This overall distribution of $^{35}$S was no longer seen after demineralization. Instead, only a thin band of $^{35}$S-labeled material was discerned at the junction of the predentine and dentine. Four days later, the $^{35}$S-labeled band was seen to be clearly displaced toward the dentino-enamel junction. Here, as in the case of the shafts of long bones, the apparent displacement was conceived as due to further additions of unlabeled material at the predentinal-dentinal junction. Although the demineralized dentine stained metachromatically with toluidine blue the $^{35}$S-labeled material was resistant to hyaluronidase (Bélanger 1952).

$^{35}$S-sulfate accumulates in preenamel in a band within a few hours after administration to very young rats or hamsters (Bélanger 1955). In teeth in early stages of development the band of $^{35}$S was seen extending over a considerable length of the enamel. On the other hand, in more fully developed teeth the band was short and limited to the more basal portion. The $^{35}$S-labeled material in the preenamel was not removed by decalcification and subsequent to decalcification was not noticeably affected on incubation with hyaluronidase.

**Brain**

A qualitative estimate of the relative concentrations of bound $^{35}$S-sulfate in various tissues of the fetal rat indicated that at 24 hr after the administration of inorganic $^{35}$S-sulfate to pregnant rats, the concentration of the $^{35}$S in the fetal brains was relatively insignificant when compared with cartilages (Dziewiatkowski 1953). In a study of the distribution of intraperitoneally-injected inorganic $^{35}$S-sulfate as a function of time in adult rats, it was found that the concentration of $^{35}S$ in the brain was less than that in blood or liver up to 24 hr after injection. Thereafter, its concentration was comparable in these three tissues. In fact, the data suggested that the concentration of $^{35}S$ in the brain after 24 hr was slightly greater than in the blood (Dziewiatkowski 1949A). Although low in concentration, the bound $^{35}S$-sulfate in brains of adult rats was shown autoradiographically by Boström and Odeblad (1953) to be present at different concentrations in the cerebrum, cerebellum, and spinal cord; the concentration of $^{35}S$ in the grey matter was greater than that in the white matter.

Intrigued by these observations, a study on the distribution of $^{35}S$-sulfate and the kinds of $^{35}S$-labeled components in the brains of suckling rats was undertaken (Dziewiatkowski and Slayman unpublished). Into each of 123 suckling rats at 8 days postnatally 250 microcuries of carrier-free inorganic $^{35}S$-sulfate were injected intraperitoneally. The rats were sacrificed after intervals of time by decapitation. Blood samples were collected, weighed, and then diluted with water. They were stored at $-15°C$ until

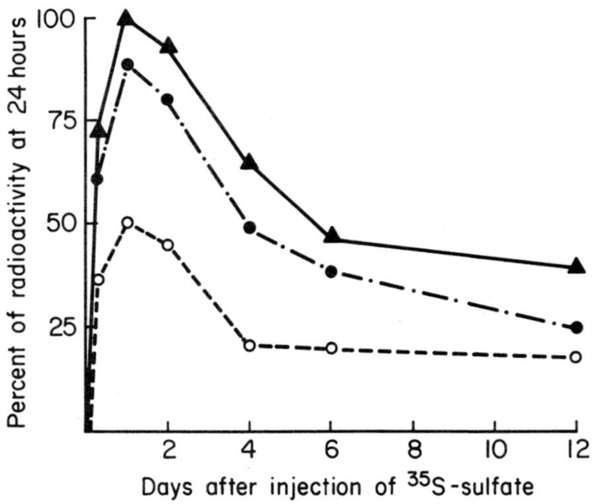


FIG. 8.6. EFFECT OF HYALURONIDASE ON $^{35}S$-LABELED SECTIONS OF BRAINS OF SUCKLING RATS

Brains removed from suckling rats after intervals of time following intraperitoneal injection of $^{35}S$-sulfate were fixed in formalin and dehydrated in ethanol. They were embedded in paraffin after clearing in xylol. The paraffin was removed before the sections were counted, ▲----▲, and then incubated in either a 1% solution of NaCl, o----o, or in a 1% solution of NaCl containing testicular hyaluronidase before they were recounted. The mean value for unincubated sections at 24 hr was taken as 100%. Each point is the mean value on eight sections.

analyzed. Within 5 min of sacrifice, the brains were removed and slit in half longitudinally. One-half of each was fixed for 72 hr in 10% formalin, which was buffered to pH 7.0 with phosphate. The tissue was dehydrated in graded concentrations of ethanol. After the tissue was cleared in xylol it was embedded in paraffin and sections, 7 $\mu$ in thickness, were prepared.

Representative sections of the brains, affixed to glass slides with the paraffin removed, were counted. Some of the sections were then incubated for 3 hr at 37°C in a 1% solution of sodium chloride, with or without hyaluronidase (Pearse 1960). The sections were rinsed in water and dried; then they were recounted. It can be seen from the results shown in Fig. 8.6 that, although there were small losses of $^{35}$S from the sections incubated in the saline solution, the losses from the sections incubated with hyaluronidase were substantial; they accounted for approximately 50% of the radioactivity at each point in time. These data suggest that in the connective tissues of the brain at least ½ of the $^{35}$S-sulfate may be linked to polysaccharides which in their lability to hyaluronidase resemble chondroitin-4-sulfate and/or chondroitin-6-sulfate. The remainder may well be in the form of sulfolipids (Pritchard 1966). Both types of components appear to turn over slowly.

The amount of $^{35}$S sequentially removed by the fixative solutions, alcohol solutions, and by the xylol was determined. Samples of the solutions were also analyzed by electrophoresis on paper. Thereby inorganic sulfate was separated from other $^{35}$S-labeled components. As can be seen in the autoradiogram of an electrophoretogram of the formalin solutions, Fig. 8.7, the concentration of inorganic $^{35}$S-sulfate in the brains decreased rapidly as a function of time. Furthermore, the relation between inorganic sulfate and other sulfated components of the brain as seen in electrophoretograms of such extracts suggests that inorganic sulfate is utilized in the synthesis of these numerous unidentified sulfated components of the brain, which are, incidentally, as transient as inorganic sulfate. In the alcohol solutions and in the xylol additional $^{35}$S-labeled components were detected electrophoretically and these, too, appeared to be transient residents of the brain.

Contact and coated autoradiograms of the brains were revealing. One hour after injection the highest concentration of bound $^{35}$S-sulfate was in cerebral and cerebellar meninges, in the choroid plexus of the lateral ventricles, and in the tela choroidea. The high concentration was probably associated with the connective tissue elements of the vascular system in these regions. Moreover, the cerebellum, thalamus, and hippocampus had a higher concentration of $^{35}$S than the midbrain and the diencephalon. The layers of the cerebellum were very sharply defined in the autoradiograms; the molecular layer and the central fiber tracts had higher concen-

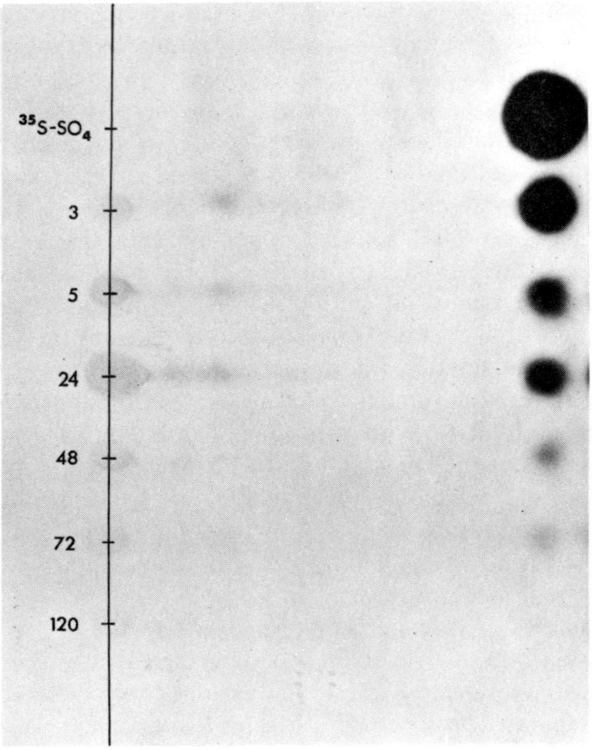


FIG. 8.7. AUTORADIOGRAM OF AN ELECTROPHORETOGRAM OF FORMALIN USED AS A FIXATIVE OF BRAINS FROM SUCKLING RATS GIVEN $^{35}$S-SULFATE

An aliquot of a 10% formalin solution, equivalent to 100 mg of brain, wet weight, was evaporated to dryness. The residue was dissolved in 1 ml of 0.03 *M* citrate buffer, pH 5. Samples of 25 $\mu$l volume were applied to Whatman 3MM paper, 23 cm × 57 cm. Electrophoresis was for 2 hr with an applied 400 volts. The dried paper was apposed to Kodak No-Screen X-Ray film. The numbers on the reproduction of the autoradiogram are times in hours after injection of the $^{35}$S-sulfate to the rats.

trations of $^{35}$S than either of the granular layers. In the cerebral cortex the molecular layer and the deep plexiform layer had less $^{35}$S than the over- and underlying cellular layer. The cerebral radiation (corona radiata) also had more $^{35}$S than either the overlying cortical layers or the underlying corpus callosum. Within the hippocampus, the alveus and central white matter were more strongly labeled than the pyramidal or granular layers.

By five hours after injection of the $^{35}$S-sulfate, the concentration of bound $^{35}S$ in the meninges and other vascular tissue was markedly decreased from that seen at 1 hr. By this time, the fiber tracts of the cerebellar gyri and medulla had more $^{35}S$ relative to the cellular layers; and in the cerebrum, the superficial molecular layer also had more bound $^{35}S$ than the underlying cellular layers. The inferior colliculus had more $^{35}S$ than the superior colliculus.

At 24 hr, the midbrain, the diencephalon, and the medulla were more radioactive than the cerebrum. The inferior olivary body had less $^{35}S$ than the surrounding grey matter.

By 60 hr postinjection marked changes in the pattern of $^{35}S$ distribution, compared to the distribution at much earlier times, had occurred. The concentration of $^{35}S$ in the cerebellum had declined below that of the forebrain, midbrain, and medulla. The concentration of $^{35}S$ in the inferior colliculus was also markedly decreased. The deep plexiform region of the cerebral cortex had more $^{35}S$ than the surrounding cellular layers.

Beyond this point in time up to 120 hr after injection there were but minor changes in the relative concentrations of $^{35}S$ in the various subdivisions of the brain, with one exception: The concentration of $^{35}S$ in the cerebellum decreased more rapidly than the concentration of $^{35}S$ in other regions. The autoradiogram Fig. 8.8 is representative of the distribution of $^{35}S$ in the various regions of the brain at 72 hr after the injection of $^{35}S$-sulfate intraperitoneally.

When the brains were removed from the animals and divided, one-half of each brain was weighed and frozen until analyzed. For analysis, the

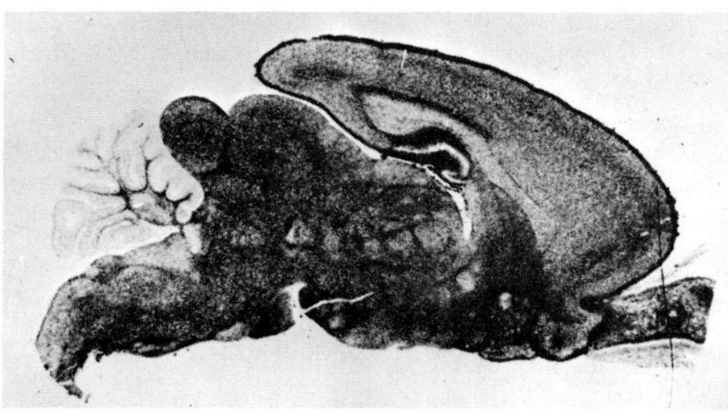

FIG. 8.8. AUTORADIOGRAM OF A SECTION OF BRAIN FROM A SUCKLING RAT GIVEN $^{35}S$-SULFATE

The brain was removed from an 11-day-old rat, 72 hr after an intraperitoneal injection of 250 microcuries of $^{35}S$-sulfate. Fixation was in 10% formalin buffered with phosphate to pH 7.0. Sections cut at 7 $\mu$ were placed in contact with strips of Kodak Contrast Process Ortho film to produce autoradiograms as reproduced here.

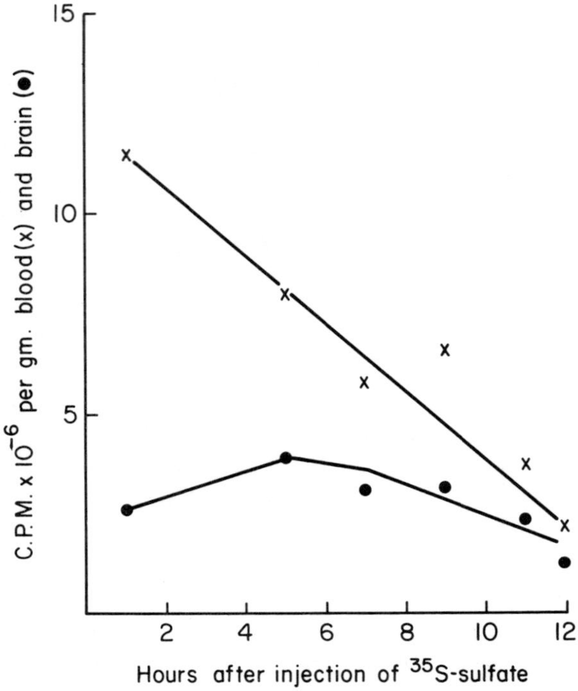


FIG. 8.9. CONCENTRATION OF $^{35}$S IN BLOOD AND IN BRAIN OF SUCKLING RATS WITH TIME AFTER INJECTION OF $^{35}$S-SULFATE

Eight-day-old rats were given 250 microcuries of $^{35}$S-sulfate as the sodium salt by intraperitoneal injection. The total concentration of $^{35}$S in the blood, x----x, and in the brain, ●----●, was determined at intervals of time. Each point is the mean value of at least three values.

brains were thawed in a refrigerator and they were then homogenized into water. An aliquot of each homogenate was analyzed for total $^{35}$S. The blood samples were analyzed similarly at the same time. The concentration of $^{35}$S in the blood was highest at the earliest time of sampling, Fig. 8.9. In the case of the brains, on the other hand, it continued to increase up to about 3 to 5 hr postinjection. At about 24 hr the concentration of $^{35}$S in both tissues was comparable. Thereafter, although the concentration of $^{35}$S in both tissues continued to decrease, it did so more rapidly in blood than in the brain, Fig. 8.10.

For comparison, 8 adult male rats (200 to 250 gm) were given 1.25 millicuries of carrier-free $^{35}$S-sulfate by intraperitoneal injection. A representative animal was sacrificed at 1, 4, 10, 24, 48, 96, 192, and 480 hr later. Their bloods and brains were analyzed for $^{35}$S as were those of the sucklings. In the blood and in the brain the highest values for $^{35}$S were observed at one hour postinjection. At this time the concentration of the label

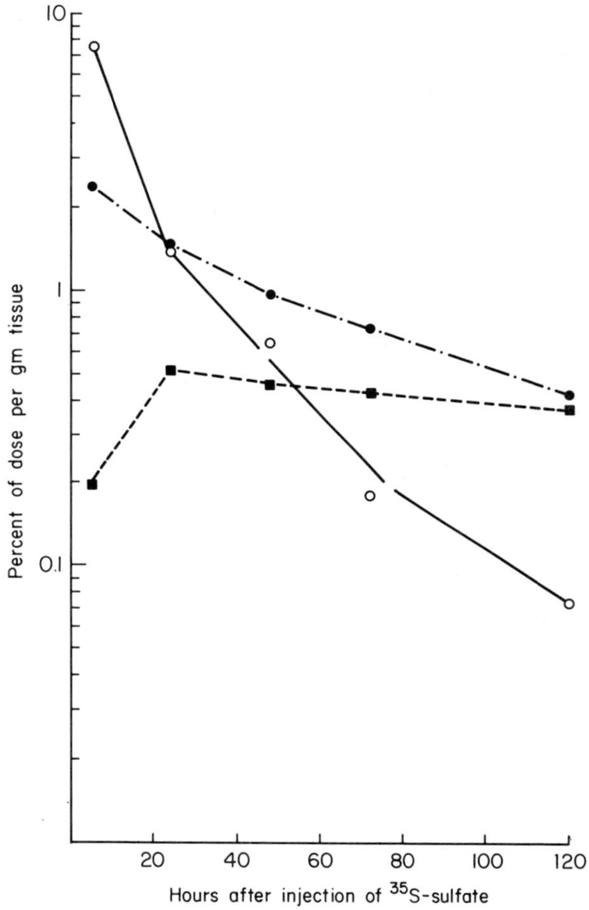


FIG. 8.10. CONCENTRATION OF $^{35}S$ IN BLOOD AND IN BRAIN OF SUCKLING RATS WITH TIME AFTER INJECTION OF $^{35}S$-SULFATE

Eight-day-old rats were given 250 microcuries of $^{35}S$-sulfate as the sodium salt by intraperitoneal injection. After extended hydrolysis of the samples with hydrochloric acid, sulfate was isolated as $BaSO_4$ for counting. The concentration in the blood, o----o, decreased more rapidly than in whole brain, ●----●. At 24 hr and thereafter the concentration of fixed $^{35}S$-sulfate in the brain, ■----■, accounted for the major fraction of the $^{35}S$-sulfate in the brain. The values for fixed $^{35}S$-sulfate were obtained on samples of brain which had been fixed in formalin and then dehydrated in increasing concentrations of ethanol and in xylol.

in the blood was ten times as great as that in the brain. Thereafter, the levels of $^{35}S$ dropped exponentially in both tissues. However, the curves crossed at about 24 hr and thereafter, as in the case of the suckling rats, the level of $^{35}S$ in the brain was maintained above the level in the blood.

As might be expected, the percentage of the dose per gram of brain in the suckling rats was greater by a factor of approximately 80 than in adult rats at equivalent time points. Some, but certainly not all, of this difference can be explained in terms of the lesser brain to body mass (1/100) in adult rats than in the suckling rats (1/15 to 1/25). More probably the difference is due to an incomplete blood-brain barrier in the developing brains of the suckling rats (Bakay 1953; Stern and Peyrot 1927). An accentuated synthesis of sulfated components in the brains of the suckling rats may also contribute to the differences observed.

In addition to the above, aliquots of the homogenates of the brains were digested with trypsin. After incubation at 37° C for 24 hr, the samples were dialyzed against running tap water for 24 hr. Insoluble materials were removed by filtration. To the filtrates cetyl trimethyl ammonium bromide was added in order to precipitate acid polysaccharides. The precipitates were recovered by centrifugation and washed 3 times in the centrifuge

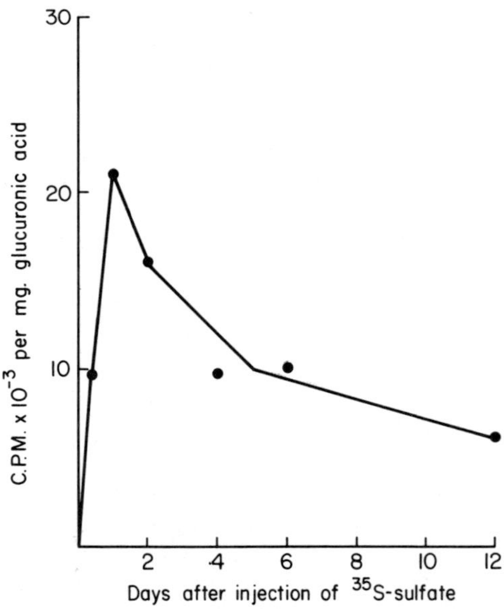


FIG. 8.11. TURNOVER OF $^{35}$S-LABELED POLYSACCHARIDES IN BRAINS OF SUCKLING RATS

Brains from suckling rats given $^{35}$S-sulfate intraperitoneally were homogenized into a 0.1 *M* $K_2HPO_4/KH_2PO_4$ buffer, pH 7.1, and then incubated with trypsin for 24 hr at 37° C. After dialysis against running tap water, insoluble materials were removed by filtration. From the filtrates, polysaccharides were precipitated by the addition of cetavalon. The precipitates after washing 3 times with 95% ethanol saturated with NaCl, were analyzed for $^{35}$S and for glucuronic acid.

with 5 ml portions of 95% ethanol saturated with sodium chloride. The washed crude precipitates of acid polysaccharides were then analyzed for $^{35}S$ and for glucuronic acid (Dische 1947). These results, Fig. 8.11, as those shown in Fig. 8.6, suggest that sulfated polysaccharides are present in brain and that, even in very young rats they do not turn over very rapidly. The results, moreover, are in agreement with the observations reported by Robinson and Green (1962) on much older rats.

The foregoing account has indicated that there are many sulfate esters in mammalian organisms. Many of them need to be fully characterized before their metabolic roles may be delineated.

## BIBLIOGRAPHY

ARNOLDT, R. I., and DE MEIO, R. H. 1941. The conjugation of phenol and its action on the respiration of tissues *in vitro*. Rev. Soc. Arg. Biol. *17*, 570–574. (Spanish)

BAKAY, L. 1953. Studies on the blood-brain barrier with radioactive phosphorus. Arch. Neurol. Psychiat. *70*, 30–39.

BALDWIN, B. C., ROBINSON, D., and WILLIAMS, R. T. 1959. Studies in detoxication. 78. Glucuronide and ethereal sulphate formation in hens. Biochem. J. *71*, 638–642.

BANDURSKI, R. S., WILSON, L. G., and SQUIRES, C. L. 1956. The mechanism of "active sulfate" formation. J. Am. Chem. Soc. *78*, 6408.

BAUMANN, E. 1876A. On the occurrence of pyrocatechol in urine. Pflügers Arch. *12*, 63–69. (German)

BAUMANN, E. 1876B. On sulfoacids in urine. Ber. Deut. Chem. Ges. *9*, 54–58. (German)

BÉLANGER, L. F. 1952. Autoradiographic visualization of the entry and transit of $^{35}S$ in cartilage, bone and dentine of young rats and the effect of hyaluronidase *in vitro*. Can. J. Biochem. Physiol. *32*, 161–169.

BÉLANGER, L. F. 1953. Autoradiographic detection of sulphur-35 synthesis by the mucous neck cells of the rat's stomach. Nature *172*, 1150.

BÉLANGER, L. F. 1954. Autoradiographic visualization of $^{35}S$ incorporation and turnover by the mucous glands of the gastrointestinal tract and other soft tissues of rat and hamster. Anat. Record *118*, 755–772.

BÉLANGER, L. F. 1955. Autoradiographic detection of radiosulfate incorporation by the growing enamel of rats and hamsters. J. Dental Res. *34*, 20–27.

BERNSTEIN, S., and MCGILVERY, R. W. 1952. The enzymatic conjugation of *m*-aminophenol. J. Biol. Chem. *198*, 195–203.

BETTELHEIM, FR. R. 1954. Tyrosine-*O*-sulphate in a peptide from fibrinogen. J. Am. Chem. Soc. *76*, 2838–2839.

BLOCK, R. J., STEKOL, J. A., and LOOSLI, J. K. 1951. Synthesis of sulphur amino acids from inorganic sulfate by ruminants. II. Synthesis of cystine and methionine from sodium sulfate by the goat and by the microorganisms of the rumen of the ewe. Arch. Biochem. Biophys. *33*, 353–363.

BLÖMBACK, B., BOSTRÖM, H., and VESTERMARK, A. 1960. On the $^{35}S$-sulfate incorporation in fibrinopeptide B from rabbit fibrinogen. Biochim. Biophys. Acta *38*, 502–512.

BORSOOK, H., KEIGHLEY, G., YOST, D. M., and MCMILLAN, E. 1937. Urinary excretion of ingested radioactive sulfur. Science *86*, 525.

BOSTRÖM, H. 1952. On the metabolism of the sulfate group of chondroitinsulfuric acid. J. Biol. Chem. *196*, 477–481.

BOSTRÖM, H. 1961. Some aspects of sulphate conjugation. Biochem. Pharmacol. *6*, 72–81.

BOSTRÖM, H., and AQVIST, S. 1952. Utilization of $^{35}$S-labelled sodium sulphate in the synthesis of chondroitin sulphuric acid, taurine, methionine, and cystine. Acta Chem. Scand. *6*, 1557–1559.

BOSTRÖM, H., and MÅNSSON, B. 1952. On the enzymatic exchange of the sulfate group of chondroitinsulfuric acid in slices of cartilage. J. Biol. Chem. *196*, 483–488.

BOSTRÖM, H., and ODEBLAD, E. 1953. Autoradiographic observations on the uptake of $^{35}$S-labelled sodium sulphate in the nervous system of the adult rat. Acta Psychiat. Neurol. Scand. *28*, 1–4.

BOSTRÖM, H., and ODEBLAD, E. 1954. An autoradiographic study on the occurrence of injected radiosulphate in the intestine. Acta Physiol. Scand. *32*, 124–128.

BOYLAND, E., MANSON, D., and ORR, S. F. D. 1957. The biochemistry of aromatic amines. II. The conversion of arylamines into arylsulphamic acids and arylamine-N-glucosiduronic acids. Biochem. J. *65*, 417–423.

CRANE, R. K. 1960. Intestinal absorption of sugar. Physiol. Rev. *40*, 789–825.

DAVIES, D. V., and YOUNG, L. 1954. Radioautographic studies of the digestive tracts of rats injected with inorganic sulphate labelled with sulphur-35. Nature *173*, 448–449.

DE MEIO, R. H., LEWYCKA, C., and WIZERKANIUK, M. 1956. Biological synthesis of sulfuric acid esters of steroid hormones. Federation Proc. *15*, 241.

DE MEIO, R. H., WIZERKANIUK, M., and SCHRIEBMAN, I. 1955. Enzymatic system synthesizing sulfuric acid esters of phenols. J. Biol. Chem. *213*, 439–443.

DISCHE, Z. 1947. A new specific color reaction of hexuronic acids. J. Biol. Chem. *167*, 189–198.

DOWDLE, E. B., SCHACHTER, D., and SCHENKER, H. 1960. Requirement for vitamin D for the active transport of calcium by the intestine. Am. J. Physiol. *198*, 269–274.

DZIEWIATKOWSKI, D. D. 1949A. Rate of excretion of radioactive sulfur and its concentration in some tissues of the rat after intraperitoneal administration of labeled sodium sulfate. J. Biol. Chem. *178*, 197–202.

DZIEWIATKOWSKI, D. D. 1949B. On the utilization of exogenous sulfate sulfur by the rat in the formation of ethereal sulfates as indicated by the use of sodium sulfate labeled with radioactive sulfur. J. Biol. Chem. *178*, 389–393.

DZIEWIATKOWSKI, D. D. 1951. Isolation of chondroitin sulfate-$^{35}$S from articular cartilage of rats. J. Biol. Chem. *189*, 187–190.

DZIEWIATKOWSKI, D. D. 1953. Sulfate-sulfur metabolism in the rat fetus as indicated by sulfur-35. J. Exptl. Med. *98*, 119–128.

DZIEWIATKOWSKI, D. D. 1954. Effect of age on some aspects of sulfate metabolism in the rat. J. Exptl. Med. *99*, 283–298.

DZIEWIATKOWSKI, D. D. 1956. Turnover of $^{35}$S-sulfate in the mucosa of the gastrointestinal tract of rats as seen in autoradiograms. J. Biophys. Biochem. Cytol. *2*, 29–32.

DZIEWIATKOWSKI, D. D. 1958. Autoradiographic studies with $^{35}$S-sulfate. Intern. Rev. Cytol. *VII*, 159–193.

DZIEWIATKOWSKI, D. D. 1962. Intracellular synthesis of chondroitin sulfate. J. Cell Biol. *13*, 359–364.

DZIEWIATOWSKI, D. D., and DI FERRANTE, N. 1957. Association of sulfate-$^{35}$S with serum proteins in the rat. J. Biol. Chem. *227*, 347–356.

DZIEWIATKOWSKI, D. D., DI FERRANTE, N., BRONNER, F., and OKINAKA, G. 1957. Turnover of $^{35}$S-sulfate in the epiphyses and diaphyses of suckling rats. Nature of the $^{35}$S-labeled compounds. J. Exptl. Med. *106*, 509–524.

DZIEWIATKOWSKI, D. D., and SLAYMAN, C. The turnover of $^{35}$S-sulfate in the brains of rats. Unpublished results.

EIBER, H. B., and DANISHEFSKY, I. 1957. The incorporation of $^{35}$S-sulfate into heparin. J. Biol. Chem. *226*, 721–724.

EMBDEN, G. 1902. Concerning the conjugation of glucuronic acid in the liver. Beitr. Chem. Physiol. Path. *2*, 591. (German)

EMBDEN, G., and GLASSNER, K. 1902. Concerning the conjugation of glucuronic acid in the liver. Beitr. Chem. Physiol. Path. *1*, 310–327. (German)

ENGFELDT, B., ENGSTRÖM, A., and BOSTRÖM, H. 1954. The localization of radiosulfate in bone tissue. Exptl. Cell Res. *6*, 251–253.

GODMAN, G. C., and LANE, N. 1964. On the site of sulfation in the chondrocyte. J. Cell Biol. *21*, 353–366.

GREENGARD, H., and WOOLEY, J. R. 1940. Studies on colloidal sulfurpolysulfide mixture; absorption and oxidation after oral administration. J. Biol. Chem. *132*, 83–89.

GREGORY, J. D., and LIPMANN, F. 1957. The transfer of sulfate among phenolic compounds with 3′, 5′-diphosphoadenosine as coenzyme. J. Biol. Chem. *229*, 1081–1090.

GREULICH, R. C. 1956. Utilization of sulfate ion by eosinophile myelocytes in the rat. Exptl. Cell Res. *11*, 225–228.

GROSS, J., BOGOROCH, R., NADLER, N. J. and LEBLOND, C. P. 1951. The theory and methods of the radioautographic localization of radioelements in tissues. Am. J. Roentgenol. Radium Therapy *65*, 420–458.

HEATLEY, N. G., JEROME, D. S., JENNINGS, M. A., and FLOREY, H. W. 1956. On the fixation of mucin in the preparation of autoradiographs. Quart. J. Exptl. Physiol. *41*, 124–130.

HELE, T. S. 1924. Studies in the sulphur metabolism of the dog. I. The synthesis of ethereal sulphate. Biochem. J. *18*, 110–119.

HELE, T. S. 1931. Studies in sulphur metabolism of dogs; synthesis of phenyl sulphate and indoxyl sulphate. Biochem. J. *25*, 1736–1744.

HENRY, W. 1829. The Elements of Experimental Chemistry, 11th Edition. Baldwin and Cradock, London.

HERTER, C. A., and WAKEMAN, A. J. 1899. The action of hepatic renal and other cells on phenol and indol, under normal and pathological conditions. J. Exptl. Med. *4*, 307–326.

HOLMGÅRD, A. 1955. On the sulfate metabolism in the sulphocerebrosides of the brain. Acta Chem. Scand. *9*, 1038.

ISSELBACHER, K. J., and MCCARTHY, E. A. 1958. Identification of a sulfate conjugate of bilirubin in bile. Biochim. Biophys. Acta *29*, 658–659.

JENNINGS, M. A., and FLOREY, H. W. 1956. Autoradiographic observations on mucous cells of stomach and intestine. Quart. J. Exptl. Physiol. *41*, 131–152.

KENT, P. W. *et al.* 1956. The deposition of $^{35}S$ in cortical bone. Biochem. J. *62*, 470–476.

LACROIX, P. 1956. The histological remodelling of adult bone. An autoradiographic study. *In* Ciba Foundation Symposium on Bone Structure and Metabolism. Little, Brown & Co., Boston.

LAIDLAW, J. C., and YOUNG, L. 1948. Studies on the synthesis of ethereal sulphates *in vivo.* Biochem. J. *42*, 1–2.

LAIDLAW, J. C., and YOUNG, L. 1953. Biochemical studies of toxic agents; study of ethereal sulphate formation *in vivo* using radioactive sulphur. Biochem. J. *54*, 142–148.

MARENZI, A. D. 1931. Is there an organ which conjugates phenols preponderantly? Compt. Rend. Soc. Biol. *107*, 743–744. (French)

MASON, M. 1970. Effects of conjugated steroids on enzymes. *In* Metabolic Conjugation and Metabolic Hydrolysis, W. H. Fishman (Editor). Academic Press, New York.

MORROW, P. C. *et al.* 1952. Gastrointestinal non-absorption of sodium cellulose sulfate labeled with $^{35}S$. J. Pharmacol. Exptl. Therap. *105*, 273–281.

ODEBLAD, E., and BOSTRÖM, H. 1952. An autoradiographic study of the incorporation of $^{35}S$-labeled sodium sulfate in different organs of adult rats and rabbits. Acta Pathol. Microbiol. Scand. *31*, 339–344.

ODELL, T. T., Jr., TAUSCHE, F. G., and GUDE, W. D. 1955. Uptake of radioactive sulfate by elements of the blood and the bone marrow of rats. Am. J. Physiol. *180*, 491–494.

PASTERNAK, C. A., and KENT, P. W. 1958. Biosynthesis of intestinal mucins. 2. Incorporation of $^{35}S$-sulphate by guinea-pig colon *in vitro.* Biochem. J. *68*, 452–457.

PEARSE, A. G. E. 1960. Histochemistry, 2nd Edition. Little, Brown & Co., Boston.

PRITCHARD, E. T. 1966. *In vivo* labelling of sulphatides from $^{35}S$-sulphate in rat brain during early growth. J. Neurochem. *13*, 13–21.

ROBBINS, P. W., and LIPMANN, F. 1956A. The enzymatic sequence in the biosynthesis of active sulfate. J. Am. Chem. Soc. *78*, 6409–6410.

ROBBINS, P. W., and LIPMANN, F. 1956B. Identification of enzymatically active sulfate as adenosine-3′-phosphate-5′-phosphosulfate. J. Am. Chem. Soc. *78*, 2652.

ROBINSON, J. D., and GREEN, J. P. 1962. Sulfomucopolysaccharides in brain. Yale J. Biol. Med. *35*, 248–257.

SATO, T., SUZUKI, T., FUKUYAMA, T., and YOSHIKAWA, H. 1956. Studies on conjugation of $^{35}S$-sulfate with phenolic compounds. III. Metabolism of *p*-cresol, *p*-hydroxybenzaldehyde and *p*-hydroxybenzoic acid in rat liver. J. Biochem. (Tokyo). *43*, 413–421.

SCHACHTER, D., DOWDLE, E. B., and SCHENKER, H. 1960. Active transport of calcium by the small intestine of the rat. Am. J. Physiol. *198*, 263–268.

SCHACHTER, D., and ROSEN, S. M. 1959. Active transport of $^{45}Ca$ by the small intestine and its dependence on vitamin D. Am. J. Physiol. *196*, 357–362.

SCHILLER, S., and DEWEY, K. F. 1956. Isolation of chondroitin sulfuric acid from normal human plasma. Federation Proc. *15*, 348.

SCHMID, K. 1953. Preparation and properties of serum and plasma proteins. XXIX. Separation from human plasma of polysaccharides, peptides, and proteins of low molecular weight. Crystallization of an acid glycoprotein. J. Am. Chem. Soc. *75*, 60–68.

SCHNEIDER, J. J., and LEWBART, M. L. 1956. Enzymatic synthesis of steroid sulfates. J. Biol. Chem. *222*, 787–794.

SCHUBERT, M., and HAMERMAN, D. 1968. A Primer on Connective Tissues. Lea and Febiger, Philadelphia.

SEGAL, H. L. 1956. Sulfate-dependent exchange of pyrophosphate with nucleotide phosphate. Biochim. Biophys. Acta *21*, 194–195.

SMITH, L. H., ANDERSON, B., and ODELL, T. T., JR. 1955. Binding of $^{35}$S-labeled sulfate in serum of rats as studied by filter-paper electrophoresis. Proc. Soc. Exptl. Biol. Med. *90*, 360–364.

STERN, L., and PEYROT, P. 1927. The function of the blood-brain barrier in various animal species at different developmental stages. Compt. Rend. Soc. Biol. *96*, 1124–1126. (French)

SUZUKI, S., and STROMINGER, J. L. 1960. Enzymatic sulfation of mucopolysaccharides in hen oviduct. I. Transfer of sulfate from 3′-phosphoadenosine-5′-phosphosulfate to mucopolysaccharides. II. Mechanism of the reaction studied with oligosaccharides and monosaccharides as acceptors. III. Mechanism of sulfation of chondroitin and chondroitin sulfate A. J. Biol. Chem. *235*, 257–266, 267–273, 274–276.

TARVER, H., and SCHMIDT, C. L. A. 1939. The conversion of methionine to cystine: experiments with radioactive sulfur. J. Biol. Chem. *130*, 67–80.

VERNE, J., WEILL, R., CACCALDI, P. F., and DE CHARPAL, O. 1952. Studies on the metabolism of radioactive sulfur in teeth of young rats. Compt. Rend. Soc. Biol. *146*, 1558–1560. (French)

VESTERMARK, A., and BOSTRÖM, H. 1959. Studies on ester sulfates. V. On the enzymatic formation of ester sulfates of primary aliphatic alcohols. Exptl. Cell Res. *18*, 174–177.

VINCENT, J. 1955. Studies on the constitution of adult bone. Editions Arscia, Brussels. (French)

WEIMER, H. E., MEHL, J. W., and WINZLER, R. J. 1950. Studies on the mucoproteins of human plasma. V. Isolation and characterization of a homogenous mucoprotein. J. Biol. Chem. *185*, 561–568.

WILSON, L. G., and BANDURSKI, R. S. 1956. An enzymatic reaction involving adenosine triphosphate and selenate. Arch. Biochem. Biophys. *62*, 503–506.

WILSON, T. H., and WISEMAN, G. 1954. The use of sacs of everted small intestine for the study of the transference of substances from the mucosal to the serosal surface. J. Physiol. *123*, 116–125.

WINZLER, R. J., DEVOR, A. W., MEHL, J. W., and SMYTH, I. M. 1948. Studies on the mucoproteins of human plasma. 1. Determination and isolation. J. Clin. Invest. *27*, 609–616.

CHAPTER 9

U. S. Garrigus

# The Need for Sulfur in the Diet of Ruminants

## INTRODUCTION

This section of the Symposium was designated by the planning committee for consideration of sulfur in relation to animal nutrition. The author, as a member of that committee, has carried out the directive to interpret this assignment relating to the ruminant animal as a charge to provide perspective by presenting a broad background of the many factors involved, to accompany the more specific discussions.

Those who have the responsibility for translating and implementing basic knowledge to research with the large animal must direct attention to a variety of purposes, including a concern for relevance to the practical.

This manuscript, at the risk of seeming elemental, will attempt to present a generalized outline of interest to those involved with sulfur and nutrition from either a more basic or a more applied orientation. Research with the large animal will necessarily be stressed.

Particularly, an effort will be made to enhance appreciation for the relevant role of the ruminant animal in the production of protein and for the significance of sulfur as an integral part of such protein for worldwide human well-being.

Mankind is indebted to all food and fiber producing animals and particularly to ruminants. The ruminant, as will be reviewed shortly, is unique in its ability to convert low-quality feed, or forage from nonarable areas of the world to high-quality protein for human consumption. There is 80% more land area of the world adaptable for use by ruminants than land suitable for cultivation (Garrigus 1968).

Since the time of Justus von Liebig in the first half of the last century it has been recognized that sulfur is a part of living matter. The role of sulfur in metabolism drew increasing interest with du Vigneaud's demonstration with labeled cystathionine that mammals could convert methionine to cystine.

Tissues of ruminants (with the possible exception of wool on sheep) do not, as presently understood, differ qualitatively from tissues of other mammals in their nutritive requirements (Black *et al.* 1952). However, it should be noted that the ruminant is not just a rat with hooves. Ruminants, through the medium of the reticulo-rumen and its symbiotic microbes, may satisfy their metabolic requirements with dietary inorganic nitrogen and sulfur compounds and cellulosic substances that are nutritionally inadequate for nonruminant mammals.

Mitchell (1963) has provided an effective reminder of the remarkable digestive adaptation of ruminants:

"Subsisting upon bulky rations of plant foods, the ruminant has developed an efficient method of meeting this difficult situation. Possessing no cellulases to dissolve the celluloses contained in plant-cell walls, the ruminant harbors vast hordes of symbiotic microorganisms and delegates this task to them. He accommodates them in specialized dilated segments of the alimentary canal, preeminently the reticulo-rumen. Besides yielding large volumes of gases ($CO_2$ and $CH_4$), together with short-chain fatty acids (2 to 4 carbon atoms) in amounts to provide more than 50% of the energy needed by the host animal for equilibrium, these symbionts also perform many syntheses of nutrients, such as vitamins and amino acids, of benefit to the host.

"The digestion of dietary protein in the reticulo-rumen involves proteolysis by bacterial proteases, yielding amino acids and other nitrogenous compounds. Depending upon conditions, a considerable but variable proportion (63 to 82% of the amino acids and ammonia) thus produced is synthesized into bacterial proteins, which are then available to the host after hydrolysis by the host or by bacterial proteases."

The metabolism of sulfur and nitrogen is closely integrated in ruminants as in other animal life; the relationship commonly expressed as nitrogen to sulfur ratios (N:S) which will be discussed in detail elsewhere in these proceedings.

At this particular point in ruminant nutrition research, where traditional techniques meld with newer approaches and the imperative for the constructive use of world resources elicits the cooperative blending of all forces, it is appropriate to assess both past and projected developments.

The author has taken the occasion of this Symposium as an opportunity to assemble a resume of research reported over many years, representing progress and contributions relating sulfur and ruminants which have led to present understanding. This manuscript incorporates references to certain studies. Space prevents note of other relevant work.

There have been a number of extensive and intensive reviews of nitrogen metabolism and sulfur, by association (Reid 1953; Chalmers and Synge 1954; Annison and Lewis 1959; Lewis 1961; Mitchell 1963; McLaren 1964; Blackburn 1965; Hungate 1965; Cocimano and Leng 1967; Farlin 1967; Chalupa 1968; Garrigus 1968; Loosli and McDonald 1968; Schelling 1968; and Tillman and Sidhu 1969). All of these reports must be analyzed carefully to determine the conditions under which the particular study was conducted to avoid the potential hazard of extrapolating nutritional data out of context of the environment in which the results are relevant.

To be of maximum use the report must differentiate the age and stage in growth or physiological process of the animal. Also, in ruminant nutrition it must be remembered that the newborn animal is not an accomplished ruminant. While function of the ruminant digestive pattern is achieved at an early age, studies prior to that achievement should be recognized as such.

Numerous studies of sulfur supplementation are of limited value because interpretation is not accurate in the absence of the reporting of sulfur forms and levels in the basal diet.

A chemically-definable reference or basal diet permits the adjustment of each nutrient independent of the others and facilitates the identification of the most limiting. Significant progress toward this end, particularly in relation to sulfur utilization, has been made by Thomas *et al.* (1951), Starks *et al.* (1953, 1954), Albert *et al.* (1956), Matrone *et al.* (1964), Virtanen (1966), Schelling and Hatfield (1968), and Oltjen *et al.* (1969).

It is the author's intent to survey the need for sulfur in the diet of ruminants by considering the following interrelated factors, all of which are necessary to provide a proper context for this discussion:

- Sulfur in Ruminant Products
  - Beef, lamb, milk
  - Products of conception
  - Wool and hair
- Sulfur in Natural or Metabolic Products Supplied to the Ruminant
  - Plant products
  - Water
  - Recycled
  - Microbial material
- Deficiencies and Excesses of Sulfur
- Interrelationship of Sulfur with Copper, Molybdenum and Selenium
- Sulfur and Disease
- Qualitative and Quantitative Requirements for Sulfur
  - Supplemental forms of sulfur
    - Methionine and cystine
    - Hydroxy analogue of methionine
    - Sulfates
    - Elemental sulfur
  - Sulfur and urea
  - Quantitative estimates
  - Absorption sites
- Concluding Comments

Each topic can be treated only briefly at best and some to be considered in depth by other authors are listed for perspective only. Emphasis is placed on dietary sources of sulfur, metabolizing these sources into desired products, and the requirements for sulfur recognizing that dietary requirements may vary measurably from requirements at absorption sites and in wool growth.

## SULFUR IN RUMINANT PRODUCTS

### Beef, Lamb, Milk

Beef, lamb and milk have similar ratios of nitrogen to sulfur (N:S) at about 15:1 (Table 9.1). Their composition remains remarkably homeostatic in N:S and amino acid ratios in the presence of widely varying dietary ratios of N:S. Extensive amino acid analyses of ruminant products are available in Bailey (1937), Masters and McCance (1939), Block and Weiss (1951), Orr and Watt (1957), American Meat Institute Foundation (1960), Mitchell (1962), Götz (1968), and others.

### Products of Conception

Sulfur is not distributed differentially in the products of conception, except in keratins. Data of Hansard and Mohammed (1968, 1969) quantitate the partition of dietary sulfur between the maternal tissues, fetus, placenta, and placental fluids in an observation of dietary $^{35}$S-sulfate absorption and excretion by gravid ewes and heifers. The $^{35}$S transversed the placenta freely at each of the three trimesters and maternal-fetal tissue concentrations were similar seven days after ingestion of $^{35}$S-sulfate. Partition of absorbed and retained $^{35}$S during the third trimester in ewes was 55.7% in maternal tissues and 44.3% in the products of conception (77% in the fetus, 16% in the placenta, and 7% in the placental fluids).

Karakul pelts are a particularly interesting product of conception. Paduceva *et al.* (1966) reported an increased number of what were graded as "first class pelts" from lambs born to ewes that were fed supplemental sodium sulfate.

### Wool and Hair

Ruminant products which deviate widely from the N:S ratio of about 15.1 are those high in keratins; wool and hair. Wool and hair contain about 13% cystine (Rimington 1929; Reis 1965A).

Studies of variability of wools with regard to sulfur are numerous. Reis (1965B) observed variations of up to 0.6% sulfur in wool grown by the same sheep under two different nutritional regimens. Different sheep under the same nutritional regimen produced wool varying by as much as 0.9% in sulfur content. Sulfur content of wool ranged from 2.7 to 4.1% due to

TABLE 9.1

EXAMPLES OF NITROGEN TO SULFUR RATIOS OF (1) POTENTIAL FEEDS FOR RUMINANTS AND (2) RUMINANT PRODUCTS PRODUCED; ALSO (3) ESTIMATED SULFUR INCORPORATED IN WORLD PRODUCTION OF RUMINANT PRODUCTS ANNUALLY (METRIC TONS)

| Item | N:S | Item | N:S |
|---|---|---|---|
| (1) Potential Feeds for Ruminants | | | |
| Alfalfa hay[4] | 7.4:1 | Maize, dent, yellow, grain[2] | 11.6 |
| Alfalfa hay, 2nd cut, early bloom[2] | 9.8 | Maize[7] | 11.0 |
| Bacteria[3] | 15.0 | Maize, stover (no ears)[4] | 6.3 |
| Barley[7] | 9.8 | Oats[7] | 8.5 |
| Barley straw[4] | 4.2 | Oat straw[2] | 2.9 |
| Bromegrass hay[4] | 17.6 | Oat straw[4] | 3.0 |
| Clover, alsike, hay, 2nd cut[2] | 5.1 | Pasture plants[12] | 23.0 |
| Clover, crimson, hay, 2nd cut[2] | 9.6 | Pasture plants[12] | 10.0 |
| Clover, ladino, hay, 2nd cut[2] | 16.7 | Peanut hulls[4] | 6.3 |
| Clover, mammoth red, hay, 2nd cut[2] | 14.0 | Potato tubers, cooked[2] | 17.8 |
| Clover, red, hay[4] | 13.5 | Rye straw[4] | 5.6 |
| Clover, silage[4] | 7.0 | Sudangrass hay[4] | 28.2 |
| Cowpea, hay, 2nd cut[2] | 8.4 | Timothy hay[4] | 8.5 |
| Grains, distil., dehy.[2] | 10.3 | Vetch hay, 2nd cut[2] | 21.2 |
| Maize, cobs, ground[2] | 1.0 | Wheat bran[7] | 12.4 |
| Maize, aerial pt., ensiled[2] | 12.1 | Wheat straw[2] | 3.1 |
| Maize silage[4] | 16.0 | Wheat straw[4] | 3.9 |
| Maize grits, by-prod., min. 5% fat[2] | 6.3 | | |
| (2) Ruminant Products Produced | | | |
| Beef, raw, frozen[6] | 15.1:1 | Milk, goat[7] | 15.7 |
| Beef, topside, roast, lean and fat[6] | 15.3 | Mutton, leg, roast[6] | 14.8 |
| Milk, fresh, whole[6] | 18.3 | Wool, clean[1] | 4.9 |
| Milk, cow[5] | 17.3 | Wool, clean[8] | 4.4 |
| Milk, goat[5] | 17.9 | | |

(3) Estimated Sulfur Incorporated in World Production of Ruminant Products Annually (Metric Tons)

| Product | | Sulfur |
|---|---|---|
| Beef[10, 11] | 32,100,000 | 81,000[6] |
| Lamb and mutton[10, 11] | 6,000,000 | 16,300[6] |
| Milk[10, 11] | 359,000,000 | 239,000[5] |
| Wool[9] | 1,560,000 | 48,600[1] |

[1]Burns *et al.* (1964).
[2]Crampton and Harris (1969).
[3]Johanson *et al.* (1949).
[4]Loosli (1952).
[5]Macy *et al.* (1953).
[6]McCance and Widdowson (1940).
[7]Mitchell and McClure (1937)
[8]Starks *et al.* (1953).
[9]Anon. (1968).
[10]Garrigus (1968).
[11]FAO (1965).
[12]Begg and Freney (1960).

the combined effects of variations in diet and in individual sheep. Wool appears to be less homeostatic in content of sulfur than is muscle tissue.

Reis *et al.* (1967) and Reis and Tunks (1968) in subsequent studies reported that the sulfur content of wools varied inversely to the rate of wool production. "High wool producers" grew wool with a mean sulfur content of 3.13%, whereas "low wool producers" grew wool with a mean sulfur content of 3.55%. Observations in these studies suggest genetic differences in the ability of sheep to incorporate sulfur into wool. Detailed studies of the wool fiber have been made by Downes *et al.* (1966), Frater (1966),

Gillespie and Reis (1966), and Fraser (1969) showing that the nonmicrofibrillar low-sulfur protein contains submicrofibrillar proteins that may polymerize to form the three-dimensional helical structure of the microfibrillar proteins within cells. The incorporation of matrix proteins into protein-synthesizing cells may be the means by which higher sulfur diets influence the total sulfur content of wools since these matrix proteins may contain 8.3% sulfur versus about 3.5% for total wool fiber.

Nutritional levels relative to wool production were observed by Mitchell *et al.* (1928). He noted that sheep on a submaintenance diet (losing 20% initial weight, 71% initial fat, 48% gross energy and 6.6% nitrogen) still continued to grow wool at a "normal" rate of 61 gm of protein and 633 kcal of gross energy per day per 453.6 kg of live weight. He observed that much of this keratin nitrogen had to come from "raided" protoplasmic tissue. He assumed the textile qualities of the wool must have been impaired.

Physical properties of wool in relation to sulfur content have been considered. Bonsma and Joubert (1934) stated: "It seems logical to conclude that the chemical composition of any substance, wool not excepted, is fundamental to its physical properties and consequently differences in chemical composition of wool will be reflected in a modification of its physical characteristics and more particularly in its tensile strength, elasticity and plasticity." Bonsma and Joubert did not find a significant correlation between fiber fineness and sulfur content of the wools they studied; however, medulated fibers were lower in sulfur. Clark and Buhrke (1954) demonstrated, using a hysteresis loop technique, that wool fibers differing in sulfur content did respond differently to load stress and relaxation tests.

Hair differs little in sulfur content from Merino wool (Reis 1965B). Differences between types of cattle (except Brahman) and between sheep and cattle are largely in the relative amount of keratin produced per unit of body weight. Springell and Leach (1965) and Springell (1966) indicated that Brahman cattle grew only about one-half as much hair per unit of surface area as did British breeds of cattle and that the Brahman hair was of lower sulfur content. It may be assumed from these data that the sulfur requirement of Brahman cattle is less than that of British breeds.

While for the purposes of this manuscript interest is in the sulfur content of ruminant products in relationship to metabolism, relative *total* amounts are interesting to note (Table 9.1).

## SULFUR IN NATURAL AND METABOLIC PRODUCTS SUPPLIED TO THE RUMINANT

Since wool contains relatively high levels of sulfur and since animal proteins contain levels of sulfur considerably higher than that of many of the feeds (particularly urea-supplemented diets) consumed by ruminants producing these proteins (Masters and McCance 1939; Block and Bolling 1951), it is logical that the adequacy of dietary sulfur for ruminants would be questioned.

**Plant Products**

Marston and Robertson renewed concern for this inadequacy when they commented in 1928 that if sheep could not synthesize cystine better than a dog, many feeds in Australia would be borderline in cystine content for high wool-producing sheep. Rimington (1929), Bonsma (1931), Rimington and Bekker (1932), Woodman and Evans (1932), Fraser and Fraser-Roberts (1932), and Aston and Lyons (1932) all raised questions about the relatively low cystine content (about 1 to 3%) of forages relative to the cystine content of wool (about 13%).

In more recent studies Hilder (1954) and Begg and Freney (1960) in Australia, and Martin (1958) in California identified wide variations in the sulfur content of forages (N:S ratios ranging from 10:1 to 23:1) (Table 9.1).

Mertz and Matsumoto (1956) reported that sulfur deficient alfalfa plants have poor protein synthesis which results in increased content of water-soluble, nonprotein nitrogen compounds within these plants.

The sulfur content of plants tends to be positively related to the sulfur content of both the soil and air environment in which they are grown (Allaway and Thompson 1966). Soils deficient in sulfur for optimum crop production are not uncommon in humid areas of granitic and basaltic origins commonly used for forage purposes. Superphosphate (11 to 13% S), ammonium sulfate (24% S) or other sulfur fertilization can remedy the deficiency. It is also interesting to note that the areas surrounding an industrial plant which burns sulfur-containing fuels gain in sulfur. Hart and Peterson as early as 1911 estimated the annual amount of sulfur from the air in the Madison, Wisconsin area at 6.8 to 9.1 kg per acre. Coleman (1966) discussed many aspects of this occurrence, and Tisdale and Nelson (1966) stated that the sulfur available from the air annually ranges from 0.45 to 45 kg per acre, the highest levels being near industrial areas. Concern for fallout and pollution may alter this occurrence.

**Water**

Because of their solubility, sulfates are present in most water supplies. Larson (1951) reported the sulfate content of public ground-water supplies in Illinois to range from 0 to 1059 ppm with a median of about 350 ppm. Altman and Dittmer (1964, 1968) reported sulfate in various waters (Table 9.2). Bray (1965) indicated the possibility that sheep drinking highly sulfated waters could consume up to 1.5 gm of sulfur per day. This appears conservative based on the level of sulfated waters reported in Table 9.2.

TABLE 9.2

SULFUR AS SULFATE[1] IN VARIOUS WATERS

| Water Source | Sulfate Mg/Liter |
|---|---|
| Swiss Alpine lakes | 25.9 |
| World river water[2] | 11.2 |
| Carbonated ground water | 10.1 |
| Chloride ground water | 11.1 |
| Siliceous ground water | 11.1 |
| Sulfate ground water[2] | 2327 |
| Seawater[2] | 2688 |

[1]Altman and Dittmer (1964, 1968).
[2]Sulfur as sulfate constitutes 70.5% of the total solids in sulfate ground water, 9.3% of the world river waters, and 7.7% of sea water.

## Recycled Sulfur

Recycled sulfur is considered elsewhere in these proceedings. In summary it might be said that Bray (1965) demonstrated the recycling of sulfur in sheep. However the amount of recycled sulfur was negligible relative to the amount of recycled nitrogen.

## Microbial Material

The microbial protein passing to the abomasum of a ruminant is presented to the host in an altered form from that originally consumed by the ruminant.

Numerous studies of the biological value of microbial proteins indicate clearly that high-quality dietary proteins are reduced in nutritive value in the rumen while low-quality proteins and inorganic nitrogen and sulfur are enhanced in nutritive value by rumen microbial metabolism, resulting in narrowed, more-nearly constant biological values of the rumen ingesta presented to the abomasum (Smuts *et al.* 1941; Johnson *et al.* 1942, 1944; Williams and Moir 1951; Hungate 1966; Purser and Buechler 1966; Bergen *et al.* 1967).

The rate of ruminal passage and ruminal syntheses under varying conditions is receiving increasing attention relative to efforts to differentially protect nutritive components from rumen degradation. Conrad *et al.* (1967A, B) studied the rate of ruminal synthesis of methionine in cows using $^{35}S$-sodium or barium sulfide and found that the level of dry matter, the level of nitrogen intake, and the grams of alfalfa nitrogen significantly affected methionine synthesis which increased at a rate of 1.5 gm per kg of feed consumed and 6.7 gm per 100 gm of alfalfa nitrogen consumed.

Other authors (Bergen *et al.* 1967) suggest that "modification of the bacterial population may be an important factor with respect to the nitrogen status of the animal and its response to dietary change."

In addition to synthesizing sulfur-containing amino acids, microorganisms also synthesize sulfur-containing thiamine and biotin which have nutritional value for the host (Kon and Porter 1954). Since these vitamins are produced in trace amounts relative to the synthesis of protein, they do not quantitatively, in and of themselves play a large role in the sulfur balance of the producing ruminant. However, thiamine and biotin levels were increased 33 and 77%, respectively, in pooled rumen samples taken from ewe lambs fed 1.6 gm of elemental sulfur daily in comparison to levels on the control diet (Albert 1954).

Deficiencies of thiamine in ruminants under field conditions have not been thought to exist, but recent reports from England and the United States note clinical symptoms of thiamine deficiency in young cattle and sheep, cured by thiamine therapy (Nutrition Foundation, Inc. 1969).

## DEFICIENCIES AND EXCESSES OF SULFUR

Cognizance of evidences of deficiencies or excesses of sulfur in ruminant metabolism is essential to the identification of optimum levels for specific physiological processes.

Sulfur status has been measured by balance studies (Warth 1932; Warth and Krishnan 1935A, B; Kellerman 1935; Starks 1953; Albert 1954; Barrow and Lambourne 1962; Bray 1964, 1965; Walker and Cook 1967), by circulating serum sulfate levels (Weir and Rendig 1954; Bray 1965), by relative serum amino acid levels (Zimmerman and Scott 1965; Schelling and Hatfield 1968), or by observation of other symptoms which have come to be recognized as associated with imbalance or toxicity.

Fecal excretion of sulfur per unit of fecal dry matter by ruminants is fairly constant in contrast to urinary excretion which tends to be much more responsive to level of intake. It is accepted that sulfate is reduced to sulfide in the rumen, absorbed and oxidized to sulfate before being excreted (Lewis 1954; Anderson 1956; Henderickx 1961).

## INTERRELATIONSHIP OF SULFUR WITH COPPER, MOLYBDENUM, AND SELENIUM

Evidence points to an inverse relationship between sulfur and both molybdenum and selenium. The dietary presence of excessive amounts of any of these minerals may alter the requirement for each of the others.

An appropriate balance of levels of copper, molybdenum and sulfur is necessary in the diet of the ruminant (Vanderveen and Keener 1964; Goodrich and Tillman 1966). The interrelationships of these three elements have been studied from a variety of approaches. Copper deficiency or liver retention relative to varying amounts of molybdenum and sulfur in pasture, forage, and in hay diets have been reported by Hogan *et al.* (1968), Beck (1962), Spais *et al.* (1968), and Dick (1953) respectively.

Sulfur added to the substrate both *in vivo* and *in vitro* reduced depression in cellulose digestion caused by copper (Evans and Davis 1966).

The relationship of sulfate to selenium in the treatment of muscular dystrophy in lambs has been reported by Hintz and Hogue (1964), Boyazoglu *et al.* (1967), and Pope *et al.* (1968).

### SULFUR AND DISEASE

Reminiscent of the beneficial effects attributed to the old "sulfur and molasses" treatments, sulfur has been used to combat digestive disorder in lambs. Christensen (1944) and Christensen *et al.* (1947) demonstrated the effectiveness of added dietary elemental sulfur in the control of enterotoxemia in feeder lambs. The significance of this treatment has been superseded by the use of bacterins.

### QUALITATIVE AND QUANTITATIVE REQUIREMENTS FOR SULFUR

The primary goal in establishing dietary requirements both qualitative and quantitative for any component is the achievement of a balanced diet which demands optimum amounts of all nutrient requirements to achieve the optimum in desired effects.

Requirements for sulfur as well as other nutrients are not as easily determined nor as precisely definable for the ruminant as for monogastric animals. This is due, as has been stated, to the flexibility of the ruminant digestive system which permits varying proportions of microbial material to reach the absorption points. Varied forms of sulfur are utilized with different efficiencies and at different sites.

#### Supplemental Forms of Sulfur

The traditional approach to understanding sulfur requirements has been through investigations involving some supplemental form of sulfur, as methionine or cystine, as sulfates, or as elemental sulfur in balance studies and radioisotope studies. With the larger animal as indicated earlier, growth and/or production measures are also commonly used.

**Methionine and Cystine.**—Various early attempts were made to supplement ruminant diets with sulfur-bearing amino acids (Anon. 1913). These and later attempts were generally inconclusive because only trace amounts of amino acids were used and/or frequently the diet supplemented already contained adequate or near-adequate levels of sulfur (Murari 1935; Du Toit *et al.* 1935, 1936; Swart 1936; Schmidt and Kliesch 1939; Smuts and Du Toit 1941; Lofgreen *et al.* 1953; Noble *et al.* 1955; Coetzee *et al.* 1967; Tomov 1967).

Loosli and Harris (1945) improved a basal diet for growth in lambs by raising the crude protein level from 6.55 to 10.28% with urea or a supplement of urea and sodium sulfate, or urea and methionine, or with linseed meal. The urea with 1.11% methionine greatly enhanced the percentage of dietary nitrogen retained, essentially equaling the linseed meal-supplemented control ration. Lofgreen *et al.* (1947) significantly increased nitrogen retention when 0.2% methionine was added to a 10% crude protein basal ration in which urea furnished 40% of the nitrogen. McLaren *et al.* (1965) and Ustinova (1968) improved nitrogen retention and wool production, respectively, with sheep receiving diets supplemented with methionine or methionine and cystine.

Initial understanding of these studies has been altered. It now appears that, unless protected from deamination in the rumen, supplemental dietary amino acids are largely or totally deaminated and as a result serve mainly to supply sulfur for anabolism by rumen microbes.

**Hydroxy Analogue of Methionine.**—Manufactured analogue of methionine appears less biodegradable in the rumen than is DL-methionine.

Brown *et al.* (1960) reported improved weight gain in Holstein and Brown Swiss dairy heifers receiving a growing diet with a high-urea supplement. Weight gains were increased ($P < .05$) with urea and either sodium sulfate or hydroxy analogue of methionine. Griel *et al.* (1968) obtained a significant ($P < .01$) increase in milk production when feeding either 40 or 80 gm of methionine hydroxy analogue daily to dairy cattle.

**Sulfates.**—Warth (1932) in cattle and Warth and Krishnan (1935A, B) in sheep observed increases in organic sulfur with increased ingestion of sulfate, suggesting conversion by rumen bacteria.

Loosli *et al.* (1949) demonstrated this synthesis when rumen material contained 9 to 20 times more amino acids than did the diet fed. They further noted (Thomas *et al.* 1951) that urea nitrogen in the absence of dietary sulfur was not utilized and deficiency resulted, corrected by the addition of inorganic sulfates.

Palfij (1963) and Palfij *et al.* (1962) increased milk production in cows fed silage containing 2 to 3 kg sodium sulfate per ton.

Ewan (1957) evaluated calcium sulfate as a source of both minerals for lambs and found both were utilizable.

The incorporation of dietary sulfate sulfur into protein products was substantiated with $^{35}S$-labeled sulfate. Block and Stekol (1950), Block *et al.* (1951), Hale and Garrigus (1953), and Kulwich *et al.* (1957) using $^{35}S$-labeled sulfate identified $^{35}S$ in amino acids of milk, wool, blood plasma, and many protein tissues.

**Elemental Sulfur.**—Steyn (1932), administering elemental sulfur to sheep to protect them from potassium cyanide poisoning noted increased weight gains and wool yield. Unfortunately similar studies with natural feeds (which were probably already adequate in sulfur) failed to yield positive results and the question of the nutritional utilization of elemental forms of sulfur by ruminants remained unanswered for some time.

At Illinois a series of increasingly refined sulfur studies were conducted over a period of time.

Garrigus *et al.* (1950) equal-fed lambs by trios a low sulfur basal control diet supplemented with either 0.5% elemental sulfur or with 0.5% methionine. Gains were significantly increased by methionine supplementation. Wool production (length and clean yield) was significantly improved by both supplements, supporting the observations of Steyn (1932) that elemental sulfur supplementation appeared to increase wool production. Starks *et al.* (1953), Garrigus *et al.* (1953), and Starks *et al.* (1954) supported these findings with nitrogen and sulfur balance studies. Sulfate sulfur appeared to be more efficiently utilized than was elemental sulfur. Albert *et al.* (1955) supplemented corn silage with urea for wintering ewe lambs. The addition of 0.2% elemental sulfur or silage preserved with sulfur dioxide to a modified urea supplement significantly increased weight gains. Sokolowski *et al.* (1969) observed that raising the basal diet (0.09% S) to the elemental sulfur equivalent of 0.5% appeared to increase the utilization of added nitrate nitrogen by lambs.

**Sulfur and Urea**

The use of dietary urea dictates special attention to sulfur requirements. Urea has been used commonly to formulate low-sulfur basal diets for experimental purposes (Loosli *et al.* 1949; Starks *et al.* 1953; Starks *et al.* 1954; Albert *et al.* 1955; Brown *et al.* 1960; McLaren *et al.* 1965).

A commercial-type diet of corn silage and a urea-containing concentrate reduced milk production in dairy cattle after 9 weeks compared to production on concentrate supplemented with 0.9% $Na_2SO_4 10H_2O$ (Jacobsen *et al.* 1967).

Progressive livestock producers cannot afford and will not accept, and the world-wide human population which needs sustenance should not be expected to accept less than optimum levels of production. The problem of sulfur in the diet of ruminants can no longer be confined to the laboratory.

TABLE 9.3

INDUSTRIAL CAPACITY FOR PRODUCTION OF UREA

| | 1959 | | 1966 | |
|---|---|---|---|---|
| Nation | Tons | Total Production % | Tons | Total Production % |
| United States | 814,800 | 38.8 | 1,445,400 | 14.6 |
| Japan | 779,100 | 37.1 | 1,494,900 | 15.1 |
| Western Europe | 459,900 | 21.9 | 1,890,900 | 19.1 |
| Other nations | 46,200 | 2.2 | 5,068,800 | 51.2 |

With the large tonnages of urea being produced throughout the world (Table 9.3), and increasing amounts being used to supplement ruminant feeds (Garrigus 1968), sulfur levels must be actively evaluated in formulating high-urea diets.

**Quantitative Estimates**

One means of estimating sulfur requirements might be by use of such established tables as those of the National Research Council. The National Research Council at this time does not attempt to define the sulfur requirements for any of the ruminant species. It is expected that factorial or other approaches will be used to make specific estimates.

Probably the most effective approach to establishing quantitative estimates of sulfur requirements in ruminants begins with the basically accepted premise that in essence the sulfur requirement is a function of the protein requirement.

Block in 1951 obtained a United States patent for the use of a mixture of nonprotein nitrogen with nonprotein sulfur at ratios varying between 10:1 and 25:1 with 16:1 being listed as optimum. Presumably the latter ratio was based upon analyses of tissue proteins and assumed a similar dietary requirement. This approach was in accord with the factorial approach to determine nutrient requirements; however, the N:S ratio arrived at by Block was too wide based upon later experimental evidence.

Later Whiting *et al.* (1954) fed mature range ewes a pelleted basal ration (B) containing 0.084% sulfur, or (B) supplemented with inorganic sulfates to either 0.128% or 0.171% total S or (B) supplemented with elemental sulfur to either 0.128% or 0.170% total S or (B) with methionine to 0.129% total S. After a 9-month experimental period, they concluded that the sulfur requirement was less than 0.1% of the total ration. This sulfur level also appears low in comparison with subsequent values in the literature unless rather low levels of production and/or unusually efficient metabolism occurred.

Subsequent direct observations have provided some remarkably similar reference points from a variety of locations, conditions and sulfur sources.

Albert *et al.* (1956) in a series of experiments self-fed young, growing-fattening lambs a purified diet (2.02% N, 92% of which came from urea) supplemented with various levels of elemental sulfur, sodium sulfate, or methionine. It was concluded on the basis of fitted curves that the dietary requirements under these conditions were $0.47 \pm 0.028\%$ elemental sulfur, $1.27 \pm 0.44\%$ sodium sulfate (0.285% S) or $0.64 \pm 0.048\%$ methionine (0.138% S). (See Table 9.4.)

Rendig and Weir (1957) fed yearling wethers alfalfa grown on soils of varied sulfur content and concluded that 0.155% total dietary sulfur resulted in good gains. Presumably much of this sulfur would have been in the

TABLE 9.4

ESTIMATED DIETARY REQUIREMENT FOR SULFUR FROM DIFFERENT SOURCES BY RUMINANTS

| Animal | Basal Diet | Supplemental Sulfur Source | Supplemental Levels (%) | Estimated Requirement (%) | Requirement as S (%) |
|---|---|---|---|---|---|
| Lambs[1] | Purified 0.048% S | Elemental S | 0.10 to 0.70 | 0.47 ± 0.028 | 0.47 |
| Lambs[1] | Purified 0.048% S | $Na_2SO_4$ | 0.44 to 3.08 | 1.27 ± 0.44 | 0.285 |
| Lambs[1] | Purified 0.048% S | DL-methionine | 0.12 to 0.82 | 0.64 ± 0.048 | 0.138 |
| Yearling wethers[4] | Alfalfa 0.155% S | varied | varied | ... | 0.155 |
| Mature wethers[2] | Oaten hulls and Urea 0.058% S | $Na_2SO_4$ | 0.377 to 1.157 | N:S = 10:1 | 0.14 |
| Jersey steers[3] | Purified High cellulose 0.15% S | $Na_2SO_4$ | varied | ... | 0.29 |

[1]Albert *et al.* (1956).
[2]Bray (1965).
[3]Evans and Davis (1966).
[4]Rendig and Weir (1957).

form of plant methionine. This compares with Albert *et al.* (1956) (0.138% sulfur as methionine). (See Table 9.4.)

Evans and Davis (1966) using purified diets found the optimum level of sulfur for maximum cellulose digestion in fistulated steers and in intact lambs to be 65 μg S/ml in the rumen fluid. This level could be obtained by feeding 0.29% S as sodium sulfate. (See Table 9.4.)

Bray (1965) and Moir *et al.* (1967–1968) have concluded that the optimum level of sulfur in the diet of sheep if maximum nitrogen is to be retained is 0.14% and that the optimum ratio of nitrogen to dietary sulfur is about 10:1. (See Table 9.4.)

The best evidence presently available in the opinion of this author indicates ruminant requirements for various forms of sulfur to be at the level shown in Table 9.4. However, research developments continually modify understandings. These levels, especially for methionine, will be subject to revision as research altering and quantifying the sulfur presented to absorption points is developed.

**Absorption Sites**

Traditional research techniques dealing directly with the large animal are being supplemented and perhaps supplanted by newer approaches which focus attention on the actual absorption points in the ruminant system. Such studies consider: (1) the ruminants' requirements at the absorption sites; (2) supplying the appropriate forms and levels of nutrients to these sites; and (3) methods of monitoring the relative nutrient status of the ruminant.

Relatively newer metabolic-research techniques with the ruminant involve postruminal infusions or intravenous, intradermal, intraperitoneal injections or infusions. Serum levels of amino acids are being observed to evaluate current nutritional status.

Application of the understandings thus gained involves chemical binding of specific amino acids, notably methionine, or encapsulating with materials which resist rumenal degradation but permit postruminal digestion.

Marsten, in 1935, anticipated these more advanced techniques in his studies of the incorporation of sulfur-bearing amino acids into wool. He reported responses in wool growth to the subcutaneous injection of cystine in a Merino ewe on a protein-poor ration.

More recently Downes (1961A) and Downes *et al.* (1964, 1965) studied the incorporation in intradermally or intravenously administered $^{35}$S-labeled cystine into the wool of sheep.

Cuthbertson and Chalmers (1950) found that ewes receiving a casein supplement through duodenal fistulae retained more nitrogen than did similar ewes receiving like supplements of casein through rumen fistulae.

Reis and Schinckel (1963) demonstrated that abomasal infusion of casein substantially increased wool growth and its sulfur content.

Numerous postruminal infusion studies have followed observation of improved utilization of high-quality protein by ruminants when received postruminally. Reis and Schinckel (1964) and Gillespie and Reis (1966) noted that relatively small amounts of L-cystine infused into the abomasum were optimum for stimulating wool growth (123 to 181%) while 3.0 gm per day resulted in less stimulation of wool growth than did a 1.5 gm level per day. The sulfur content of the wool increased 9 to 19% with casein administration but did not increase with gelatin administration.

Reis (1967) administered abomasally various amounts of amino acids by aqueous drips. Small equimolar amounts 0.5 to 2.0 gm per day of either L-cysteine or DL-methionine increased wool growth as much as 100%. With relatively high levels of 6 to 8 gm per day of L-cysteine, rate of wool growth was reduced slightly below the maximum response. Administration of equimolar amounts of methionine caused a substantial reduction in wool growth to near pretreatment levels. These relatively high levels seem to have caused a nutrient imbalance unfavorable to the sheep. Methionine hydroxy analog was as effective as DL-methionine.

Schelling and Hatfield (1968) and Hatfield (1970) in an extended series of abomasal infusions of growing lambs receiving a purified diet ad libitum, demonstrated that a mixture of amino acids designed to provide the same relative mixture of the ten essential amino acids comparable to hydrolyzed casein, resulted in improved nitrogen balance. When methionine-supplemented casein was infused, the voluntary feed intake increased 15% indicating improved nutrient balance. They also adapted the Zimmerman and Scott (1965) technique of monitoring serum amino acid levels relative to treatment.

Wright (1969) using either intraperitoneal injections of 1.5, 3.0, or 4.5 gm methionine or dietary addition to a 0.15% sulfur basal diet found that sulfur-bearing amino acids were limiting for maximum gain in weight on an 8% C.P. diet and for wool production on both the 8% and a 12% crude protein diet.

Appropriate sulfur-bearing amino acids or intact protein administered by intraperitoneal injection or postruminally to the digestive tract, then, give promising results.

More dietary high-quality nutrients might be supplied to absorption points by increasing their rate of passage through the rumen.

Reduction of time spent in the rumen does not appear to be very effective because better quality proteins tend to be most soluble and hence to be acted upon more rapidly by rumen microflora while synthesis of microbial proteins from inorganic compounds and the degradation of cellulose to obtain energy proceed best at a moderate rate of passage.

However, supplying nutrients to absorption points in optimum levels and forms is being accomplished by chemically binding or by encapsulating, thus enabling them to escape rumen degradation to a large degree while remaining available for abomasal digestion.

Chemical binding of high-quality meals with tannic acid or formaldehyde has been reported by Leroy *et al.* (1964), Ferguson *et al.* (1967), Delort-Laval and Zelter (1969), and Driedger *et al.* (1969). Some possible problems related to chemically binding proteins were identified by Oslage and Becker (1958) and Tagari *et al.* (1965).

Linton *et al.* (1968) and Sibbald *et al.* (1969), improved levels of plasma methionine and gains in steers fed a diet supplemented with "a core of methionine, kaolin and triglycerides enveloped in a continuous film of triglycerides" which was relatively insoluble at rumen pH but soluble at abomasal pH.

## CONCLUSION

The preceding resume depicts through a review of the literature and delineation of the factors involved the development of a general understanding of sulfur in relation to the diet of ruminants.

It has attempted to provide an outline of interest to those concerned with sulfur metabolism from either a more basic or a more applied orientation.

An effort has been made to identify differences in nutrition research between the laboratory animal and the large animal and between the monogastric and the ruminant.

Particularly it is hoped that an appreciation has been aroused for the relevant role of the ruminant animal in the production of protein and for the significance of sulfur as an integral part of such protein for world-wide human well-being.

ACKNOWLEDGMENT

The author is grateful to Miss Marjorie Edman for translation of references.

## BIBLIOGRAPHY

ALBERT, W. W. 1954. The utilization of varied sulfur sources in urea containing rations fed to lambs. Ph.D. Thesis, Univ. Illinois, Urbana, Ill.

ALBERT, W. W., GARRIGUS, U. S., FORBES, R. M., and HALE, W. H. 1955. Modified urea supplements with corn silage for wintering ewe lambs. J. Animal Sci. *14*, 143–152.

ALBERT, W. W., GARRIGUS, U. S., FORBES, R. M., and NORTON, H. W. 1956. The sulfur requirement of growing-fattening lambs in terms of methionine, sodium sulfate, and elemental sulfur. J. Animal Sci. *15*, 559–569.

ALLAWAY, W. H., and THOMPSON, J. F. 1966. Sulfur in the nutrition of plants and animals. Soil Sci. *101*, 240–247.

ALTMAN, P. L., and DITTMER, DOROTHY S. 1964. Biology Data Book. Federation Am. Soc. Exptl. Biol., Bethesda, Md.

ALTMAN, P. L., and DITTMER, DOROTHY S. 1968. Metabolism. Federation Am. Soc. Exptl. Biol., Bethesda, Md.

AMERICAN MEAT INSTITUTE FOUNDATION. 1960. The Science of Meat and Meat Products. W. H. Freeman & Co., San Francisco.

ANDERSON, C. M. 1956. The metabolism of sulfur in the rumen of sheep. New Zealand J. Sci. Technol. *37*, Sec. A, 379–394.

ANNISON, E. F., and LEWIS, D. 1959. Nitrogen metabolism. *In* Metabolism in the Rumen. Methuen and Co., London.

ANON. 1913. Influence of sulfur in feeds upon wool production. Univ. Wisconsin Bull. *228*, 37–38.

ANON. 1968. World wool production, 1968–69. World Wool Dig. *19*, 185–186.

ASTON, B. C., and LYONS, J. 1932. Stock licks and their composition. New Zealand J. Agr. *44*, 305–308.

BAILEY, K. 1937. The sulphur distribution of proteins. Biochem. J. *31*, 1396–1405.

BARROW, N. J., and LAMBOURNE, L. J. 1962. Partition of excreted nitrogen, sulfur and phosphorus between feces and urine of sheep being fed pasture. Australian J. Agr. Res. *13*, 461–471.

BECK, A. B. 1962. The levels of copper, molybdenum and inorganic sulphate in some Western Australian pastures. Australian J. Exptl. Agr. Animal Husbandry *2*, 40–45.

BEGG, J. E., and FRENEY, J. R. 1960. Vegetation survey in the Macquarie region New South Wales. Commonwealth Sci. Ind. Res. Organ., Div. Plant Ind. Rept. *18*, 1–26.

BENESCH, R. *et al.* (Editors) 1959. Sulfur in Proteins, Symp. Academic Press, New York.

BERGEN, W. G. 1968. Determination of limiting amino acids of rumen-isolated microbial proteins fed to rat. J. Dairy Sci. *51*, 1698–1700.

BERGEN, W. G., PURSER, D. B., and CLINE, J. H. 1967. Enzymatic determination of the protein quality of individual rumen bacteria. J. Nutr. *92*, 357–364.

BLACK, A. L., KLEIBER, M., and SMITH, A. H. 1952. Carbonate and fatty acids as precursors of amino acids in casein. J. Biol. Chem. *197*, 365–370.

BLACKBURN, T. H. 1965. Nitrogen metabolism in the rumen. *In* Physiology of Digestion in the Ruminant, R. W. Dougherty *et al.* (Editors). Butterworths, Washington, D.C.

BLACKBURN, T. H., and HOBSON, P. N. 1960. Proteolysis in the sheep rumen by whole and fractionated rumen contents. J. Gen. Microbiol. *22*, 272–281.

BLAXTER, K. L., and WOOD, W. A. 1952. The nutrition of the young Ayrshire calf. 1. The biological value of gelatin and of casein when given as the sole source of protein. British J. Nutr. *6*, 56–71.

BLOCK, R. J. 1951. Livestock Feed. U.S. Pat. 2,569,282. Sept. 25.

BLOCK, R. J., and BOLLING, DIANA. 1951. The Amino Acid Composition of Proteins and Foods, 2nd Edition. Charles C. Thomas, Springfield, Ill.

BLOCK, R. J., and STEKOL, J. A. 1950. Synthesis of sulfur amino acids from inorganic sulfate by ruminants. Proc. Soc. Exptl. Biol. Med. *73*, 391–394.

BLOCK, R. J., STEKOL, J. A., and LOOSLI, J. K. 1951. Synthesis of sulfur amino acids from inorganic sulfate by ruminants. II. Synthesis of cystine and methionine from sodium sulfate by the goat and by the microorganisms of the rumen of the ewe. Arch. Biochem. Biophys. *33*, 353–363.

BLOCK, R. J., and WEISS, KATHRYN W. 1951. Amino Acid Handbook, 2nd Edition. Charles C. Thomas, Springfield, Ill.

BONSMA, F. N. 1931. The sulfur content of some South African wool. J. Textile Inst. *22*, 305–313.

BONSMA, F. N., and JOUBERT, P. J. 1934. The sulfur content of Merino wool: Its distribution and relation to fineness and quality. S. African J. Sci. *31*, 347–370.

BOYAZOGLU, P. A., JORDAN, R. M., and MEADE, R. J. 1967. Sulfur-selenium-vitamin E interrelations in ovine nutrition. J. Animal Sci. *26*, 1390–1396.

BRAY, A. C. 1964. The recycling and excretion of sulphur in sheep. Proc. Australian Soc. Animal Prod. *5*, 336–344.

BRAY, A. C. 1965. Studies on sulfur metabolism in sheep. Ph.D. Thesis, Univ. Western Australia, Nedlands.

BROWN, L. D., EMERY, R. S., BENNE, E. J., and LASSITER, C. A. 1960. Effect of level and source of sulfur on the utilization of urea by dairy heifers. J. Dairy Sci. *43*, 890. (Abstr.)

BURNS, R. H. *et al.* 1964. Minerals in domestic wools. J. Animal Sci. *23*, 5–11.

CHALMERS, MARGARET I. 1961. Protein synthesis in the rumen. *In* Digestive Physiology and Nutrition of the Ruminant. D. Lewis (Editor). Butterworths, London.

CHALMERS, MARGARET I., JAYSINGHE, J. B., and MARSHALL, S. B. M. 1964. The effect of heat treatment in the processing of ground nut meal on the value of the protein for ruminants with some additional experiments on copra. J. Agr. Sci. *63*, 283–288.

CHALMERS, MARGARET I., and SYNGE, R. L. M. 1954. The digestion of protein and nitrogenous compounds in ruminants. Advan. Protein Chem. *9*, 93–120.

CHALUPA, W. 1968. Problems in feeding urea to ruminants. J. Animal Sci. *27*, 207–219.

CHRISTENSEN, J. 1944. Sulfur prophylaxis of coccidiosis of feeder lambs. Am. J. Vet. Res. *5*, 341–345.

CHRISTENSEN, J. F., DEEM, A. W., ESPLIN, A. L., and CROSS, F. 1947. The use of sulfur in reducing death losses from enterotoxemia in feeder lambs. J. Am. Vet. Med. Assoc. *111*, 144–148.

CLARK, G. L., and BUHRKE, V. E. 1954. Effect of elemental sulfur in the diet on load-extension hysteresis in single wool fibers. Science *120*, 40.

COCIMANO, M. R., and LENG, R. A. 1967. Metabolism of urea in sheep. British J. Nutr. *21*, 353–371.

COETZEE, C. G., LESCH, S. F., and NEL, J. W. 1967. Yellow maize as a supplement to winter veld grass with reference to ruminal utilization of fish meal and urea. S. African J. Agr. Sci. *10*, No. 4, 1037–1048.

COLEMAN, R. 1966. The importance of sulfur as a plant nutrient in world crop production. Soil Sci. *101*, 230–239.

CONRAD, H. R., HIBBS, J. W., and PRATT, A. D. 1967A. Effect of plane of nutrition and source of nitrogen on methionine synthesis in cows. J. Nutr. *91*, 343–352.

CONRAD, H. R., MILES, R. C., and BUTDORF, J. 1967B. Estimation of methionine synthesis in intact cows after administering sulfide-$^{35}$S. J. Nutr. *91*, 337–342.

CRAMPTON, E. W., and HARRIS, L. E. 1969. United States—Canadian tables of feed composition. Natl. Acad. Sci., Natl. Res. Council Publ. *1684*.

CUTHBERTSON, D. P., and CHALMERS, M. I. 1950. Utilization of a casein supplement administered to ewes by ruminal and duodenal fistulae. Biochem. J. *46*, 17–18.

DANKE, R. J., SHERROD, L. B., NELSON, E. C., and TILLMAN, A. D. 1966. Effect of autoclaving time on the utilization of solvent extracted cottonseed meal by sheep. J. Animal Sci. *25*, 181–184.

DELORT-LAVAL, J., and ZELTER, S. Z. 1969. Improving the nutritive value of proteins by tanning process. *In* Proc. Second World Conf. Am. Prod., Am. Dairy Sci. Assoc., Urbana, Ill.

DICK, A. T. 1953. Influence of inorganic sulfate on the copper-molybdenum interrelationship in sheep. Nature *172*, 637–638.

DICKINSON, E. M., and SCOFIELD, R. H. 1939. The effect of sulfur against artificial infection with *Eimeria aceroulina* and *Eimeria tenella*. Poultry Sci. *18*, 419–431.

DOWNES, A. M. 1961A. On the mechanism of incorporation of ($^{35}$S) cystine into wool. Australian J. Biol. Sci. *14*, 109–119.

DOWNES, A. M. 1961B. On the amino acids essential for the tissues of the sheep. Australian J. Biol. Sci. *14*, 254–259.

DOWNES, A. M., FERGUSON, K. A., GILLESPIE, J. M., and HARRAP, B. S. 1966. A study of the proteins of the wool follicle. Australian J. Biol. Sci. *19*, 319–333.

DOWNES, A. M., SHARRY, L. F., and TILL, A. R. 1964. The fate of intradermal doses of labeled amino acids in sheep. Australian J. Biol. Sci. *17*, 945–959.

DOWNES, A. M., SHARRY, L. F., and TILL, A. R. 1965. Further studies of the metabolism of cystine in sheep. Australian J. Biol. Sci. *18*, 140–153.

DRIEDGER, A., HATFIELD, E. E., and GARRIGUS, U. S. 1969. Effect of tannic acid treated soybean meal on growth and nitrogen balance. J. Animal Sci. *29*, 156–157.

DU TOIT, P. J., MALAN, A. I., GROENEWALD, J. W., and BOTHA, M. L. 1935. The effect of different forms of sulphur in the diet upon the growth and wool production of sheep. Onderstepoort J. Vet. Sci. Animal Ind. *4*, 229–239.

DU TOIT, P. J., MALAN, A. I., GROENEWALD, J. W., and BOTHA, M. L. 1936. Studies in mineral metabolism. XXXIV. The effect of sulphur upon the weight and wool production of sheep when food intake is not limited. Onderstepoort J. Vet. Sci. Animal Ind., *7*, 181–188.

DUVIGNEAUD, V. 1952. A trail of research in sulfur chemistry and metabolism and related fields. Cornell Univ. Press, Ithaca, N.Y.

EMERY, R. S., SMITH, C. K., and FAI TO, L. 1957. Utilization of inorganic sulfate by rumen microorganisms. II. The ability of single strains of rumen bacteria to utilize inorganic sulfate. Appl. Microbiol. *5*, 363–366.

EMERY, R. S., SMITH, C. K., and HUFFMAN, C. F. 1957. Utilization of inorganic sulfate by rumen microorganisms. I. Incorporation of inorganic sulfate into amino acids. Appl. Microbiol. *5*, 360–362.

EVANS, J. L., and DAVIS, G. K. 1966. Influence of sulfur, molybdenum, phosphorus and copper interrelationships in cattle upon cellulose digestion *in vivo* and *in vitro*. J. Animal Sci. *25*, 1014–1018.

EWAN, R. C. 1957. A study of the utilization of calcium and nitrogen by growing lambs. M.S. Thesis, Univ. Illinois, Urbana, Ill.

FAO 1965. Food and Agriculture Organization Yearbook, U.N., Rome.

FARLIN, S. D. 1967. Metabolism of urea by ruminant and nonruminant animals. Ph.D. Thesis, Univ. Illinois, Urbana, Ill.

FERGUSON, K. A., HEMSLEY, J. A., and REIS, P. J. 1967. Nutrition and wool growth. The effect of protecting dietary protein from microbial degradation in the rumen. Australian J. Sci. *30*, 215–217.

FRASER, A. H., and FRASER-ROBERTS, J. A. 1932. Cystine and wool production. Nature *130*, 473–474.

FRASER, I. E. B. 1969. Proteins of keratin and their synthesis. II. Incorporation of ($^{35}$S) cystine into prekeratin and keratin proteins. Australian J. Biol. Sci. *22*, 231–238.

FRATER, R. 1966. Comparison of similar protein components isolated from wool and wool roots. Australian J. Biol. Sci. *19*, 699-710.

GARRIGUS, U. S. 1968. Conversion of nonprotein nitrogen to animal protein for human consumption. Univ. Illinois Coll. Agr. Special Publ. *12*, 19–32.

GARRIGUS, U. S., KAMMLADE, W. G., BOLEY, L. E., and HARDENBROOK, H., Jr. 1953. A study of elemental sulfur and of some other suggested preventives of enterotoxemia in a lamb fattening ration. J. Animal Sci. *12*, 474–479.

GARRIGUS, U. S., MITCHELL, H. H., HALE, W. H., and ALBIN, J. S. 1950. The value of elemental sulfur in a methionine deficient sheep ration. J. Animal Sci. *9*, 656–657.

GILLESPIE, J. M., and REIS, P. J. 1966. The dietary-regulated biosynthesis of high-sulphur wool proteins. Biochem. J. *98*, 669–677.

GILLESPIE, J. M., REIS, P. J., and SCHINCKEL, P. G. 1964. The isolation and properties of some soluble proteins from wool. IX. The proteins in wool of increased sulphur content. Australian J. Biol. Sci. *17*, 548–560.

GOODRICH, R. D. 1966. Interrelationships of copper, sulfur and nitrogen in ruminants. Dissertation Abstr. (B), 27:9–B.

GOODRICH, R. D., and TILLMAN, A. D. 1966. Copper, sulfate and molybdenum interrelationships in sheep. J. Nutr. *90*, 76–80.

GÖTZ, GERTRUD. 1968. The comparative amino acid content of wool fibers containing varying amounts of crimp. M.S. Thesis, Univ. Illinois, Urbana, Ill.

GRIEL, L. C., JR., PATTON, R. A., MCCARTHY, R. D., and CHANDLER, P. T. 1968. Milk production response to feeding methionine hydroxy analog to lactating dairy cows. J. Dairy Sci. *51*, 1866–1868.

HALE, W. H., and GARRIGUS, U. S. 1953. Synthesis of cystine in wool from elemental sulfur and sulfate sulfur. J. Animal Sci. *12*, 492–496.

HANSARD, S. L., and MOHAMMED, A. S. 1968. Maternal-fetal utilization of sulfate sulfur by the gravid ewes. J. Nutr. *96*, 247–254.

HANSARD, S. L., and MOHAMMED, A. S. 1969. Absorption, excretion and maternal-fetal utilization of sulfate by the bovine. J. Animal Sci. *28*, 283–287.

HART, E. B., and PETERSON, W. H. 1911. Sulfur requirements of farm crops in relation to the soil and air supply. Wisconsin Agr. Expt. Sta. Res. Bull. *14*, 1-21.

HATFIELD, E. E. 1970. Selected topics related to amino acid nutrition of the growing ruminant. Proc. Federation Am. Soc. Exptl. Biol. *29*, 44–50.

HATFIELD, E. E. *et al.* 1959. Biuret—a source of NPN for ruminants. J. Animal Sci. *18*, 1208–1219.

HENDERICKX, H. 1961. The incorporation of sulfate in the ruminal proteins. Arch. Intern. Physiol. Biochim. *69*, 449–458.

HILDER, E. J. 1954. Some aspects of sulfur as a nutrient for pastures in New England soils. Australian J. Agr. Res. *5*, 39–54.

HINTZ, H. F., and HOGUE, D. E. 1964. Effect of selenium, sulfur and sulfur amino acids on nutritional muscular dystrophy in the lamb. J. Nutr. *82*, 495–498.

HOGAN, K. G., MONEY, D. F. L., and BLAYNEY, A. 1968. The effect of a molybdate and sulphate supplement on the accumulation of copper in the livers of penned sheep. New Zealand J. Agr. Res. *11*, 435–444.

HUNGATE, R. E. 1965. Quantitative aspects of the rumen fermentation. *In* Physiology of Digestion in the Ruminant, R. W. Dougherty *et al.* (Editors). Butterworths, Washington, D.C.

HUNGATE, R. E. 1966. The Rumen and Its Microbes. Academic Press, New York.

JACOBSON, D. R., BARNETT, J. W., CARR, S. B., and HATTON, R. H. 1967. Voluntary feed intake, milk production, rumen content and plasma-free amino acid levels of lactating cows on low sulfur and sulfur-supplemented diets. J. Dairy Sci. *50*, 1248–1254.

JOHANSON, R., MOIR, R. J., and UNDERWOOD, E. J. 1949. Sulfur-containing amino acids in the rumen bacteria of sheep. Nature *163*, 101.

JOHNSON, B. C., HAMILTON, T. S., MITCHELL, H. H., and ROBINSON, W. B. 1942. The relative efficiency of urea as a protein substitute in the ration of ruminants. J. Animal Sci. *1*., 236–245.

JOHNSON, B. C., HAMILTON, T. S., ROBINSON, W. B., and GARVEY, J. C. 1944.

On the mechanism of nonprotein nitrogen utilization by ruminants. J. Animal Sci. *3*, 287–298.

KELLERMAN, J. H. 1935. Sulfur metabolism. 1. The absorption and excretion of flowers of sulfur. Onderstepoort J. Vet. Sci. *4*, 199–228.

KON, S. K., and PORTER, J. W. G. 1954. The intestinal synthesis of vitamins in the ruminant. *In* Vitamins and Hormones, Vol. 12, R. S. Harris, and K. V. Thimann (Editors), Academic Press, New York.

KOVACS, A. 1936. Cystine content of the skin and hair of some domestic animals. Thesis, Univ. Budapest. (German)

KULWICH, R., STRUGLIA, L., and PEARSON, P. B. 1957. The metabolic fate of $^{35}S$ in sheep. J. Nutr. *61*, 113–126.

LARSON, T. E. 1951. Mineral content of public groundwater supplies in Illinois. Circ. *31*. State of Illinois, Water Survey Div., Urbana.

LEROY, F., ZELTER, S. Z., and FRANCOIS, A. C. 1964. Protection of dietary proteins from bacterial deamination in the environment of the rumen. Studies in an artificial rumen. Compt. Rend. Acad. Sci. *259*, 1592–1595. (French)

LEWIS, D. 1954. Reduction of sulphate in the rumen of the sheep. Biochem. J. *56*, 391–399.

LEWIS, D. 1961. The fate of nitrogenous compounds in the rumen. *In* Digestive Physiology and Nutrition of the Ruminant, D. Lewis (Editor). Butterworths, London.

LINTON, J. H., LOUGHHEAD, T. C., and SIBBALD, I. R. 1968. Elevation of free methionine in bovine plasma. J. Animal Sci. *27*, 1168.

LITTLE, O. C., and MITCHELL, G. E., Jr. 1967. Abomasal versus oral administration of proteins to wethers. J. Animal Sci. *26*, 411–413.

LOFGREEN, G. P., LOOSLI, J. K., and MAYNARD, K. 1947. The influence of protein source upon nitrogen retention by sheep. J. Animal Sci. *6*, 343–347.

LOFGREEN, G. P., WEIR, W. C., and WILSON, J. F. 1953. Gains in weight, nitrogen retention and wool growth of lambs fed a ration containing urea supplemented with sodium sulfate. J. Animal Sci. *12*, 347–352.

LOOSLI, J. K. 1952. Meeting the sulfur requirements of ruminants. Feed Age *44*, 45–58.

LOOSLI, J. K., and HARRIS, L. E. 1945. Methionine increases the value of urea for lambs. J. Animal Sci. *4*, 435–437.

LOOSLI, J. K., and MCDONALD, I. W. 1968. Nonprotein nitrogen in the nutrition of ruminants. Food Agr. Organ., U.N., FAO Agr. Studies Bull. *75*.

LOOSLI, J. K. *et al.* 1949. Synthesis of amino acids in the rumen. Science *110*, 144–145.

MACY, ICIE G., KELLY, HARRIET J., and SLOAN, R. E. 1953. The composition of milks. A compilation of the comparative composition and properties of human, cow, and goat milk, colostrum and transitional milk. Natl. Res. Council Publ.

MARTIN, W. E. 1958. Sulfur deficiency widespread. California Agr. *11*, 10–12.

MARSTON, H. R., and ROBERTSON, B. 1928. The utilization of sulfur by animals with especial reference to wool production. Australian Council Sci. Ind. Res. Bull. *39*.

MASTERS, MARGERY, and MCCANCE, R. A. 1939. The sulfur content of foods. Biochem. J. *33*, 1304–1312.

MATRONE, G., BUNN, C. R., and MCNEILL, J. J. 1964. Investigation of dietary factors in purified diets for ruminants. J. Nutr. *84*, 215–219.

McCANCE, R. A., and WIDDOWSON, E. M. 1940. The chemical composition of foods. Privy Council Med. Res. Council, His Majesty's Stationery Office, London.

McDONALD, I. W. 1968. The nutrition of grazing ruminants. Nutr. Abstr. Rev. *38*, 381–400.

McLAREN, G. A. 1964. Symposium on microbial digestion in ruminants: Nitrogen metabolism in the rumen. J. Animal Sci. *23*, 577–590.

McLAREN, G. A., ANDERSON, G. C., and BARTH, K. M. 1965. Influence of methionine and tryptophan on nitrogen utilization by lambs fed high levels of nonprotein nitrogen. J. Animal Sci. *24*, 231–234.

MERTZ, E. T., and MATSUMOTO, H. 1956. Further studies on the amino acids and proteins of sulfur-deficient alfalfa. Arch. Biophys. *63*, 50–63.

MITCHELL, H. H. 1962. Comparative Nutrition of Man and Domestic Animals, Vol. 1. Academic Press, New York.

MITCHELL, H. H. 1963. Some comparative studies in protein nutrition. *In* Symposium on Protein Nutrition and Metabolism, Joseph Kastelic *et al.* (Editors). Univ. Illinois, Coll. Agr. Special Publ. *4*.

MITCHELL, H. H., KAMMLADE, W. G., and HAMILTON, T. S. 1928. Relative energy value of alfalfa, clover and timothy hay for the maintenance of sheep. Illinois Agr. Expt. Sta. Bull. *317*, 127–167.

MITCHELL, H. H., and McCLURE, F. J. 1937. Mineral nutrition of farm animals. Natl. Res. Council Bull *99*.

MOIR, R. J., SOMERS, M., and BRAY, A. C. 1967–1968. Utilization of dietary sulphur and nitrogen by ruminants. Sulphur Inst. J. *3*, 15–18.

MURARI, T. 1935. The effect of sulfur on Bellary sheep. Madras Agr. J. *23*, 268–269.

NEHRING, K., and SCHRAMM, W. 1939. The influence of urea and glycine on the protein exchange and progress of growing sheep. VII. The question of substituting amides for proteins. Z. Tierernahr. Futtermittelk. *2*, 201–238. (German)

NOBEL, R. L., POPE, L. S., and GALLUP, W. D. 1955. Urea and methionine in fattening rations for lambs. J. Animal Sci. *14*, 132–136.

NUTRITION FOUNDATION, INC. 1969. Thiamine deficiency in ruminants? Nutr. Rev. *27*, 176–178.

OLTJEN, R. R., BOND, J., and RICHARDSON, G. V. 1969. Growth and reproductive performance of bulls and heifers fed purified and natural diets. III. Blood proteins, glucose, amino acids and hair amino acid analysis. J. Animal Sci. *29*, 81–87.

ORR, M. L., and WATT, B. K. 1957. Amino acid content of foods. U.S. Dept. Agr., Agr. Res. Serv., Home Econ. Res. Rept. *4*.

OSLAGE, H. J., and BECKER, M. 1958. Studies on the nutritive value of carob beans for ruminants, particularly on the detrimental effect of the tannic acid (content) of the feeds on the digestibility of protein. Arch. Tierernahr *8*, 271–277. (German)

PADUCEVA, A. L., KALANTAROV, S. S., and NIJAZOV, G. 1966. Effect of sodium sulfate on productivity of Karakul ewes. Ovcevodstvo *10*, 27–28. (Russian); Nutr. Abstr. Rev. *37*, 1247.

PADUCEVA, A. L., VOLKOVA, Z. I., and FEOFILOVA, Z. A. 1963. Rate of formation of sulphur-containing wool keratin precursors in the skin of sheep after administration of sulphate and methionine. Dokl. Akad. Nauk SSSR *148*, 1205–1206. (Russian); Nutr. Abstr. Rev. *33*, 775.

PALFIJ, F. YU. 1963. S-containing compounds in metabolic process in animal organism. Zhivotnovodstvo *25*, No. 7, 20–23. (Russian); Chem. Abstr. *59*, 15686–15687.

PALFIJ, F. YU., NAKONECNAJA, JA. JA., and MAKAR, G. N. 1962. Enriching maize silage with S- and other amino acids. Zhivotnovodstvo 1962, No. 8, 26–27. (Russian); Nutr. Abstr. Rev. *33*, 578.

POPE, A. L., MOIR, R. J., SOMERS, M., and UNDERWOOD, E. J. 1968. Effect of sulfur on $^{75}$selenium absorption in sheep. J. Animal Sci. *27*, 1771.

PURSER, D. B., and BUECHLER, SUZANNE M. 1966. Amino acid composition of rumen organisms. J. Dairy Sci. *49*, 81–84.

REED, F. M., MOIR, R. J., and UNDERWOOD, E. J. 1949. Rumenal flora studies in sheep. 1. The nutritive value of rumen bacterial protein. Australian J. Sci. Res. *B2*, 304–317.

REID, J. T. 1953. Urea as a protein replacement. Review. J. Dairy Sci. *36*, 955–996.

REIS, P. J. 1965A. Variations in the sulphur content of wool. *In* Biology of the Skin and Hair Growth, A. G. Lyne, and B. F. Short (Editors). American Elsevier Publishing Co., New York.

REIS, P. J. 1965B. The growth and composition of wool. Australian J. Biol. Sci. *18*, 671–687.

REIS, P. J. 1967. The growth and composition of wool. IV. The differential response of growth and of sulphur of wool to the level of sulphur-containing amino acids given per abomasum. Australian J. Biol. Sci. *20*, 809–825.

REIS, P. J., and SCHINCKEL, P. G. 1963. Some effects of sulphur-containing amino acids on the growth and composition of wool. Australian J. Biol. Sci. *16*, 218–230.

REIS, P. J., and SCHINCKEL, P. G. 1964. The growth and composition of wool. II. The effect of casein, gelatin and sulphur-containing amino acids given per abomasum. Australian J. Biol. Sci. *17*, 532–547.

REIS, P. J., and TUNKS, D. A. 1968. Reproducibility of individual differences in the sulphur content of wool grown by Merino sheep. Proc. Australian Soc. Animal Prod. 7, 402–406.

REIS, P. J., TUNKS, D. A., WILLIAMS, O. B., and WILLIAMS, A. J. 1967. A relationship between sulphur content of wool and wool production by Merino sheep. Australian J. Biol. Sci. *20*, 153–163.

REIS, P. J., and WILLIAMS, O. B. 1965. Variations in the sulphur content of wool from Merino sheep on two semi-arid grasslands. Australian J. Agr. Res. *16*, 1011–1020.

RENDIG, V. V., and WEIR, W. C. 1957. Evaluation by lamb feeding tests on alfalfa hay grown on low-sulphur soil. J. Animal Sci. *16*, 451–461.

RIMINGTON, C. 1929. VII. The relation between cystine yield and total sulfur in wool. Biochem. J. *23*, 41–46.

RIMINGTON, C., and BEKKER, J. G. 1932. Is preformed cystine essential to sheep for wool production? Nature *129*, 687–688.

ROSS, D. A. 1964. Wool production and sulphur content. New Zealand J. Agr. Res. *7*, 663–665.

SCHELLING, G. T. 1968. Amino acid nutrition of the ruminant. Ph.D. Thesis, Univ. Illinois, Urbana, Ill.

SCHELLING, G. T., and HATFIELD, E. E. 1968. Effect of abomasally infused nitrogen sources on nitrogen retention of growing lambs. J. Nutr. *96*, 319–326.

SCHELLING, G. T., HINDS, F. C., and HATFIELD, E. E. 1967. Effect of dietary protein levels, amino acid supplementation and nitrogen source upon the plasma free amino acid concentrations in growing lambs. J. Nutr. *92*, 339–347.

SCHINCKEL, P. G. 1963. The potential for increasing efficiency of feed utilization through newer knowledge of animal nutrition: Sheep and goat. Proc. World Conf. Animal Prod. *1*, 199–239.

SCHMIDT, J., and KLIESCH, J. 1939. The effect of feeding "Eskinoll," organic sulfur, to milk cows. Zeit. Tierernahr. Futtermittelkunde *3*, 159–161. (German)

SIBBALD, R., LOUGHHEED, T. C., and LINTON, J. H. 1969. A methionine supplement for ruminants. *In* Proc. Second World Conf. Am. Prod. Am. Dairy Sci. Assoc., Urbana, Ill.

SMUTS, D. B., and DU TOIT, B. A. 1941. The milk-producing qualities of certain proteins: The special role played by amino-acid cystine. Farming S. Africa *16*, 49–50.

SMUTS, D. B., DU TOIT, B. A., and WATH, J. G. V.D. 1941. A study on the possibility of cystine synthesis in the rumen of sheep together with the effect of cystine supplementation on the nitrogen utilization of lucerne in young stock. Onderstepoort J. Vet. Sci. *16*, 181–190.

SOKOLOWSKI, J. H., HATFIELD, E. E., and GARRIGUS, U. S. 1969. Effects of inorganic sulfur on potassium nitrate utilization by lambs. J. Animal Sci. *28*, 391–396.

SPAIS, A. G., LAZARIDIS, T. K., and AGIANNIDIS, A. K. 1968. Studies on sulphur metabolism in sheep in association with copper deficiency. Res. Vet. Sci. *9*, 337–344.

SPRINGELL, P. H. 1966. Sulphur requirements for cattle hair growth. Proc. Australian Soc. Ann. Prod. *6*, 399–402.

SPRINGELL, P. H., and LEACH, S. J. 1965. Cystine content of cattle hair. Nature *208*, 1326–1327.

STARKS, P. B. 1953. Studies on the value of urea and certain sulfur compounds in the nutrition of growing lambs. Ph.D. Thesis, Univ. Illinois, Urbana, Ill.

STARKS, P. B., HALE, W. H., GARRIGUS, U. S., and FORBES, R. M. 1953. The utilization of feed nitrogen by lambs as affected by elemental sulfur. J. Animal Sci. *12*, 480–491.

STARKS, P. B. *et al.* 1954. Response of lambs fed varied levels of elemental sulfur, sulfate and methionine. J. Animal Sci. *13*, 249–257.

STEYN, D. G. 1931. The effects of sulfur on Merino sheep, 17th Rept. Dir. Vet. Serv. Animal Ind., Union South Africa.

STEYN, D. G. 1932. The effects of sulfur on Merino sheep and their resistance to potassium cyanide poisoning, 18th Rept. Dir. Vet. Serv. Animal Ind., Union South Africa.

STREETER, C. L., LITTLE, C. O., and MITCHELL, G. E. 1969. $^{35}S$ incorporation as a measure of ruminal protein synthesis. J. Animal Sci. *29*, 172.

SWART, J. C. 1936. The influence of sulfur in a ration on the growth and properties of wool. Ziet. f. Zuchtung, B. Tierzucht. n. Zuchtungs Biol. *36*, 307–336. (German)

TAGARI, H., HENIS, Y., TAMIR, M., and VOLCANI, R. 1965. Effect of carob pod extract on cellulolysis, proteolysis, deamination and protein biosynthesis in an artificial rumen. Appl. Microbiol. *13*, 437–442.

THOMAS, W. E., LOOSLI, J. K., WILLIAMS, H. H., and MAYNARD, L. A. 1951. The utilization of inorganic sulfates and urea nitrogen by lambs. J. Nutr. *43*, 515–523.

TILLMAN, A. D., and SIDHU, K. S. 1969. Nitrogen metabolism in ruminants: Rate of ruminal ammonia production and nitrogen utilization by ruminants. Review. J. Animal Sci. *28*, 689–697.

TISDALE, S. L., and NELSON, W. L. 1966. Soil Fertility and Fertilizers. The Macmillan Co., New York.

TOBISKA, J. W., and JENSEN, R. 1946. Livestock poisoning caused by sulfur in combination with protein concentrates. Colorado A & M News, Sept.

TOMOV, I. 1967. Effect of silage rich in nitrogen and sulphur on amount and quality of wool. Zivotnovadni Nauki *4*, 57–63, (Russian); Nutr. Abstr. Rev. *33*, 578.

TRENKLE, A., CHENG, E., and BURROUGHS, W. 1958. Availability of different sulfur sources for rumen microorganisms in *in vitro* cellulose digestion. J. Animal Sci. *17*, 1191.

USTINOVA, M. P. 1968. Effect of synthetic methionine on wool production of sheep. Zivotnovodstvo *6*, 85–89, (Russian); Nutr. Abstr. Rev. *39*, 265.

VANDERVEEN, J. E., and KEENER, H. A. 1964. Effects of molybdenum and sulfate sulfur on metabolism of copper in dairy cows. J. Dairy Sci. *47*, 1224–1230.

VICKERY, H. B. 1942. Liebig and the chemistry of proteins. *In* A Century of Progress in Agricultural Chemistry. Am. Assoc. Advan. Sci. Publ. *16*, 19–29.

VIRTANEN, A. I. 1966. Milk production of cows on protein-free feed. Science *153*, 1603–1614.

WALKER, D. M., and COOK, L. J. 1967. Nitrogen balance studies with the milk-fed lamb. 4. Effect of different nitrogen and sulphur intakes on live-weight gain and wool growth and on nitrogen and sulphur balances. British J. Nutr. *21*, 237–256.

WARNER, A. C. I. 1956. Proteolysis by rumen micro-organisms. J. Gen. Microbiol. *14*, 747–762.

WARTH, F. J. 1932. Sulphur and sulphate balance experiments with cattle. Indian J. Vet Sci. Animal Husbandry *2*, 225–241.

WARTH, F. J., and KRISHNAN, T. S. 1935A. The determination of sulphur and sulphur balance experiments with sheep. Indian J. Vet. Sci. Animal Husbandry *5*, 210–216.

WARTH, J., and KRISHNAN, T. S. 1935B. Sulfur and sulfate balance experiments with sheep. Indian J. Vet. Sci. Animal Husbandry *5*, 319–331.

WEIR, W. C., and RENDIG, V. V. 1954. Serum inorganic sulfate as a measure of the sulfur intake of the sheep. J. Nutr. *54*, 87–96.

WHITING, F., SLEN, S. B., BEZEAU, L. M., and CLARK, R. D. 1954. The sulfur requirements of mature range ewes. J. Animal Sci. *12*, 936.

WILLIAMS, V. J., and MOIR, R. J. 1951. Ruminal flora studies in sheep. III. The influence of different sources of nitrogen upon nitrogen retention and upon the total number of free micro-organisms in the rumen. Australian J. Sci. Res. *B4*, 377–387.

Woodman, H. E., and Evans, R. E. 1932. Cystine and wool production. Nature *130*, 1001.

Woodman, H. E., and Evans, R. E. 1933. The determination of total sulfur and sulfate sulfur in feeding stuffs. J. Agr. Sci. *23*, 459–462.

Wright, P. L. 1969. Body weight gain and wool growth response to casein and sulfur amino acid supplementation. J. Animal Sci. *29*, 177.

Zimmerman, R. A., and Scott, H. M. 1965. Interrelationship of plasma amino acid levels and weight gain in the chick as influenced by suboptimal and superoptimal dietary concentrations of single amino acids. J. Nutr. *87*, 13–18.

CHAPTER 10

P. D. Whanger
G. Matrone

# Effects of Sulfur Deficiency on Metabolism in Sheep[1]

This study is one of several (Van Campen and Matrone, 1960, 1964; Peters and Matrone, 1967) conducted recently concerning the metabolic pathways utilized by the mixed population of ruminal microorganisms for the formation of energy substrates. The roughage-free, purified diet (Matrone *et al.* 1959) developed for ruminants is particularly useful in the investigation of the metabolic consequences of dietary deficiencies. A diet composed similarly to that used previously was employed in the present study to investigate certain consequences of sulfur deficiency in sheep.

Synthesis of long-chain fatty acids in the rumen is accomplished by the condensation of the shorter chain acids with 2-carbon units (Elsden *et al.* 1951; Gray *et al.* 1952; Ladd 1959). Several sulfur-containing compounds such as biotin, coenzyme A, lipoic acid and thiamine are required for fatty acid metabolism (Wakil, 1960). In addition, there are many other sulfur compounds as well as sulfur-containing enzymes required for fatty acid metabolism. Therefore, changes in the organic acids in the rumen as well as some related effects were investigated in sulfur deficient sheep.

Purified diets with and without sulfur, containing urea as the sole source of nitrogen (Matrone *et al.* 1959; Whanger and Matrone 1965) were fed to sheep. Since sulfate has been shown to be utilized by rumen microorganisms (Block *et al.* 1951; Lewis 1954; Anderson 1956), sodium sulfate was added to one of the diets at the recommended level (Albert *et al.* 1956).

During the 8 weeks that the first 2 sheep were on experiment, the sheep fed sulfur lost 3 lb, whereas the sulfur-deficient animal lost 25 lb. The sulfur-supplemented animal consumed its alloted ration of 2 lb of feed daily, but the sheep fed the low-sulfur diet would eat only ¾ lb feed daily. The general appearance of the sulfur-supplemented sheep was excellent whereas emaciation, dullness, poor appetite, and loss of wool were signs observed in the sulfur-deficient sheep. Similar observations have been made previously on sheep fed sulfur-deficient diets (Thomas *et al.* 1951; Starks *et al.* 1954).

Since the sheep fed the low-sulfur diet became so emaciated after 8 weeks, the sheep in subsequent studies were maintained on this diet for only

[1]This work was supported by a grant from the Herman Frasch Foundation.

6 weeks. A second sheep was placed on the low-sulfur diet so that it would have been fed this diet for 2 weeks by the time the first one was removed from the low-sulfur diet at 6 weeks. A third sheep was treated likewise. Subsequently, the low-sulfur sheep were placed on a sulfur diet to regain health after removal from the sulfur-deficient diet. Thus, by this rotational plan sulfur-fed and sulfur-deficient sheep were constantly available for studies. Most of the changes due to sulfur deficiency are evident by two weeks (Whanger 1965).

Examination of the rumen microorganisms revealed that those from the sulfur-supplemented sheep were predominantly Gram positive, whereas those from the sulfur-deficient sheep were mostly Gram negative (Whanger 1965; Whanger and Matrone 1965). The bacterial population was greater in the rumen fluid from the sulfur-fed animal. The viscosity of the rumen fluid was about twice that of the fluid from the sulfur-deficient sheep. These observations are in agreement with Gall *et al.* (1951) who

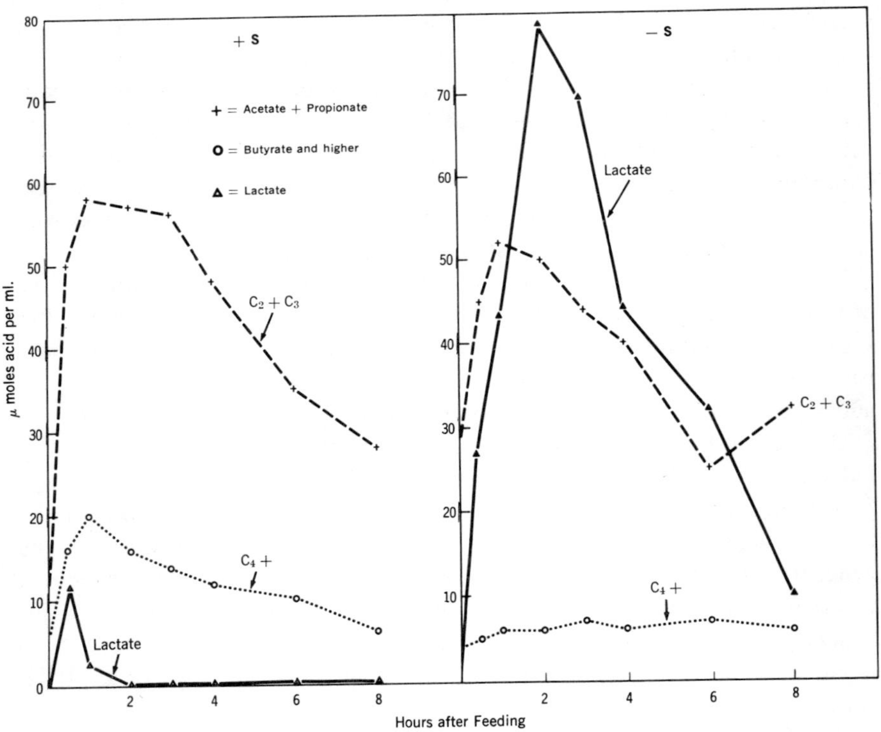


Fig. 10.1. EFFECT OF SULFUR DEFICIENCY ON VOLATILE FATTY ACIDS AND LACTIC ACID CONTENT OF RUMEN FLUID IN SHEEP

The sheep were starved for 36 hr and fed 1/2 lb of their respective diets.

reported that a ration of urea plus sulfur supported a bacterial population about double that found in the rumen contents of animals fed the urea ration without added sulfur.

Figure 10.1 indicates that the volatile fatty acids and lactic acid content of rumen fluid from the experimental sheep were different. The acetate plus propionate levels in the rumen fluid did not differ significantly between the sulfur-fed and sulfur-deficient sheep. However, the concentration of butyric and higher acids was 2 to 3 times greater in the rumen fluid from the sulfur-fed sheep than from the sulfur-deficient sheep. The total fatty acid concentration in the rumen fluid averaged about 20% higher in the sulfur-fed sheep. The lactate concentration in the rumen fluid of the sulfur-deficient animal reached a peak in about 2 hr after feeding and then gradually diminished. In contrast, the concentration of lactate reached a peak after ½ hr and then almost disappeared after 1 hr in the rumen fluid of the sulfur-fed sheep. Even with this high level of ruminal lactate, the pH of the rumen fluid did not drop significantly (Whanger and Matrone 1965) which strongly supports the postulated role of bicarbonates in rumen buffering of animals fed purified diets (Bunn and Matrone 1968). With sheep fed natural diets, the accumulation of lactate in the rumen has been shown to drastically reduce the pH of the ruminal fluid (Briggs *et al.* 1957).

*In vitro* studies with a washed cell suspension of rumen microorganisms indicate that the sulfur-deficient microorganisms produced large amounts of lactate (Fig. 10.2). Incubation of the sulfur-sufficient organisms with the diet produced a peak in lactate production at 4 hr which then disappeared almost completely by the end of 8 hr. In contrast, the sulfur-deficient microorganisms produced large amounts of lactate very rapidly up to about 8 hr and then the level increased slowly over the rest of the incubation. The lactate apparently was not converted to volatile fatty acids to any appreciable extent. The majority at the early stages of incubation was D-lactate, but the proportion of L-lactate slowly increased with time (Whanger and Matrone 1966). Other work (Bunn and Matrone 1968) has indicated that D-lactate accounted for the majority of lactate present in the rumen of sheep fed purified diets and that, under limited feeding conditions, the addition of both cations and alfalfa to these diets were required to decrease ruminal lactate to low levels.

Since the accumulated lactate did not appear to be converted to volatile fatty acids to any significant extent by the sulfur-deficient microorganisms, this was suggestive that it may be absorbed as such from the rumen of the sulfur-deficient sheep. The experimental data in Table 10.1 support this proposal. The total blood lactate levels in the sulfur-deficient sheep were significantly higher ($P < 0.05$) than in the sulfur-fed sheep, which presumably resulted in higher blood reducing sugars in the former animal

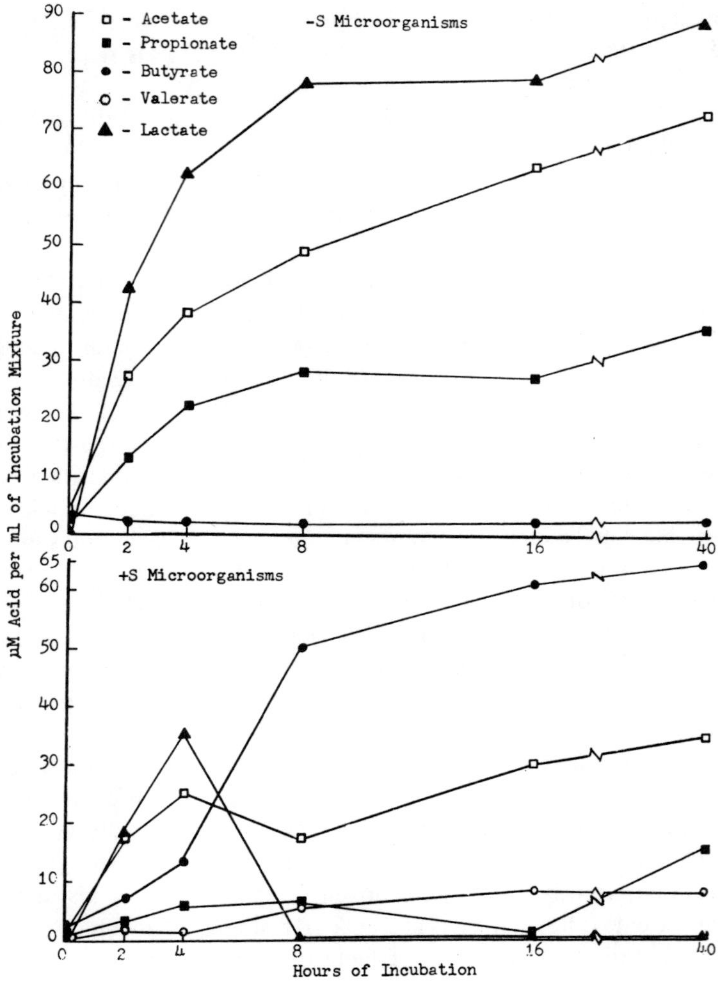


Fig. 10.2. ORGANIC ACID PRODUCTION FROM THE INCUBATION OF SULFUR-SUFFICIENT AND SULFUR-DEFICIENT MICROORGANISMS WITH THE DIET

The respective washed cell suspensions were incubated with 2 gm of sulfur-deficient or sulfur-sufficient diet in 125 ml, sidearmed, (closed with a serum stopper) Erlenmeyer flasks. The incubation mixture contained 30 ml cell suspension and 24 ml buffer with the diet. Nine milliliters was withdrawn with a needle and syringe via the sidearm at various intervals for analysis. The contents of the flasks were incubated under 95% $N_2$ plus 5% $CO_2$.

TABLE 10.1

THE STEREO FORMS OF LACTATE IN THE BLOOD OF THE SULFUR-FED AND SULFUR-DEFICIENT SHEEP[1]

| Hours After Feeding | Blood Lactate (Mg/100 Ml) | | | |
|---|---|---|---|---|
| | Sulfur-fed | | Sulfur-deficient | |
| | Total | L-lactate | Total | L-lactate |
| 0 | 7.8 | 7.6 | 8.2 | 8.0 |
| 1/4 | 12.4 | 12.2 | 20.0 | 20.5 |
| 1/2 | 10.3 | 10.5 | 19.8 | 19.5 |
| 1 | 7.9 | 8.0 | 18.7 | 15.3 |
| 2 | 5.6 | 5.4 | 18.4 | 15.5 |
| 4 | 3.7 | 3.5 | 9.9 | 8.2 |
| 8 | 9.8 | 9.9 | 24.6 | 22.0 |

[1]The sheep were starved for 36 hr and fed ½ lb of their respective diets.

(Whanger and Matrone 1966). There was good agreement between the values for total and L-lactate in the blood of the sulfur-fed sheep. In contrast, there was a discrepancy between the 2 values for blood lactate of the sulfur-deficient sheep, beginning with the 1-hr sample. The difference between total and L-lactate presumably represents D-lactate. There were high levels of lactate in the urine of sheep fed the low-sulfur diet, whereas only traces of it were found in urine of the sulfur-fed sheep (Whanger and Matrone 1966). Since it was shown that the D form of lactate was produced by rumen microorganisms and that tissues apparently do not metabolize D-lactate, one can assume that this stereo form of lactate in the blood was of ruminal origin. The presence of high levels of lactate in the urine of the sulfur-deficient sheep also supports this reasoning. Dunlop and Hammond (1965) have presented evidence for absorption of lactate in cattle after force feeding excessive levels of starchy feeds. The appearance of lactate in the urine of sheep after interruminal infusion of this acid indicates lactate absorption and excretion in sheep (Armstrong 1961).

Other *in vitro* work supports the results obtained *in vivo.* Incubation of the respective microorganisms with lactate, glucose, or the combination of these compounds resulted in the production of about 8 times the amount of butyrate and about 6 times the amount of valerate by the sulfur-sufficient microorganisms as compared to the sulfur-deficient microorganisms (Fig. 10.3). With the sulfur-deficient microorganisms there was a greater production of acetate and propionate. Thus, the sulfur-deficient microorganisms converted more of the substrates to acetate and propionate whereas the sulfur-sufficient organisms converted more of them to butyric and higher acids.

Since these studies suggested that the microorganisms may be using different metabolic pathways for the conversion of lactate to propionate, the respective microorganisms were incubated with lactate-2-$^{14}$C. If lac-

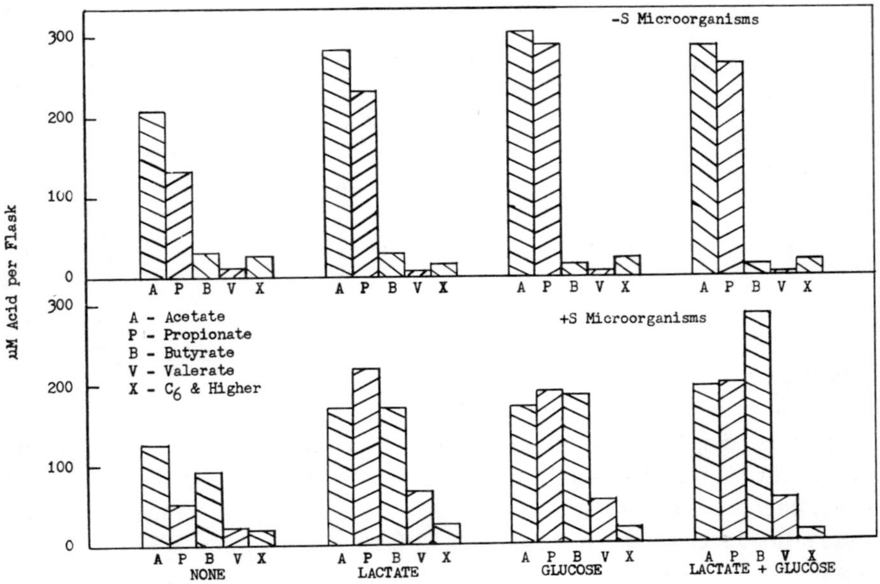


Fig. 10.3. PRODUCTION OF VOLATILE FATTY ACIDS FROM GLUCOSE OR LACTATE BY SULFUR-DEFICIENT AND SULFUR-SUFFICIENT MICROORGANISMS

The microorganisms were incubated for 4 hr under 95% $N_2$ + 5% $CO_2$ with 500 $\mu$M lactate, 500 $\mu$M glucose, or 250 $\mu$M lactate + 250 $\mu$M glucose. The incubation mixture contained 10 ml cell suspension plus 10 ml buffer.

tate is converted to propionate by the direct route (acrylate), the radioactivity should be in the methylene carbon of propionate. However, the conversion of lactate to propionate by the randomizing pathway (succinate) would produce propionate with the label in the methylene and methyl carbons. The propionate that was produced by the microorganisms from the sulfur-fed sheep was labeled as though 30% of it was formed by the acrylate route, but little of the propionate was formed via this route by the microorganisms from the sulfur-deficient sheep (Whanger and Matrone 1967).

Metabolism of intermediates of these two pathways by the respective microbes supports the lactate-2-$^{14}$C study. The sulfur-deficient microorganisms decarboxylated succinate more rapidly than the sulfur-adequate organisms (Table 10.2). The specific activity of the propionate produced from succinate by the sulfur-deficient microorganisms remained fairly constant, suggesting that the only pathway for the formation of propionate was from succinate. In contrast, the specific activity of propionate produced by the sulfur-sufficient microorganisms gradually increased during the incubation. This suggests that another pathway was operating, at least

TABLE 10.2

DECARBOXYLATION OF SUCCINATE BY SULFUR-SUFFICIENT AND SULFUR-DEFICIENT MICROORGANISMS[1]

| Hours of Incubation | μM Propionate[2] +S | −S | Cpm/μM Propionate +S | −S |
|---|---|---|---|---|
| ¼ | 122.0 | 119.6 | 254 | 1,178 |
| ½ | 123.6 | 182.8 | 273 | 991 |
| 1 | 189.6 | 286.8 | 307 | 1,109 |
| 2 | 143.2 | 398.8 | 434 | 1,139 |
| 4 | 211.2 | 559.6 | 554 | 1,154 |

[1]The cells were incubated in 125 ml side arm flasks, which contained 20 ml of cell suspension, 16 ml buffer, 1 ml succinate-2-[14]C (2 microcuries) and 3 ml of sodium succinate (1000 μM) for a total of 40 ml. Eight milliliters were withdrawn at ¼, ½, 1, 2, and 4 hr of incubation through the side arm with a needle and syringe. The cells were incubated under an atmosphere of $N_2$.

[2]μM propionate per flask.

during the early part of incubation, in addition to the one for conversion of succinate to propionate.

The sulfur-sufficient microorganisms converted higher levels of acrylate to propionate than did the sulfur-deficient microbes (Table 10.3). The

TABLE 10.3

PRODUCTION OF PROPIONATE FROM VARIOUS LEVELS OF ACRYLATE BY THE SULFUR-SUFFICIENT AND SULFUR-DEFICIENT MICROORGANISMS[1]

| Level of Acrylate (μm/Flask) | Propionate Present (μm/Flask) Total +S | −S | Endogenous +S | −S | Net +S | −S |
|---|---|---|---|---|---|---|
| 200 | 133 | 112 | 19 | 46 | 114 | 66 |
| 500 | 347 | 42 | 54 | 46 | 293 | −4 |
| 1,000 | 100 | — | 54 | — | 46 | — |
| 2,000 | 39 | — | 54 | — | −15 | — |

[1]The microbes were incubated in 25 ml flasks which contained 5 ml cell suspension and 5 ml buffer with substrates to make a total of 10 ml. The cells were incubated under an atmosphere of hydrogen.

optimum level of acrylate for the sulfur-sufficient microbes was 500 μm, whereas the sulfur-deficient organisms exhibited a small amount of activity at the 200 μm level. Presumably there would be high acrylate levels in rumen microorganisms when lactate accumulates in the rumen. If the sulfur-deficient microorganisms can metabolize acrylate only at low levels, then this is of limited benefit under a lactate load. The factors affecting acrylate metabolism by rumen microorganisms have been discussed (Whanger 1965; Whanger and Matrone 1967).

The acrylate pathway has been shown to operate at the maximum when high carbohydrate diets are fed (Baldwin *et al.* 1962, 1963), and these are conditions in which lactate accumulation is most likely to occur. Therefore, it has been proposed that this pathway acts as a "safety valve"

(Whanger and Matrone 1967). From a biochemical standpoint, the diminution of the acrylate pathway by the sulfur-deficient microorganisms could be one of the factors contributing to the accumulation of lactate. From the species standpoint, the accumulation of lactate in the rumen of the sulfur-deficient sheep could be due to a paucity of lactic acid-fermenting microorganisms. Since many of the lactate-utilizing organisms are Gram positive (Huhtanen and Gall 1953; Hungate *et al.* 1952), the Gram stains of the sulfur-deficient and sulfur-sufficient microorganisms support such a proposal.

Enzymatic studies with cell-free extracts from the respective microorganisms revealed differences. Acid and alkaline phosphatases, aryl sulfatase, and β-glucuronidase activities were significantly higher in the sulfur-deficient microorganisms (Table 10.4). This suggests that those organisms

TABLE 10.4

ACTIVITY OF ACID PHOSPHATASE, ALKALINE PHOSPHATASE, ARYL SULFATASE AND β-GLUCURONIDASE IN CELL-FREE EXTRACT OF SULFUR-DEFICIENT AND SULFUR-SUFFICIENT MICROORGANISMS

| Enzyme[1] | Microorganisms +S | −S |
|---|---|---|
| Acid phosphatase | $134 \pm 21^{a}$ | $418 \pm 84^{b}$ |
| Alkaline phosphatase | $134 \pm 31^{c}$ | $553 \pm 93^{d}$ |
| Aryl sulfatase | $24.5 \pm 3.5^{a}$ | $49.6 \pm 6.5^{b}$ |
| β-glucuronidase | $9.1 \pm 1.0^{a}$ | $23.9 \pm 4.6^{b}$ |

[1]Activity is expressed as μmoles product per 30 min per mg protein. Each value represents means ± standard error of 3 samples. a and b are significantly different at 5% level. c and d are significantly different at 1% level.

which could not exist under sulfur-deficiency did not possess these enzymes in significant amounts. The higher level of aryl sulfatase in the sulfur-deficient microbes is interesting in light of the greater incorporation of $^{35}S$-sulfate into bacterial protein by the sulfur-deficient microorganisms (Whanger 1965).

Except for tyrosine and phenylalanine transaminases, all of the transaminases tested were higher in the sulfur-sufficient microorganisms (Table 10.5). Unpublished work indicates that the nitrogen content is greater in sulfur-adequate organisms than in sulfur-deficient microorganisms (5.8% nitrogen for +S versus 4.7% for −S organisms). Sulfur deficiency has been shown to limit nonprotein nitrogen utilization (Thomas *et al.* 1951; Moir *et al.* 1967–1968) and the ratio of these two elements for maximum nitrogen utilization is critical (Moir *et al.* 1967–1968). The reduced transaminase activity in sulfur-deficient microorganisms could contribute to the inefficiency of nitrogen utilization during sulfur deficiency.

TABLE 10.5

TRANSAMINASE ACTIVITY IN CELL-FREE EXTRACT OF SULFUR-SUFFICIENT AND SULFUR-DEFICIENT MICROORGANISMS

| | Microorganisms | |
|---|---|---|
| Enzyme[1] | +S | −S |
| Tyrosine transaminase | 1.01 ± 0.15 | 1.26 ± 0.20 |
| Ornithine transaminase | 0.098 ± 0.035 | 0.048 ± 0.013 |
| Ornithine transcarbamylase | 0.56 ± 0.13 | 0.37 ± 0.17 |
| Glutamic oxalacetic transaminase | 0.921 ± 0.102 | 0.59 ± 0.056 |
| Glutamic pyruvic transaminase | 0.646 ± 0.10 | 0.258 ± 0.005 |
| Tryptophan transaminase | 0.26 ± 0.02 | 0.17 ± 0.01 |
| Phenylalanine transaminase | 0.90 ± 0.08 | 1.02 ± 0.06 |

[1]Activity is expressed as micromoles product per hour per milligram protein. Each value represents means ± standard error of 3 to 6 samples.

Lactyl Co A dehydrase, succinic dehydrogenase, malic dehydrogenase and 2-keto-3-deoxy-6-phosphogluconate aldolase were higher in sulfur-deficient microorganisms, whereas isocitrate dehydrogenase (both NAD and NADP linked) was higher in the sulfur-sufficient microorganisms (Table 10.6). Lactyl Co A dehydrase, malic dehydrogenase and succinic dehydrogenase have been reported to be higher in rumen microorganisms from animals fed concentrate rations (Baldwin and Palmquist 1965; Palmquist and Baldwin 1966), whereas 2-keto-3-deoxy-6-phosphogluconate aldolase (Whanger *et al.* 1967) is higher in microorganisms from animals fed alfalfa diets. Evidence has been presented for a metabolic role for the last enzyme in uronic acid metabolism in the rumen. The higher succinic and malic dehydrogenases in sulfur-deficient microorganisms support the metabolic studies, but the higher lactyl Co A dehydrase is in conflict with the metabolic results. These results suggest that the diminution of the

TABLE 10.6

SOME SELECTED ENZYMATIC ACTIVITY OF THE CELL FREE EXTRACT FROM SULFUR-SUFFICIENT AND SULFUR-DEFICIENT MICROORGANISMS

| | Microorganisms | |
|---|---|---|
| Enzyme[1] | +S | −S |
| 2-keto-3-deoxy-6-phosphogluconate aldolase | 19.2 ± 1.2 | 34.2 ± 3.0 |
| Lactyl Co A dehydrase | 1.9 ± 0.1 | 2.8 ± 0.20 |
| Succinic dehydrogenase | 1.04 ± 0.43 | 4.87 ± 1.34 |
| Malic dehydrogenase | 4.23 ± 1.73 | 7.49 ± 0.13 |
| Isocitrate dehydrogenase (NADP-linked) | 15.8 ± 1.62 | 10.8 ± 1.02 |
| Isocitrate dehydrogenase (NAD-linked) | 4.7 ± 0.46 | 1.1 ± 0.08 |

[1]Activity is expressed as micromoles product formed per minute per milligram protein times $10^3$. Each value represents means ± standard error of 3 to 6 samples.

acrylate pathway in the sulfur-deficient microorganism is not due to the absence of lactyl Co A dehydrase. Previous work (Baldwin and Palmquist 1965; Palmquist and Baldwin 1966) and the results of Table 10.6 indicate that caution must be used to assess the contribution of various metabolic pathways by enzymatic techniques alone. For example, an inverse relationship was observed between the activities of enzymes in the rumen microorganisms which synthesize butyrate and the amount of this acid found in the rumen contents (Palmquist and Baldwin 1966).

## SUMMARY

Gram positive organisms were predominant in rumen of sheep fed the sulfur-adequate diet, but Gram negative organisms were predominant in the rumen of the sulfur-deficient sheep. *In vitro* and *in vivo* studies indicate that the microorganisms from sheep fed the sulfur-adequate diet formed more butyric and higher acids, whereas the microorganisms from the sulfur-deficient sheep formed more acetate and propionate. There was an accumulation of lactate in the rumen of sheep fed the low-sulfur diet, but only traces of it were found in the rumen of sulfur-fed sheep. Incubation of these two respective groups of microorganisms with lactate-2-$^{14}$C produced propionate labeled as though very little of it was formed by the acrylate pathway by the sulfur-deficient microorganisms, whereas as much as 30% of it was formed by this route by the sulfur-sufficient microorganisms. The sulfur-deficient microorganisms metabolized succinate faster, whereas the sulfur-adequate organisms utilized acrylate faster than the deficient microbes. Most of the lactate which accumulated in the rumen of the deficient animals was the D form, and evidence was obtained for its absorption and excretion in the urine. Most of the transaminases were higher in the sulfur-sufficient microbes, whereas the phosphatases, sulfatase, and $\beta$-glucuronidase were higher in the deficient organisms. No consistent trend was noted for the enzymes involved in energy metabolism between the adequate and deficient microorganisms.

## BIBLIOGRAPHY

ALBERT, W. W., GARRIGUS, U. S., FORBES, R. M., and NORTON, H. W. 1956. The sulfur requirement of growing-fattening lambs in terms of methionine, sodium sulfate, and element sulfur. J. Animal Sci. *15*, 559–569.

ANDERSON, E. M. 1956. The metabolism of sulfur in the rumen of the sheep. New Zealand J. Sci. Technol. Sec. A. *37*, 379–394.

ARMSTRONG, D. G. 1961. *In* Digestive Physiology and Nutrition of the Ruminant, D. Lewis (Editor). Butterworths, London.

BALDWIN, R. L., and PALMQUIST, D. L. 1965. Effect of diet on the activity of several enzymes in extracts of rumen microorganisms. Appl. Microbiol. *13*, 194–200.

BALDWIN, R. L., WOOD, W. A., and EMERY, R. S. 1962. Conversion of lactate-2-$^{14}$C to propionate by the rumen microflora. J. Bacteriol. *831*, 907–913.

BALDWIN, R. L., WOOD, W. A., and EMERY, R. S. 1963. Conversion of glucose-$^{14}$C to propionate by the rumen microbiota. J. Bacteriol. *85*, 1346–1349.

BLOCK, J. R., STEKOL, J. A., and LOOSLI, J. K. 1951. Synthesis of sulfur amino acids from inorganic sulfate by ruminants. Arch. Biochem. Biophys. *33*, 353–363.

BRIGGS, P. K., HOGAN, J. P., and REID, R. L. 1957. The effect of volatile fatty acids, lactic acid and ammonia on rumen pH in sheep. Australian J. Agr. Res. *8*, 674–690.

BUNN, C. R., and MATRONE, G. 1968. Dietary factors affecting utilization of urea nitrogen by sheep fed purified diets. J. Nutr. *95*, 122–128.

DUNLOP, R. H., and HAMMOND, P. B. 1965. D-lactic acidosis of ruminants. Ann. N.Y. Acad. Sci. *119*, 1109–1132.

ELSDEN, S. R., GILCHRIST, F. M. C., LEWIS, D., and VALCONI, B. E. 1951. The formation of fatty acids by a Gram-negative coccus. Biochem. J. *49*, IXiX.

GALL, L. S., THOMAS, W. E., LOOSLI, J. K., and HUHTANEN, C. N. 1951. The effect of purified diets upon rumen flora. J. Nutr. *44*, 113–122.

GRAY, F. V., PULGRIM, A. F., RODDA, H. J., and WELLER, R. A. 1952. Fermentation in the rumen of the sheep. IV. The nature and origin of the volatile fatty acids in the rumen of the sheep. J. Exptl. Biol. *29*, 57–65.

HUHTANEN, C. N., and GALL, L. S. 1953. Rumen organisms. II. Two lactate utilizers and six miscellaneous types. J. Bacteriol. *65*, 554–559.

HUNGATE, R. E., DOUGHERTY, R. W., BRYANT, M. P., and CELLO, R. M. 1952. Microbiological and physiological changes associated with acute indigestion in sheep. Cornell Vet. *42*, 423–449.

LADD, J. N. 1959. The fermentation of lactic acid by a Gram negative coccus. Biochem. J. *71*, 16–22.

LEWIS, D. 1954. The reduction of sulfate in the rumen of the sheep. Biochem. J. *56*, 391–399.

MATRONE, G., RAMSEY, H. A., and WISE, G. H. 1959. Effect of volatile fatty acids, sodium and potassium bicarbonate in purified diets for ruminants. Proc. Soc. Exptl. Biol. Med. *100*, 8–11.

MOIR, R. J., SOMERS, M., and BRAY, A. C. Winter 1967–1968. Utilization of dietary sulphur and nitrogen by ruminants. Sulphur Inst. J. *3*, 15–18.

PALMQUIST, D. L., and BALDWIN, R. L. 1966. Enzymatic techniques for the study of pathways of carbohydrate utilization in the rumen. Appl. Microbiol. *14*, 60–69.

PETERS, D. S., and MATRONE, G. 1967. The propionate-activating system of rumen bacteria. Biochim. Biophys. Acta *137*, 478–483.

STARKS *et al.* 1954. Response of lambs fed varied levels of elemental sulfur, sulfate sulfur and methionine. J. Animal Sci. *13*, 249–257.

THOMAS, W. E., LOOSLI, J. K., WILLIAMS, H. H., and MAYNARD, L. A. 1951. The utilization of inorganic sulfate and urea nitrogen by lambs. J. Nutr. *43*, 515–523.

VAN CAMPEN, D. R., and MATRONE, G. 1960. Investigation of precursors of ruminal fatty acids of sheep fed purified diets. J. Nutr. *72*, 277–282.

VAN CAMPEN, D. R., and MATRONE, G. 1964. An acetate-activating system from rumen microorganisms. I. Presence of phosphotransacetylase and effect of cations on its activity. II. Presence of acetokinase and effect of cations on its activity. Biochim. Biophys. Acta *85*, 400–419.

WAKIL, S. J. 1960. The mechanism of fatty acid synthesis. Am. J. Clin. Nutr. *8*, 630–644.

WHANGER, P. D. 1965. The effect of sulfur depletion on organic acid production and some related effects in the rumen of sheep. Ph.D. Thesis. North Carolina State Univ., Raleigh.

WHANGER, P. D., BEITZ, D. C., WOOD, W. A., and THOMAS, J. W. 1967. A role for 2-keto-3-deoxy-6-phosphogluconic aldolase in rumen microbes. J. Animal Sci. *26*, 932.

WHANGER, P. D., and MATRONE, G. 1965. Effect of dietary sulfur upon the fatty acid production in the rumen. Biochim. Biophys. Acta *98*, 454–461.

WHANGER, P. D., and MATRONE, G. 1966. Effect of dietary sulfur upon the production and absorption of lactate in sheep. Biochim. Biophys. Acta *124*, 273–279.

WHANGER, P. D., and MATRONE, G. 1967. Metabolism of lactic, succinic and acrylic acids by rumen microorganisms from sheep fed sulfur-adequate and sulfur-deficient diets. Biochim. Biophys. Acta *136*, 27–35.

CHAPTER 11

R. J. Moir

# Implications of the N:S Ratio and Differential Recycling

The newborn ruminant is a conventional mammal; with the onset of the ruminating habit and the adoption of its associated ruminal microbial population it develops into the complex ruminant system. While the mammalian moiety of this system remains specifically mammalian with all its limited capacities and fastidious requirements, the microbial portion, with its vast range of metabolic activities, and its capacity for adaptation to new chemical environments, enlarges the nutritional possibilities of the ruminant system far beyond that of the conventional mammal. The sulfur metabolism of the ruminant, particularly when considered in association with its nitrogen metabolism, is an excellent example of this point. Satisfying the simpler sulfur and nitrogen needs of the microbial system will, in fact, satisfy the basic needs of the ruminant animal. However, it is possible to further the productive capacity of the animal beyond that inherent in the bacterial system by dietary manipulation.

Appreciation of the sulfur metabolism of the ruminant system is only a recent phenomenon, now properly based upon the separate requirements of microbial and animal components. It is clear that in the previous two decades, the animal's requirement for methionine overshadowed, and probably diverted the attention of animal nutritionists away from the sulfur metabolism of the bacteria in which methionine plays a secondary role.

McDonald's (1948, 1952) studies on protein metabolism in the rumen have established the central role of ammonia-nitrogen in this process. The ammonia derived from the rapid microbial fermentation of dietary proteins, together with that derived from other nonprotein-nitrogen sources, serves as the major source of nitrogen for the growth of microorganisms. However, ammonia-nitrogen is also absorbed directly through the rumen wall.

The rate of absorption of ammonia depends largely upon its concentration (Hogan 1961), but as pH rises above seven, the loss-rate increases. This is due to the increasing proportion of unionized $NH_3$ which is absorbed more rapidly across the ruminal epithelium (Coombe *et al.* 1960). The loss of ammonia-nitrogen through the omasum depends upon the level within the rumen and the rate of flow through the omasum. With excess nitrogen intake (20 gm/day), Hume (1970) has established that up to 2 gm/day of ammonia-N may flow into the omasum. While very low levels of ammonia-nitrogen are found normally in blood, the level does not rise appreciably until absorption exceeds the liver's capacity to convert ammonia to urea (Lewis *et al.* 1957).

Many authors have shown a strong relationship between ruminal ammonia, blood urea and, ultimately, urinary urea. However, competition exists between urinary excretion and the urea free space of the gut. McDonald (1948) first proposed a urea cycle; urea-nitrogen returned to the rumen through the saliva enters the ammonia-nitrogen pool of the rumen and becomes available for microbial use. More than ½ gm a day may be added in this manner through the saliva of the sheep; the amount depends on blood urea concentration and total flow (Somers 1961). Simonnet *et al.* (1957) and Houpt (1959) extended McDonald's hypothesis by demonstrating diffusion of urea from the blood through the rumen wall. Recent quantitative flow studies have shown that the net amount of N added by these routes in sheep may be as great as 3–7.5 gm/day (Weston and Hogan 1968A, B; Hume *et al.* 1970; Hume 1970). In cattle the net amount of urea returned to the rumen may be 17–20 gm/day, with the possibility of some 60% coming from saliva (Vercoe 1969).

Although the use of ammonia-nitrogen is widespread among the rumen microorganisms, and it is indeed obligatory to many (Bryant and Robinson 1961), other nitrogenous compounds may be stimulatory or even essential. Thus methionine and cysteine are stimulatory, or essential, to some rumen microorganisms, while the growth of others is stimulated by peptides (Pittman and Bryant 1964; Pittman *et al.* 1967). Recently, Hume (1970) has found significant increases in the total rumen microbial protein synthesis where some protein is added to urea-containing diets. In this work the nature of the protein was important, but the nature and the absolute requirement for these protein-derived substances is not yet clear. While these may be important where large amounts of urea are fed, they are unlikely to be limiting under normal conditions, because of dietary protein intake, the inflow of salivary protein, and the lysis and turnover of microbial protein within the rumen (Hoogenraad *et al.* 1967).

The major portion of plant nitrogen is in the form of protein, together with up to 30% of NPN (Ferguson and Terry 1954) as free amino acids, glutamine and asparagine together with peptides, nucleic acids, purine, and pyrimidines and occasionally alkaloids (Synge 1952). Occasionally, however, nitrates accumulate in plants and the concentration of nitrate may constitute a significant proportion of the total nitrogen. Although nitrate is readily reduced to nitrite in the rumen, and then to ammonia, the intermediate, nitrite, may accumulate to significant concentrations in the rumen (Lewis 1951; Jamieson 1959). Neither nitrate nor nitrite is absorbed directly from the rumen (Pfander *et al.* 1956) and the time relationship for the formation of methhemoglobinemia following nitrate or nitrite ingestion is consistent with absorption from the lower gastro-intestinal tract. Whether nitrite inhibits or competes with sulfite metabolism, remains to be determined.

Thompson (1967) in his review of sulfur metabolism in plants states that the study of this subject has lagged behind that of carbon and nitrogen metabolism, "no doubt, due to lesser intrinsic interest as well as such practical considerations as the instability of sulfur compounds." The proportion of sulfur to nitrogen in plants varies very widely. Stewart and Porter (1969) found ratios of N:S ranging from 4 to 1 to as much as 55 to 1. With limited sulfur, accumulation of nitrates, amides, and amino acid nitrogen accounted for over 75% of the nitrogen and gave wide N:S ratios. Conversely, when nitrogen was deficient and sulfur adequate, total sulfur far exceeded protein sulfur and inorganic sulfate accumulated. This study, and many others (Eaton 1966), indicates that the plant leaf protein sulfur content maintains a constant relationship to the nitrogen content during the vegetative stages of about 1 to 15.

From a review of the nitrogen-sulfur ratio of the organic matter of leaves, Dijkshoorn and van Wijk (1967) concluded that the ratio varied with species, gramineous plants approximated 14 to 1 whereas leguminous species ranged up to 17 to 1. Due to the variable proportions of protein to inorganic sulfur in plants, the ruminal population is presented with a variable sulfur environment. As the population is not homogeneous, the metabolism of sulfur in the rumen is complex. Clearly, microbial cystine and methionine synthesis and incorporation into protein are the major end-products of the metabolism and the efficiency of these processes is basic to the maintenance of the rumen system and to productivity from ruminants.

The metabolism of sulfur in the rumen closely parallels that of nitrogen with some notable exceptions. Based on known pathways, rumen microorganisms may potentially obtain their essential sulfur supplies (a) entirely from cyst(e)ine directly; (b) partially from methionine directly; (c) partially or wholly from inorganic sulfates by assimilatory reduction (see Peck 1962); (d) partially or wholly from sulfide accumulating from dissimilatory reduction of inorganic sulfates; or (e) from sulfide originating from cystine, cysteine or other organic substances. The metabolic pathways involved in reduction and incorporation of sulfur into plants and microorganisms has been reviewed by Peck (1962) and Thompson (1967) and will not be considered further.

There is a developing consensus of opinion that sulfide is the central source of sulfur for microbial protein synthesis, and that sulfur amino acid synthesis proceeds from this substrate *de novo* in most circumstances. The recent results of Walker and Nader (1968) and of Hume and Bird (1970) are consistent with this view. Despite the few rumen organisms known to require some cysteine or methionine, there is little evidence of direct incorporation of sulfur amino acids by ruminal microorganisms although this process cannot be disregarded. Emery *et al.* (1957A) found that, in the presence of protein-containing substrates, the incorporation of $^{35}S$ from sulfate into cysteine was twice as fast as into methionine. Microradioautographs, however, indicated that the majority of organisms did not incorporate radioactive sulfate in that work. In a further study (Emery *et al.* 1957B) sulfate incorporation by pure cultures of rumen organisms was apparently inhibited by the presence of cysteine in even smaller quantities. This may be expected in assimilatory sulfate utilization where more reduced sulfur compounds generally inhibit the process. In an experiment in our own laboratories the total growth of microorganisms fermenting cellulose *in vitro* was not influenced by the source of sulfur over 24 hr, although cystine and methionine both gave significantly higher values for TCA-N during the first 12 hr. In the presence of $^{35}S$-sulfate, methionine and cysteine accounted for 28 and 34% of TCA-S produced, but sulfide accounted for 88%, that is, virtually suppressing sulfate incorporation.

The reduction of sulfate to sulfide in the rumen is well-established (Lewis 1954; Anderson 1956; Henderickx 1961; Halverson *et al.* 1968; Spais *et al.* 1968; Bray 1969A; Hume and Bird 1970), as is the desulfuration of cystine and cysteine both *in vitro* (Lewis 1954; Anderson 1956; Sirotnak *et al.* 1953) and *in vivo* (Hume and Bird 1970; Bird 1969). In the latter work, whereas cysteine and cystine were rapidly broken down, the rise in $H_2S$ in the rumen following administration of methionine was very small; Bird considers that methylmercaptan is also a degradation product and has recently identified this substance in the rumen. Despite the slow rate of methionine degradation in the rumen (Sirotnak *et al.* 1953; Lewis 1955; Lewis and Emery 1962), the amino acid in a free state has only been identified once in rumen fluid (Lewis 1955) although Spais *et al.* (1968) showed levels of up to $12.0 \pm 4.4$ mg, and Anderson (1956) less than 4 mg per 100 ml of rumen fluid of free amino acid sulfur.

The metabolism of methionine in the rumen has not yet been clarified; it may serve as a partial, but not an only, source of sulfur for rumen microorganisms (Prescott 1961). The use of methionine as a sulfur supplement has been extensively investigated (Loosli and Harris 1945; Albert *et al.* 1956; Lofgreen *et al.* 1947; Williams and Moir 1951; Moir *et al.* 1967–1968). It is perhaps significant that where comparisons have been made with other sulfur sources, or where methionine has been added to sulfur adequate diets, no outstanding growth or production advantage has been obtained. Rendig and Weir (1957) found that the increase in urinary sulfate from lambs fed alfalfa and methionine equalled the methionine sulfur added. Precise quantitative evidence of sulfur amino acid metabolism, particularly methionine metabolism in the rumen is still required.

There is now substantial evidence that sulfate sulfur, when adaptation to give reduction is established, is an adequate source of sulfur for the rumen microbiota. With minor exceptions, sulfide is probably the central metabolite in rumen sulfur metabolism, as is ammonia in nitrogen metabolism. It follows that the utilization of dietary sulfur will depend not only upon the quantity and nature of available sulfur, but upon (1) the rate of sulfide production, (2) the uptake of sulfide by the microorganisms, and (3) the loss of sulfide from the rumen. The rate of sulfide production is rapid whether from sulfate, from more reduced inorganic sulfur substrates, or from cysteine and cystine (Lewis 1954; Anderson 1956; Bray 1964; Halverson *et al.* 1968; Spais *et al.* 1968). Sulfide concentration rises steeply and usually reaches peak values with 2 hr of feeding. In Bray's (1964) experiment, radioactivity could not be demonstrated in the sulfide 1 hr following a single dose of $^{35}$S-sulfate, although sulfide remained elevated for some hours after feeding as did the radioactivity of the rumen digesta. The rate and extent of reduction of sulfate is modified by adaptation (Lewis 1954; Halverson *et al.* 1968; Spais *et al.* 1968; Bird and Moir 1967. Occasionally individual animals do not develop a vigorous sulfur dissimilating population. Cystine yields sulfide even more rapidly than sulfate (Halverson *et al.* 1968; Hume and Bird 1970; Bird 1969). It is probably that the cyst(e)ine desulfurase systems are widespread and systematic, paralleling the amino acid deaminase systems. The author is unaware of studies on these desulfurase systems in ruminal bacteria.

The level of sulfide in the rumen after dosing sheep with sulfate or S-containing amino acids is shown in Table 11.1 with the time taken to reach that level.

The very high levels of sulfide in the work of Lewis (1954) and Anderson (1956) did not result in any toxic symptoms, nor did the addition of sulfide to give rumen concentrations of 330 $\mu$g $S^{2-}$ per ml (Spais *et al.* 1968) or 113 $\mu$g per ml (Anderson 1956). Anderson records recovered sulfide from jugular blood. Bray (1969B), however, observed respiratory difficulty in sheep when sulfide levels of approximately 330 and 266 $\mu$g per ml were added to buffers replacing rumen contents, and sulfide could be smelled on the animals' breath in 10 to 15 min. As rumen concentration fell, the animals recovered without incident. The loss of sulfide from the rumen followed a log linear course and the half-life values obtained in three experiments were 21.5, 15.5, and 10.7 min. This compares with values of approximately 20 min derived from Anderson's (1956) data and 35 min from that of Spais *et al.* (1968).

The conditions in these two experiments differed from those of Bray's;

TABLE 11.1

ACCUMULATION OF SULFIDE IN RUMEN LIQUOR AFTER DOSING WITH SULFATE

| Feeding Conditions | Dietary S% | Sulfur Given | Rumen Sulfide μg/Ml | Time (Hr.) | Reference |
|---|---|---|---|---|---|
| Concentrates + hay | — | 80 gm $Na_2SO_4 \cdot 10H_2O$ | 160 | 4 | Lewis (1954) |
| Concentrates + hay[1] | — | 80 gm $Na_2SO_4 \cdot 10H_2O$ | 288 | 4–6 | Lewis (1954) |
| Lucerne hay | 0.28 | 90 gm $Na_2SO_4 \cdot 10H_2O$ | 97 | 5 | Anderson (1956) |
| ditto | | 60 gm $Na_2SO_4 \cdot 10H_2O$ | 72 | 3 | ditto |
| ditto | | 30 gm $Na_2SO_4 \cdot 10H_2O$ | 38 | 2 | ditto |
| ditto | | nil | 10 | 0–7 | ditto |
| Halophytes | 0.6DM | nil | 7–8 | 2–4 | Spais *et al.* (1968) |
| Lucerne | 0.146 | nil | 5 | 4 | |
| ditto | 0.146 | 10 gm $SO_4$ | 9.8 ± 2.2 | 6 | |
| Lucerne fasted | 0.146 | nil | 1.7 | constant 0–12 | |
| ditto | 0.146 | 10 gm $SO_4$ | 10.5 ± 2.4 | 6 | |
| Oat hulls + lucerne | — | 3 gm cystine-S | 44 | 2 | Bird (1969) |
| | — | 3 gm cystine-S | 47 | 2 | ditto |
| | — | 3 gm meth.-S | 2 | 35 | ditto |
| | — | 3 gm meth.-S[2] | 5 | 6 | ditto |
| Oat hulls, starch, urea | 0.075 | nil | 0.54 | constant | Hume and Bird (1970) |
| | 0.237 | 1.6 gm $SO_4$-S | 1.9 | ditto | ditto |
| | 0.231 | 1.6 gm cystine-S | 4.3 | ditto | ditto |
| | 0.423 | 1.6 gm $SO_4$-S + 1.6 gm cystine-S | 3.5 | ditto | ditto |
| Oat hulls, urea (800 gm) | 0.058 | nil | 0.00 | 4 | Bray and Hemsley (1969) |
| | 0.143 | 0.38% $Na_2SO_4$ | 2 | 2 | ditto |
| | 0.318 | 1.16% $Na_2SO_4$ | 6 | 8 | ditto |

[1]+40 gm $Na_2SO_4 \cdot 10H_2O$ for 15 days.
[2]Fed for 20 days.

sulfide was added to existing rumen contents in which sulfide production was proceeding but maintaining relatively constant basal levels, whereas Bray used buffer solutions in the rumen. The longer half-life calculated from the data of Spais *et al.* (1968) is significant. Whereas the pH of the rumen contents was above 7 (7.1 to 7.7) in this work, it was approximately 6.5 in the other experiments. Clearly, the amount of sulfide absorbed depends on the concentration in the rumen contents, but contrary to the situation with ammonia, the rate of absorption is apparently reduced at elevated pH.

Sulfate, like nitrate, is not absorbed to any measurable extent from the rumen. Added $^{35}$S-sulfate was not detected in the blood during a 3 hr study period whereas $^{35}$S-sulfide-S was found within 5 min (Bray 1969A). Bray reports that sulfide is also rapidly absorbed from the duodenum, 40 to 90% being absorbed in 60 min but again sulfate absorption is much slower, 25% or less being absorbed in the same period. Nevertheless apparent absorption of sulfate from the small intestine may account for 80 to 90% or more, of up to 6 gm of sulfate-sulfur administered into the duodenum (Bird and Moir 1968). Normally little sulfate passes from the rumen with the ingesta and only a small proportion of the sulfide. On present evidence it may be stated that sulfur not in organic form is lost largely by absorption of sulfide directly through the rumen wall. Whether significant absorption of the sulfur amino acids occurs by this route remains to be demonstrated.

The data in Table 11.2 illustrate the flow pattern of sulfur products from the rumen through the omasum (Bird and Hume 1969).

TABLE 11.2

RUMEN SULFUR DATA[1]

| | Treatments | | | |
|---|---|---|---|---|
| | A | B | C | D |
| | Basal | $+SO_4$ | +Cystine | $+SO_4$ + Cystine |
| S intake mg | 610 | 1940 | 1950 | 3420 |
| S outflow total mg[2] | 1072 | 1227 | 1260 | 1505 |
| Inorganic $SO_4$-S outflow mg[2] | 28 | 65 | 49 | 81 |
| (Sheep No. 67) | 31 | 848 | 132 | 966 |
| Sulfide-S flow mg[2] | 4 | 21 | 37 | 44 |
| Sulfur gain or loss mg | +460 | −713 | −690 | −1915 |

[1] Bird and Hume (1969).
[2] Excludes data for sheep No. 67 for which cystine but not sulfide was reduced.

Net loss by absorption was 36.7% from treatment B, 35.4% from C, and 56% from D. On the other hand, there was a net gain of sulfur in the rumen system in the sulfur deficient treatment A of almost half a gram. This most significant addition confirms the presence of a sulfur recycling system, first proposed by Bray (1964). The quantity shown in Table 11.2 indicates that there is a greater return of sulfur to the rumen than originally calculated by Bray.

By analogy with the nitrogen recycling mechanisms only three sources are possible. These are: (1) the desquamation of ruminal epithelium, (2) entry through the rumen wall, and (3) entry through salivary secretions.

No data are available for the quantitative loss of epithelial constituents into the rumen; as hypertrophy and hyperplasia of the ruminal epithelium occurs in response to nutritional condition, perhaps confounded by the physical nature of the diet, even primary estimates are difficult to arrive at, and further discussion at this stage is pointless.

Sulfate absorption from the rumen is negligible (Bray 1969B). Unlike urea which moves readily from the blood into the urea-free space of the rumen in significant amounts, the movement of sulfate into the rumen also appears to be negligible. With hypertonic, isotonic, and hypotonic buffers in the rumen, only 1.4, 0.5 and 0.5% (mean 0.7%) of intravenous doses of $^{35}$S-sulfate accumulated in the rumen buffer during 4 hr experimental periods (Bray 1969C). The higher figure for the hypertonic buffer condition was associated with an increased water flow into the rumen. Bray calculated that the amount of sulfate-sulfur transfer for a 45 kg sheep with a sulfate space of 27% of body weight and a passage of 0.7% of available sulfate would be only 20.4 mg per day with a blood level of 40 $\mu$g sulfate-sulfur per milliliter and only 5.1 mg at 10 $\mu$g/ml. Houpt (1959) using a similar technique found $4.9 \pm 0.9$ mM/hr of urea N passed through the rumen wall of sheep. This would yield an input of approximately 1.6 gm urea-nitrogen per day, or an N:S ratio of approximately 80:1. These values for sulfate and nitrogen flows have been obtained using washed out rumen preparations with buffers and are not necessarily entirely realistic. Recent quantitative studies under physiological conditions (Weston and Hogan 1968A, B; Hume *et al.* 1970) have revealed that under some conditions net inflows into the rumen of 3 to 7.5 gm N are possible. In this case the N:S ratio would be widened even further, provided that the flow of sulfate into the rumen is not favorably influenced by metabolical activity within the rumen. Values for blood inorganic sulfur content available (Table 11.3) suggest that Bray's estimate may be quite realistic. Even if doubled, these sulfur inflows would remain nutritionally insignificant.

Values for the sulfur content of ruminant saliva are few and variable; this may well be expected in view of the nature of the several salivary glands and of their secretions, the changes in salivary composition with different flow rates, and the volumes produced from the various glands (Kay 1960).

TABLE 11.3

BLOOD INORGANIC SULFATE LEVELS

| Species | Diets | Blood Inorganic S Mg Per 100 Ml | Reference |
|---|---|---|---|
| Cattle | adequate | 3.44 | Brown and Lewis (1941) |
| Sheep | adequate<br>deficient | 2.0–4.0<br>1.0–1.5 | Weir and Rendig (1954) |
| Sheep | adequate | 3.5–6.5 | Whiting *et al.* (1954) |
| Sheep | adequate<br>adequate | 12.9–16.7<br>11.6 | Peirce (1960) |
| Cow | adequate | 14.2 | |
| Sheep | adequate<br>deficient | 1.5–6.0<br>0.3–0.8 | Bray (1964) |
| Sheep | 0.058%<br>0.143%<br>0.318% | 0.32–0.76<br>2.97–4.07<br>2.71–5.12 | Bray and Hemsley (1969) |
| Sheep | 0.4–1.4% | 6.5–10 | L'Estrange *et al.* (1969) |

Most analyses of the inorganic components of saliva show, in agreement with those of Kay, that bicarbonate, phosphate and chloride account for most of the anions in saliva; total cations, however, usually exceed total anions measured by 5–10 meq/liter. Few sulfate values are available, but if these anions were wholly sulfate the saliva could contain 80–160 mg S/liter, thus yielding up to 1.6 gm per day in the sheep. Although the sulfur concentrations reported by Thomas *et al.* (1951) of 0.5 to 2.2 mg/ml would support this view, most recent analyses do not, and suggest much lower yields even with mixed saliva with its higher ester sulfate content (Table 11.4).

TABLE 11.4

SALIVARY SULFUR CONCENTRATIONS ($\mu$g/ml)

| Species | Saliva Type | Total S | $SO_4$-S | Reference |
|---|---|---|---|---|
| Sheep | parotid | 9.1 ± 4.0 | 3.1 | Bray (1964) |
| | | 12.5 ± 4.5 | 11.5 | |
| | mixed | 30.4 ± 6.6 | 9.3 | |
| | | 38.8 ± 7.8 | 8.4 | |
| Sheep | parotid | 1.4–2.1 | 0.3–1.28 | Bray and Hemsley (1969) |
| Cattle | mixed (cardia) | — | 5–45 | Goss *et al.* (1969) |
| | mixed (mouth) | — | 147–446 | |

Bray (1964), assuming flows of 5 liters of parotid saliva and 3 liters of other saliva, calculated from the sulfur concentrations in his experiments that from 138 to 180 mg sulfur would return to the rumen in these fluids. These animals were maintained on a high protein diet containing 1.7 to 2.5% nitrogen and 0.2 to 0.3% sulfur. In a later experiment (Bray and Hemsley 1969), the dietary nitrogen level was lower (1.38%) and the sulfur content varied with added sulfate from 0.058 to 0.318%. Total collections of saliva from 1 parotid gland were made over a period of 7 days for each of 3 sheep for each sulfur treatment. Selected data are given in Table 11.5.

TABLE 11.5

BLOOD AND SALIVARY SULFATE CONCENTRATION AND OUTPUT

| Diet | | | Blood ($T_0$) | | Parotid Saliva Output/Day | | | | |
|---|---|---|---|---|---|---|---|---|---|
| Diet | % N | % S | Urea N Mg/ 100 Ml | Sulfate $\mu$g S/ Ml | l. | Total N Mg | Urea N Mg | Sulfate-S Mg | Total-S Mg |
| A | 1.38 | 0.06 | 1.5 | 3.2 | 3.13 | 373 | 336 | 1.0 | 4.4 |
| B | 1.38 | 0.14 | 7.9 | 29.7 | 2.98 | 264 | 235 | 3.8 | 6.4 |
| C | 1.37 | 0.32 | 6.7 | 27.1 | 3.18 | 243 | 216 | 4.0 | 6.7 |
| S.E. of means | — | — | 0.3[1] | 3.0[1] | 0.12 | 28 | 28 | 0.2[1] | 0.3[1] |

[1]Diets differ significantly $P < 0.05$ (Bray and Hemsley 1969).

It is noteworthy that the N:S ratio in the residual-S (protein) fraction of the saliva remains relatively constant at 11–12:1 whereas the ratio for other constituents varies with nutritional conditions. From the trends in these data it appears that urea moves more freely into the saliva than does sulfate which, nevertheless, tends to follow blood sulfate.

Parotid saliva, however, is low in mucin whereas all the other salivary glands, excluding the inferior molar, secrete a copious mucin (Kay 1960). Although the palatine, buccal and pharangeal glands maintain a slow continuous flow, the mucous producing glands generally respond to direct stimuli and their output is largely determined by the duration of eating and ruminating. Kay (1960) estimates that from 2.5 to 8 liters of this residual saliva may be secreted each day in the sheep. Lyttleton (1960) could not detect sulfur in a bovine salivary mucoprotein preparation, but McCrea (1953) gives a value of 1.4% S (and 10.1% N) for a typical preparation (C1) from sheep submandibular glands. If this is correct, then Bray and Hemsley's values for residual (protein) sulfur may be incorrect. The sulfur content of the mucins, however, appears to respond to blood sulfate level to a much greater extent than parotid saliva, presumably because of increased ester-sulfate incorporation. In a current study in our laboratory salivary sulfur secretion in relation to blood sulfur level in cattle is being examined (Goss *et al.* 1969). Whereas saliva collected at the cardia of cattle with bailed out rumens contains 5–45 μg sulfate-S/ml, that collected from the mouth contained 147–446 μg/ml. Blood sulfates ranged up to 74 μg/ml. The predominantly parotid saliva from the cardia has not shown any significant response to blood sulfate level, but the more mucid secretions obtained from the mouth show a strong positive relationship.

$$Y = -82.46 + 6.23x \; (r = 0.8173^{***})$$

where $Y$ = salivary sulfate
$x$ = blood sulfate concentration

Preliminary analyses indicate that the N to S ratio in this saliva is substantially below 10 to 1 (from 7.1 to 0.2). It has been estimated from these data that the daily sulfate-sulfur recycled to the bovine rumen is between 500 and 2000 mg. As these values were obtained without feeding and with bailed out rumens, they must be considered as minimal values. Vercoe (1969) found a limit of 17–20 gm N/day in net urea N flow into the rumen of steers, of which some 60% was derived from saliva. It is therefore, conceivable that the N to S ratio of recycled N to recycled S could approach 10 to 1.

The data of Table 11.6 show that under sulfur deficient conditions the net gain in sulfur flowing from the rumen above dietary approach 500 mg per day, that is, more than double Bray's (1964) estimates. This estimate

TABLE 11.6

N:S RATIOS IN DIET, BLOOD AND PAROTID SALIVA OF SHEEP[1]

| | Diet | Blood | Saliva | | |
|---|---|---|---|---|---|
| Diet | Dietary N:S | Urea N:$SO_4$-S | Total N:S | Urea N:$SO_4$-S | Residual N:S |
| A | 24.0 | 36 | 83 | 376 | 11 |
| B | 9.7 | 2.7 | 42 | 61 | 12 |
| C | 4.3 | 2.5 | 36 | 53 | 11 |

[1]Derived from Bray and Hemsley 1969.

purported to be sulfate-S, but is incorrect (Bray 1969A). However, the figures will still represent possible minimal total sulfur inflows.

That mechanisms exist resulting in recycling of sulfur to the rumen is now quite clear. On present evidence little sulfate is returned to the rumen by diffusion from the blood through the rumen wall, but significant amounts may be presented in the saliva depending on its mucin content and the blood sulfate levels. On quantitative values for the net return, both of sulfur and of nitrogen, to the rumen it appears that N to S ratios may be much narrower than previously calculated.

Walker and Nader's (1968) directly determined N:S ratio for ruminal bacteria is narrower than that given by Moir *et al.* (1967–1968) and much closer to their recommendation of 10 to 1 for dietary nitrogen to sulfur. This ratio is narrower than that found in many pastures. Barrow and Lambourn's (1962) data show an average N to S ratio of 13 to 1 for a range of species with sulfur contents ranging from 0.05 to 0.58%. Estimates from the results of Begg and Freney (1960) show the N to S ratio narrowed from 23 to 1 to 10 to 1 as the sulfur content increased from 0.03 to 0.29%. Nitrogen fertilizers without sulfur may extend rations even further (Stewart and Porter 1969).

The nitrogen and sulfur contents, and the N:S ratios of a number of feedstuffs are given in Table 11.7. The N:S ratios range about 10:1 except for peanut meal. The values for most cereal grains on average composition (Kent 1966) are also close to 10:1, but pure proteins, particularly legume proteins depart from this substantially. Thus, peanut meal has an N:S ratio of 18.2 to 1 (Table 11.7) and legumin 46.9 to 1 (Sherwin 1935). Other well known pure vegetable proteins, such as zein (26.9 to 1), edestin (21.2 to 1), gliadin (17.2 to 1) have wide ratios, but so, too, have the animal proteins casein (19.7 to 1), myosin (13.1 to 1), and serum globulin (14.3 to 1). Egg protein is close to 10 to 1. However, the N:S ratio of the protein moiety of the foodstuff in ruminant nutrition is irrelevant if the protein is totally degraded in the rumen; present evidence suggests that in the presence of protein the total sulfur intake in relation to nitrogen intake is the critical parameter (Hume and Bird 1970; Moir *et al.* 1967–1968; Walker

TABLE 11.7

NITROGEN AND SULFUR CONTENTS AND N:S RATIO OF FEEDSTUFFS AND PROTEIN[1]

| Material Roughages | % N | % S | N:S | Material Meals | % N | % S | N:S |
|---|---|---|---|---|---|---|---|
| Lucerne chaff (1) | 2.65 | 0.191 | 13.9 | Coconut | 3.44 | 0.283 | 12.2 |
| (2) | 2.83 | 0.293 | 9.7 | Cottonseed | 6.46 | 0.441 | 14.6 |
| (3) | 2.72 | 0.377 | 7.2 | Linseed | 5.87 | 0.452 | 12.9 |
| Wheaten chaff | 0.91 | 0.111 | 8.2 | Peanut | 7.24 | 0.396 | 18.2 |
| Oat straw | 0.25 | 0.042 | 6.0 | Safflower | 2.86 | 0.212 | 13.5 |
| Wheat | 2.18 | 0.178 | 12.2 | Meatmeal | 7.98 | 0.606 | 13.1 |
| Oats | 1.18 | 0.124 | 9.5 | Meatmeal | 7.80 | 0.506 | 15.4 |

[1]Franklin *et al.* (1956).

and Nader 1968). Where nitrogen and sulfur intakes are inherently low, and where urea supplementation is resorted to, recycling of sulfur and nitrogen would not correct a wide dietary N to S ratio. The values obtained, however, indicate that significant contributions of both nutrients to maintenance requirements may result from recycling. Bray and Hemsley (1969) using a simple oat hull, urea, and mineral diet with sheep found dry matter digestion increased from 46.6 to 51.9% by increasing sulfate-sulfur intake to narrow the N:S ratio from 24 to 9.7 to 1. Rumen microbial protein and the rate of cellulose digestion both increased with more favorable sulfur supply. Hume and Bird (1970) found a significant increase in the amount of protein synthesized in the rumen (82 gm against 91–98 gm protein/day), with either sulfate or cystine additions. As digestibilty influences intake, (Blaxter *et al.* 1961) favorable N:S ratios, in affecting digestibility, will greatly increase digestible energy intake and as more total protein is synthesized by the rumen organisms under such conditions, protein balance, as well as energy balance, is improved (Thomas *et al.* 1951; Moir *et al.* 1967–1968; Bray and Hemsley 1969; Hume and Bird 1970).

The relationship between nitrogen intake ($X_n$), sulfur intake ($X_s$), and nitrogen balance ($Y$) has been described by the regression $Y = -3.945 + 0.436 X_n + 2.669 X_s$ (Moir *et al.* 1967–1968). Maximum nitrogen balance was achieved with dietary nitrogen to sulfur ratio of 10:1.

Bray and Hemsley (1969) found a highly significant linear relationship between nitrogen balance and sulfur balance which showed an approximately 10 to 1 ratio between these parameters. Although Bird and Hume (1969) have found a rather wider ratio (18.25:1) between tungstic acid precipitate nitrogen and sulfur in the digesta flowing through the omasum, the tungstic-acid nitrogen to neutral-sulfur ratio at 12.3 was closer to Walker and Nader's (1968) figure of 10.7 to 1. There is no evidence that the amino acid composition of the rumen bacteria will be influenced by dietary conditions (Holmes *et al.* 1953; Weller 1957; Purser and Buechler 1966). With inadequate sulfur, total yield of protein will be reduced, and there remains the possibility that other nonprotein-sulfur components, such as glutathione and inorganic S will be substantially reduced. The uptake of appreciable sulfide nonenzymatically is known to occur where sulfide is available (Walker and Nader 1968). The nutritional and metabolic significance of such changes in microbial composition remain to be explored fully. The known conditions of ruminal sulfur metabolism, allowing for losses from the system and the direct requirements of the microorganisms, give general support for the suggestion that the dietary N:S ratio should be close to 10:1.

The influence of sulfur extends beyond immediate protein production and metabolic effects to many other fields. For example, Whanger and Matrone (1965) show sulfur deficiency has a profound influence on the products of rumen fermentation as well as the rate of fermentation and the crop of organisms produced. Wide sulfur to nitrogen ratios promote selenium uptake and retention, whereas narrow ratios depress retention (Pope *et al.* 1968) and may result in acidosis and depress feed intake (L'Estrange *et al.* 1969). It is probable that rumen sulfide is implicated in most of these processes, therefore the control of nutritional energy-nitrogen-sulfur relationships assumes considerable importance in animal production.

## BIBLIOGRAPHY

ALBERT, W. W., GARRIGUS, U. S., FORBES, R. M., and NORTON, H. W. 1956. The sulfur requirement of growing-fattening lambs in terms of methionine, sodium sulfate, and elemental sulfur. J. Animal Sci. *15*, 559–569.

ANDERSON, C. M. 1956. The metabolism of sulphur in the rumen of the sheep. New Zealand J. Sci. Technol. *37*, 379–394.

BARROW, N. J., and LAMBOURNE, L. J. 1962. Partition of excreted nitrogen, sulphur and phosphorus between the faeces and urine of sheep being fed pasture. Australian J. Agr. Res. *13*, 461–471.

BEGG, J. E., and FRENEY, J. R. 1960. Chemical composition of some grazed native pasture species in New England region of New South Wales. C.S.I.R.O. Australia, Div. Pl. Ind., Divl. Rept. *18*, Canberra.

BIRD, P. R. 1969. Personal communication. Nedlands, West Australia.

BIRD, P. R., and HUME, I. D. 1969. Unpublished data.

BIRD, P. R., and MOIR, R. J. 1967. Personal communication. Nedlands, West Australia.

BIRD, P. R., and MOIR, R. J. 1968. Unpublished data.

BLAXTER, K. L., WAINMAN, F. W., and WILSON, R. S. 1961. The regulation of food intake by sheep. Animal Prod. *3*, 51–61.

BRAY, A. C. 1964. The recycling and excretion of sulphur in sheep. Proc. Australian Soc. Animal Prod. *5*, 336–344.

BRAY, A. C. 1969A. Sulphur metabolism in sheep. I. Preliminary investigations on the movement of sulphur in the sheep's body. Australian J. Agr. Res. *20*, 725–737.

BRAY, A. C. 1969B. Sulphur metabolism in sheep. II. The absorption of inorganic sulphate and inorganic sulphide from the sheep's rumen. Australian J. Agr. Res. *20*, 739–748.

BRAY, A. C. 1969C. Sulphur metabolism in sheep. III. The movement of blood inorganic sulphate across the rumen wall of sheep. Australian J. Agr. Res. *20*, 749–758.

BRAY, A. C., and HEMSLEY, J. A. 1969. Sulphur metabolism in sheep. IV. The effect of a varied dietary sulphur content on some body fluid sulphate levels and on the utilization of a urea-supplemented roughage by sheep. Australian J. Agr. Res. *20*, 759–773.

BROWN, B. H., and LEWIS, H. B. 1941. The metabolism of sulphur. XXVII. The distribution of sulphur in the ultrafiltrates of blood plasma. J. Biol. Chem. *138*, 705–726.

BRYANT, M. P., and ROBINSON, I. M. 1961. Studies on nitrogen requirements of some ruminal cellulolytic bacteria. Appl. Microbiol. *9*, 96–103.

COOMBE, J. B., TRIBE, D. E., and MORRISON, J. W. C. 1960. Some experimental observations on the toxicity of urea to sheep. Australian J. Agr. Res. *11*, 247–256.

DIJKSHOORN, W., and VAN WIJK, A. L. 1967. The sulphur requirements of plants as evidenced by the sulphur-nitrogen ratio in the organic matter.—A review of published data. Plant Soil *26*, 129–157.

EATON, T. M. 1966. Sulphur. *In* Diagnostic Criteria for Plants and Soils, H. D. Chapman (Editor). Univ. California, Div. Agr. Sci., Berkeley.

EMERY, R. S., SMITH, C. K., and HUFFMAN, C. F. 1957A. Utilization of inorganic sulfate by rumen microorganisms. I. Incorporation of inorganic sulfate into amino acids. Appl. Microbiol. *5*, 360–362.

EMERY, R. S., SMITH, C. K., and HUFFMAN, C. F. 1957B. Utilization of inorganic sulfate by rumen microorganisms. II. The ability of single strains of rumen bacteria to utilize inorganic sulfate. Appl. Microbiol. *5*, 363–366.

FERGUSON, W. S., and TERRY, R. A. 1954. The fractionation of the non-protein-nitrogen of grassland herbage. J. Sci. Food Agr. *5*, 515–524.

FRANKLIN, M. C., MCCLYMONT, G. L., and BRIGGS, P. 1956. Unpublished data.

GOSS, K., THORNTON, R. F., and MOIR, R. J. 1969. Unpublished data.

HALVERSON, A. W., WILLIAMS, G. D., and PAULSON, G. D. 1968. Aspects of sulfate utilization by the microorganisms of the ovine rumen. J. Nutr. *95*, 363–368.

HENDERICKX, H. 1961. The incorporation of sulfate in the ruminal proteins. Arch. Intern. Physiol. Biochim. *69*, 449–458.

HOGAN, J. P. 1961. The absorption of ammonia through the rumen of the sheep. Australian J. Biol. Sci. *14*, 448–460.

HOLMES, P., MOIR, R. J., and UNDERWOOD, E. J. 1953. Ruminal flora studies in the sheep. V. The amino acid composition of rumen bacterial protein. Australian J. Biol. Sci. *6*, 637–644.

HOOGENRAAD, N. J., HIRD, F. J. R., HOLMES, I., and MILLIS, N. F. 1967. Bacteriophages in rumen contents of sheep. J. Gen. Virology *1*, 575–576.

HOUPT, T. R. 1959. Utilization of blood urea in ruminants. Am. J. Physiol. *197*, 115–120.

HUME, I. D. 1970. Synthesis of microbial protein in the rumen. III. The effect of dietary protein. Australian J. Agr. Res. (In press)

HUME, I. D., and BIRD, P. R. 1970. Synthesis of microbial protein in the rumen. IV. The influence of the level and form of dietary sulphur. Australian J. Agr. Res. (In press)

HUME, I. D., MOIR, R. J., and SOMERS, M. 1970. Synthesis of microbial protein in the rumen. I. Influence of the level of nitrogen intake. Australian J. Agr. Res. (In press)

JAMIESON, N. D. 1959. Nitrate reduction in the rumen of the grazing sheep. New Zealand J. Agr. Res. *2*, 96–106.

KAY, R. N. B. 1960. The rate of flow and composition of various salivary secretions in sheep and calves. J. Physiol. (London) *150*, 515–537.

KENT, N. L. 1966. Technology of Cereals. Pergamon Press, London.

L'ESTRANGE, J. L., CLARKE, J. J., and MCALEESE, D. M. 1969. Studies on high intakes of various sulphate salts and sulphuric acid in sheep. I. Effects on voluntary feed intake, digestibility and acid base balance. Irish J. Agr. Res. *8*, 133–150.

LEWIS, D. 1951. The metabolism of nitrate and nitrite in the sheep. I. The reduction of nitrate in the rumen of the sheep. Biochem. J. *48*, 175–180.

LEWIS, D. 1954. The reduction of sulphate in the rumen of the sheep. Biochem. J. *56*, 391–399.

LEWIS, D. 1955. Amino acid metabolism in the rumen of the sheep. Brit. J. Nutr. *9*, 215–230.

LEWIS, D., HILL, K. J., and ANNISON, E. F. 1957. Studies on the portal blood of sheep. I. Absorption of ammonia from the rumen of the sheep. Biochem. J. *66*, 587–592.

LEWIS, T. R., and EMERY, R. S. 1962. Relative deamination rates of amino acids by rumen microorganisms. J. Dairy Sci. *45*, 765–768.

LOFGREEN, G. P., LOOSLI, J. K., and MAYNARD, L. A. 1947. The influence of protein source upon nitrogen retention by sheep. J. Animal Sci. *6*, 343–347.

LOOSLI, J. K., and HARRIS, L. E. 1945. Methionine increases the value of urea for lambs. J. Animal Sci. *4*, 435–437.

LYTTLETON, G. W. 1960. Bloat in cattle. XVIII. The proteins of bovine saliva. New Zealand J. Agr. Res. *3*, 63–68.

MCCREA, J. F. 1953. Studies on influenza virus receptor-substance and receptor-substance analogues. I. Preparation and properties of a homogeneous mucoid from the salivary gland of sheep. Biochem. J. *55*, 132–138.

MCDONALD, I. W. 1948. The absorption of ammonia from the rumen of the sheep. Biochem. J. *42*, 584–587.

MCDONALD, I. W. 1952. The role of ammonia in ruminal digestion of protein. Biochem. J. *51*, 86–90.

MOIR, R. J., SOMERS, M., and BRAY, A. C. 1967–1968. Utilization of dietary sulphur and nitrogen by ruminants. Sulphur Inst. J. *3*, 15–18.

PECK, H. D. JR. 1962. Symposium on metabolism of inorganic compounds. V. Comparative metabolism of inorganic sulfur compounds in microorganisms. Bacteriol. Rev. *26*, 67–94.

PEIRCE, A. W. 1960. Studies of salt tolerance of sheep. III. The tolerance of sheep for mixtures of sodium chloride and sodium sulphate in the drinking water. Australian J. Agr. Res. *11*, 548–556.

PFANDER, W. H., ELLIS, W. C., GARNER, G. B., and MURKER, M. E. 1956.

Absorption of glucose, lactic acid, potassium nitrate and potassium nitrite from the rumen of mature wethers. J. Animal Sci. *15*, 1292.

Pittman, K. A., and Bryant, M. P. 1964. Peptides and other nitrogen sources for growth of *Bacteroides ruminicola.* J. Bacteriol. *88*, 401–410.

Pittman, K. A., Lakshmanan, S., and Bryant, M. P. 1967. Oligopeptide uptake by *Bacteroides ruminicola.* J. Bacteriol. *93*, 1499–1508.

Pope, A. L., Moir, R. J., Somers, M., and Underwood, E. J. 1968. Effect of sulphur on $^{75}$selenium absorption in sheep. J. Animal Sci. *27*, 1771.

Prescott, J. M. 1961. Utilization of sulfur compounds by *Streptococcus bovis.* J. Bacteriol. *82*, 724–728.

Purser, D. B., and Buechler, S. M. 1966. Amino acid composition of rumen organisms. J. Dairy Sci. *49*, 81–84.

Rendig, V. V., and Weir, W. C. 1957. Evaluation by lamb feeding tests of alfalfa hay grown on a low-sulfur soil. J. Animal Sci. *16*, 451–461.

Sherwin, H. C. 1935. Chemistry of Food and Nutrition. The Macmillan Co., New York.

Simonnet, H., Le Bars, H., and Mollé, J. 1957. The cycling of urea administered orally in ruminants. Compt. Rend. Acad. Sci. *244*, 943–945. (French)

Sirotnak, F. M., Doetsch, R. N., Brown, R. E., and Shaw, J. C. 1953. Amino acid metabolism of bovine rumen bacteria. J. Dairy Sci. *36*, 1117–1123.

Somers, M. 1961. Factors influencing the secretion of nitrogen in sheep saliva. I. The distribution of nitrogen in the mixed and parotid saliva of sheep. Australian J. Exptl. Biol. Med. Sci. *39*, 111–122.

Spais, A. G., Lazaridis, T. K., and Agiannidis, A. K. 1968. Studies on sulphur metabolism in sheep in association with copper deficiency. Res. Vet. Sci. *9*, 337–344.

Stewart, B. A., and Porter, L. K. 1969. Nitrogen-sulfur relationships in wheat (*Triticum aestivum* L.), corn (*Zea mays*), and beans (*Phaseolus vulgarus*). Agron. J. *61*, 267–271.

Synge, R. L. M. 1952. The utilization of herbage protein by animals. Brit. J. Nutr. *6*, 100–104.

Thomas, W. E., Loosli, J. K., Williams, H. H., and Maynard, L. A. 1951. The utilization of inorganic sulfates and urea nitrogen by lambs. J. Nutr. *43*, 515–523.

Thompson, J. F. 1967. Sulfur metabolism in plants. Ann. Rev. Plant Physiol. *18*, 59–84.

Vercoe, J. E. 1969. The transfer of nitrogen from the blood to the rumen in cattle. Australian J. Agr. Res. *20*, 191–197.

Walker, D. J., and Nader, C. J. 1968. Method for measuring microbial growth in rumen content. Appl. Microbiol. *16*, 1124–1131.

Weir, W. C., and Rendig, V. V. 1954. Serum inorganic sulfate sulfur as a measure of the sulfur intake of sheep. J. Nutr. *54*, 87–96.

Weller, R. A. 1957. The amino acid composition of hydrolysates of microbial protein from the rumen of sheep. Australian J. Biol. Sci. *10*, 384–389.

Weston, R. H., and Hogan, J. P. 1968A. The digestion of pasture plants by sheep. II. The digestion of ryegrass at different stages of maturity. Australian J. Agr. Res. *19*, 963–979.

Weston, R. H., and Hogan, J. P. 1968B. Factors limiting the intake of feed by sheep. IV. The intake and digestion of mature ryegrass. Australian J. Agr. Res. *19*, 567–576.

WHANGER, P. D., and MATRONE. G. 1965. Effect of dietary sulfur upon the fatty acid production in the rumen. Biochim. Biophys. Acta *98*, 454–461.

WHITING, F., SLEN, S. B., BEZEAU, L. M., and CLARK, R. D. 1954. The sulphur requirements of mature range ewes. Can. J. Agr. Sci. *34*, 261–268.

WILLIAMS, V. J., and MOIR, R. J. 1951. Ruminal flora studies in the sheep. III. The influence of different sources of nitrogen upon nitrogen retention and upon the total number of free microorganisms in the rumen. Australian J. Sci. Res., Ser. B, Biol. Sci. *4*, 377–390.

CHAPTER 12

A. R. Till
P. F. May
I. W. McDonald

# The Use of Tracer Techniques to Study the Sulfur Cycle in Grazed Pastures

## INTRODUCTION

Research into mineral nutrition of plants in agricultural and grazing systems has frequently neglected sulfur or has accorded it only minor importance compared with nitrogen, phosphorus and potassium. However, sulfur is essential for the production of proteins and is certainly a major plant nutrient.

Most of the world's population of ruminant animals obtain their feed by grazing for either part, or in many instances, the whole of their lifetime. In considerable areas of the world, each flock or herd grazes year-round upon, and obtains the whole of its food from, a fixed area of land; this is particularly true for the wool industry in Australia.

In the grazing system, the soil minerals essential for plant growth travel in a continuous cycle from the soil through the plants and then back to the soil either directly or after passage through the grazing animals. The availability to plants of these essential minerals exercises control of plant growth, within the limits of the environment, and thus of the number of animals that can be supported and their production per head.

When essential elements are deficient, considerable increases in plant growth, animal numbers and production have been obtained by the application of appropriate fertilizers to correct deficiencies. In the New England area of Australia soils are often deficient in sulfur and substantial increases in wool production have been obtained by the application of gypsum and other sulfur sources to grazed pasture; however, the responses are difficult to predict and may not be lasting.

These responses in production depend on the amounts of nutrient in the pools within the cycle, the transport rates between them, and the accession or loss of nutrient from the system. Widely used techniques in grassland research can provide information on the sizes of nutrient pools in the system but cannot be used to estimate the utilization of applied nutrient or the transport rates between pools.

A "tracer" for a particular nutrient is necessary to enable its uptake and transport rates to be studied. In the investigations described here sulfur-35 has been used to study the sulfur cycle in the grazed pasture. Sulfur-35 was chosen for the following reasons:

(1) Wool is a protein that contains about 3.5% sulfur. As wool is derived from the amino acids of plant proteins (and of the microbial proteins produced from them in the rumen), wool production per acre is necessarily related to the production of plant proteins. (2) Other sulfur studies in this region had shown a lack of correlation between sulfur applications, pool sizes, and animal production. These experiments emphasized the need for studies of the system's dynamics. (3) The radiation ($\beta$ $E_{max}$ 0.167 Mev) and half-life (87.2 days) make sulfur-35 convenient to use experimentally. (4) Research on the metabolism of sulfur in sheep had revealed efficient incorporation of ($^{35}S$) cystine into wool and had provided considerable experience in the processing and measurement of samples containing sulfur-35.

The experiments were initiated, first, to establish the feasibility of using radiotracers in grazed pastures and, second, to investigate the form of the cycling process, the sizes of the sulfur pools, and the rates of transport between pools. The results have demonstrated the usefulness of sulfur-35 for studying the system, and suggest that other nutrients could be studied by similar techniques provided that suitable nuclides are available.

## CONCEPTS AND METHODS USED TO STUDY THE GRAZED PASTURE

### Radiotracers

Radiotracers enable direct measurement of a nutrient in an otherwise undisturbed system. The pools with which the tracer mixes, the rates of movement, and the chemical forms can be used to study the dynamics and reaction mechanisms of the system. Experiments with tracers in the field may be difficult at times due to the impossibility of precisely defining the size and environment of the experimental unit (e.g., depth of soil explored by plant roots, occurrence of drought or flood, and unspecified inputs or outputs), but in most cases concurrent experiments on similar sites should at least partially resolve the difficulties.

The experiments described here have been designed to develop techniques for radiotracer studies in the field and to obtain a basic understanding of the role of a nutrient in a grazed pasture system.

Sulfur-35 has been used to study the length of time that an applied nutrient is available to plants, to estimate adsorption, leaching, and recycling.

The sizes of some sulfur pools and transport rates have been estimated and these, together with the proposed sulfur cycle (Fig. 12.1) have been used to model the system on an analog computer. The sulfur pools have

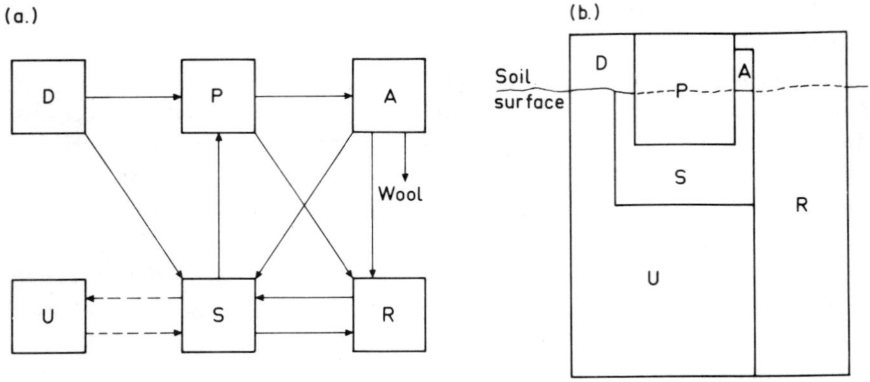


FIG. 12.1. DIAGRAM OF SULFUR CYCLE (a) AND RELATIVE SIZES OF SULFUR POOLS (b)

(a.) The rectangles represent sulfur pools and the arrows show known (——) or possible (-----) transports of sulfur. (b.) The areas of the rectangles are approximately proportional to the total sulfur in each pool and show their position relative to the soil surface. They are for a stocking rate of about 15 sheep per hectare.

LEGEND

D—Fertilizer sulfur (D = 0 at equilibrium)
P—Sulfur in entire plants
A—Sulfur in sheep
R—Residues pool which comprises organic sulfur in soil, dead plant material, organisms, excreta, etc., most of which is considered as being below ground level
S—Soil sulfur in forms suitable for immediate utilization by plants (mainly sulfates)
U—Sulfur in the soil that apparently does not enter the cycle
C—Cycling sulfur pool
$C = P + A + R + S$

been selected for convenience of study and are defined in the legend to Fig. 12.1.

**Analog Computing**

Changes in the specific activity of any pool in the cycle following the application of sulphur-35 are determined by the mixing of the sulfur applied into the cycling sulfur pool. The shape of the curve obtained by plotting the specific activity against time, here described as the transient, depends on the transport rates within the cycle and the sizes of the various compartment pools.

The rates of change in the radioactivity content of the various compartment pools in the system at any time are expressed as functions of a set of transport parameters in a series of differential equations which are used to describe the system shown in Fig. 12.1.

The solution of these equations then gives estimates of the parameters. Because theoretical solutions to such sets of equations are generally unwieldy, it is more convenient to apply numerical methods. In these studies numerical solutions were obtained with the aid of an analog computer.

Because the model fitted in these studies involves a number of component pools, the generation of the behavior of compartments corresponding to measured parts of the experimental systems incidentally generates the corresponding behavior of the associated compartments. Some parts of the system are inaccessible to experimental observation because of lack of resources, suitable techniques, or physical definition of the compartments themselves; the model enables the behavior of these parts to be studied by simulation. Further, the model provides indications of those parts of the system that are likely to be of major or minor importance, and thus helps in selecting topics for more detailed research.

## TECHNIQUES

### Preparation of Labeled Materials

Sulfur-35 was obtained as carrier-free sulfate in aqueous solution.

**Solid.**—Gypsum ($CaSO_4 2H_2O$) was prepared by mixing the carrier-free sulfate solution with sulfuric acid and then reacting the labeled acid with aqueous calcium chloride. The precipitate of gypsum was filtered off, dried, and ground to the required particle size (May *et al.* 1968).

Carrier-free sulfate was prepared for distribution by adding the sulfate solution to clean sand in a cylindrical container which then was revolved about a horizontal axis and the water removed by a stream of warm air.

Where further dilution was required the ($^{35}S$) gypsum or labeled sand was added to similar nonradioactive material and sealed in a drum. Mixing was obtained by rotating the drum on horizontal rollers.

**Liquid.**—The dose of sulfur-35 was mixed with the required amounts of carrier sulfate and water immediately before application to the paddocks.

### Application of Radioactivity to the Paddocks

**Solid.**—This was the most difficult material to distribute by machine because of lack of control of quantity, nonuniform application, and the risk of fractionation of material if particle size is not uniform. Depending upon the aim of the experiment, the nonuniform application may not be critical (see results of strip and uniform labeling). Small quantities of material were applied to small areas by distribution from bottles with caps perforated like those of pepper shakers.

**Liquid.**—The solutions were applied to whole paddocks from a boom spray mounted on a tractor. Small areas were labeled by using a small boom spray having one pair of wheels and being pushed by hand. The wheels were connected, by cycle chains and sprockets, to a double-acting piston pump so that the amount delivered was directly proportional to the distance traveled.

**Proportion of Paddock Initially Labeled.**—Three basic types of labeling techniques have been used. They are "uniform" (whole area), "strip" (about 3.3% of area), and "hot spot" (about 0.015% of area), each of which allows successively higher specific activity sulfur to be applied, while providing the same equivalent total radioactivity for the entire pasture. With the amounts of radioactivity used, the uniform labeling technique was confined to studies of the wool specific activity transient. The strip labeling gave similar wool specific activity transients but in addition provided a technique for the estimation of translocation of nutrients, direct measurement of uptake by plants, and an indirect check on adsorption and retention of nutrients by soil. The hot spot method produced similar wool transients but with much greater variation during the initial stages. The advantages were that it allowed direct measurement of sulfur fractions in the soil and their relationship to the specific activity of plant sulfur. It also provided a conclusive demonstration of the recycling of sulfur.

### Sampling

The following conventional sampling techniques were used: soil—steel coring tube; plants—hand plucking; ingesta—oesophageal fistula; wool—Oster animal clippers; and faecal excreta—sampling from the rectum.

### Sulfur Determination and Radioassay

Sulfur was estimated using a modification of the method of Garrido (1964). Samples were wet ashed using a nitric-perchloric acid mixture, with ammonium vanadate as catalyst and potassium dichromate as indicator; the digest diluted with an acetic, hydrochloric, and phosphoric acid mixture. The sulfur content of the resulting solution was determined turbidimetrically on an Auto Analyser (Technicon Corp., United States).

The sulfur contents of wool, plant material, ingesta, and excreta were determined to within 1% (95% confidence) of the observed value, but of soil to only 7% (95% confidence) for values of 100 $\mu$gS/gm or greater. It has been reported (Chapman and Pratt 1961) that the above method of oxidation may not release all the sulfur from some soils. We have not observed this, but it may partly account for the larger variations in the soil sulfur estimates.

Sulfur-35 was determined by liquid scintillation counting of the buffered digest. Samples of the digests (each 7.0 ml) were added to a toluene based scintillator and Triton x100 mixture (7:6 v/v, 13 ml) and the sulfur-35 measured with approximately 50% efficiency. The scintillator solution consisted of *p*-terphenyl (0.3% v/v) and 1,4-bis-2′-(5′ phenyloxazolyl)-benzene (0.01% w/v) in toluene.

## Calculation of the Cycling Sulfur Pool

A diagram of a sulfur cycle is presented in Fig. 12.1a; the size of the cycling sulfur pool was calculated as follows.

It was assumed that the transports shown as broken lines were of negligible magnitude and that all the applied sulfur entered the cycle and became mixed with the sulfur already in the cycle. The equilibrium specific activity of the cycling sulfur was estimated from the plateau values of the wool specific activity; an estimate of the mass of sulfur in the cycling sulfur pool was then obtained by dividing the dose of sulfur-35 applied by the equilibrium specific activity.

The operation of the transports shown as broken lines could be considered equivalent to losses of radioactivity by processes such as adsorption, leaching, and carryoff in products and/or an input of nonradioactive material from some other source. Under these circumstances additional estimates of the losses or gains would be required before the size of the cycling sulfur pool could be calculated.

## Translocation of Sulfur

When a strip labeling technique is used and all (or a known proportion) of the applied sulfur enters and remains in the cycle then, once the wool specific activity reaches a plateau, the proportion of the applied nutrient that has been translocated can be calculated as follows

$$w = fp + (1 - f)r$$

where w = specific activity of wool at the plateau
p = specific activity of plants on the originally labeled area
f = fraction of the area originally labeled
r = average specific activity of plant sulfur in the originally unlabeled part of the area

The proportion of the original dose remaining in the site will be

$$\frac{fp}{fp + (1 - f)r} = \frac{fp}{w}$$

This will be the net retention and will provide a small underestimate of total translocation due to redeposition of activity on the strips.

## Calculation of Transport Rates

The differential equations describing the model of the sulfur cycle were programmed into an analog computer and the rate constant potentiometers were adjusted until the model generated a transient closely resembling that obtained experimentally.

The transport rates were calculated from the scale values and settings of the rate constant potentiometers of the computer.

Calculation of an estimate of variance due to discrepancy between the observed specific activity of the animal sulfur pool and that generated by the model, enabled an objective assessment of the goodness of fit of the proposed cycle to observations from the experiments.

A method of obtaining conditionally optimal estimates of a subset of parameters, based on minimum discrepancy variances, was used to examine the effects of treatments on transport rates in restricted parts of the cycling system.

**Simulated Effects of Modifications to the Cycle**

The simulation of effects of modifications to the cycling process, by computer studies on the behavior of the model, have enabled an examination of the sensitivity of the recycling system to varying degrees of departure from estimated transport rates in different parts of the cycle. In extreme cases a transport pathway is eliminated entirely by setting the value of an appropriate parameter to zero.

## RESULTS

The results presented are from a series of experiments conducted at the Pastoral Research Laboratory, Armidale.

In a preliminary experiment (May *et al.* 1968) it was shown that some of the applied sulfur was available for a long period ( > 1000 days), leaching appeared to be negligible, and some of the sulfur-35 had been translocated by the sheep to previously unlabeled areas and reutilized by plants. The results also suggested that a large proportion of the soil sulfur remained isolated from the rest of the sulfur in the system and that the total amount of sulfur potentially available to plants throughout their growing period was much greater than would be expected from relationships between total soil sulfur and that estimated as "available" by conventional procedures (Ensminger and Freney 1966).

It was proposed that the utilization of sulfur in grazed pastures could be represented by a simple sulfur cycle as shown in Fig. 12.1.

Another experiment on the site used for the preliminary experiment showed that the total sulfur in the system and the cycling sulfur pool had not changed significantly during a period of about 1000 days (Till and May 1970). The wool specific activity transients and rainfall distribution for both these experiments are shown in Fig. 12.2. The initial parts of the two transients have different shapes (probably largely due to the different rainfall patterns), but the plateaus reached similar values. These results supported the concept of a simple cycle (Fig. 12.1) which was used as the basis for simulation of the system on an analog computer. However, further experiments were necessary to verify the assumptions used in the

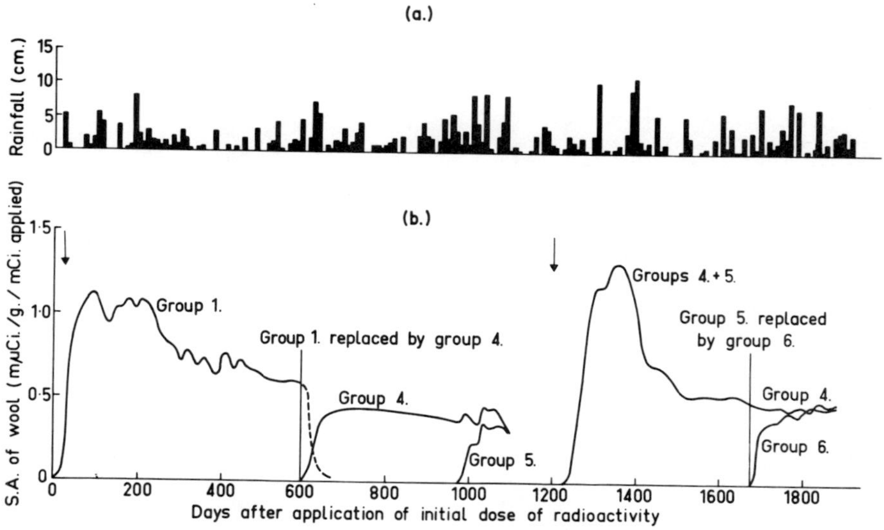


FIG. 12.2. RAINFALL DISTRIBUTION (a) AND WOOL SPECIFIC ACTIVITY TRANSIENTS (b) THROUGHOUT EXPERIMENT

(a.) Rainfall distribution presented as successive ten-day totals. (b.) Average wool specific activity per mCi applied versus time. Each curve represents a different group of sheep. The differences in the shapes of the initial parts of the transients following the application of Sulfur-35 are probably mainly due to the different rainfall patterns. The similarity of the plateau level attained after each dose shows the constancy of the size of the cycling sulfur pool. The portion of the curve shown as a broken line indicates the decline in wool specific activity when the sheep were removed from the radioactive pasture. Arrows show applications of Sulfur-35 to the pasture.

isotope dilution calculations and to test the effects and usefulness of alternative methods of applying the sulfur-35. Figure 12.3 shows the wool specific activity transients from sheep grazing eight comparable paddocks which had received similar amounts of sulfur-35 but applied in different ways as given in Table 12.1. Similar plateaus were reached after approximately the same length of time even though there were considerable differences in the early parts of the transients.

These results support the assumption that there was no adsorption of the applied sulfur and show that the alternative labeling techniques tested are satisfactory. Further support for the above is provided by considering the specific activities of sulfur in plant material harvested from the labeled strips (see Fig. 12.4). The sulfur was applied at rates of 130 mg per hectare or 7.8 kg per hectare and yet the specific activities of similar plant material at the plateaus did not differ significantly between paddocks. This would not have been expected if plant production was primarily controlled by

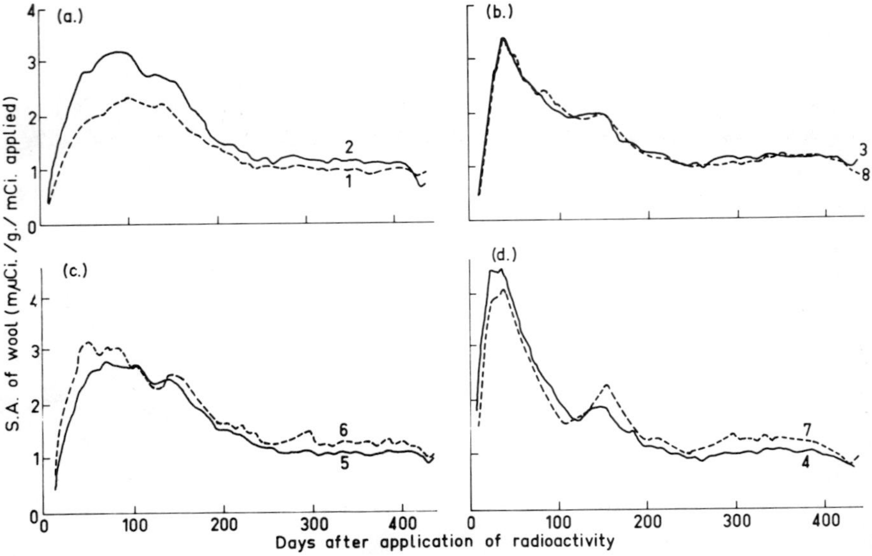


FIG. 12.3. COMPARISONS BETWEEN THE AVERAGE SPECIFIC ACTIVITIES OF WOOL GROWN BY SHEEP GRAZING STRIP (-----) AND UNIFORMLY LABELED PADDOCKS (—)

The number against each curve indicates the paddock number as shown in Table 12.1. (a.) Solid gypsum application. (b.) Sulfate on sand. (c.) Aqueous magnesium sulfate. (d.) Aqueous sulfate. The similarity between each pair of transients indicates that the incorporation of Sulfur-35 into wool was the same after either strip or uniform labeling. The different shapes of the initial portions of the transients are clearly due to the form of the material applied, but the similarity of plateau levels shows that this did not influence the estimation of the cycling sulfur pool.

TABLE 12.1

SULFUR APPLICATIONS AND POOL SIZES

| Paddock No. | Sulfur Applied | | | Specific Activity of Wool Sulfur[2] MμCi/Gm/MCi Applied | Cycling Sulfur Pool (Kg) | Sulfur in Top 10 Cm Soil (Kg) |
|---|---|---|---|---|---|---|
| | Weight[1] | MCi | Material, Form and Method | | | |
| 1 | 2.34 kg | 5.76 | $CaSO_4 2H_2O$ solid, strip | 30.8 | 32.4 | 66.2 |
| 2 | 2.34 kg | 5.76 | ditto solid, uniform | 35.8 | 27.9 | 61.6 |
| 3 | 38 mg | 6.40 | $K_2SO_4$ on sand—uniform | 32.5 | 30.8 | 55.4 |
| 4 | 930 mg | 5.76 | $CaSO_4$, aqueous—uniform | 30.8 | 32.4 | 52.6 |
| 5 | 2.34 kg | 5.76 | $MgSO_4$, aqueous—uniform | 33.9 | 29.5 | 51.5 |
| 6 | 2.34 kg | 6.40 | ditto aqueous—strip | 38.5 | 26.0 | 44.3 |
| 7 | 38 mg | 6.40 | $K_2SO_4$, aqueous—strip | 37.9 | 26.3 | 54.7 |
| 8 | 38 mg | 6.40 | ditto on sand—strip | 30.9 | 33.3 | 52.6 |

[1]Each paddock had an area of 0.3 hectares.
[2]Divided by dose applied to enable direct comparison between paddocks.

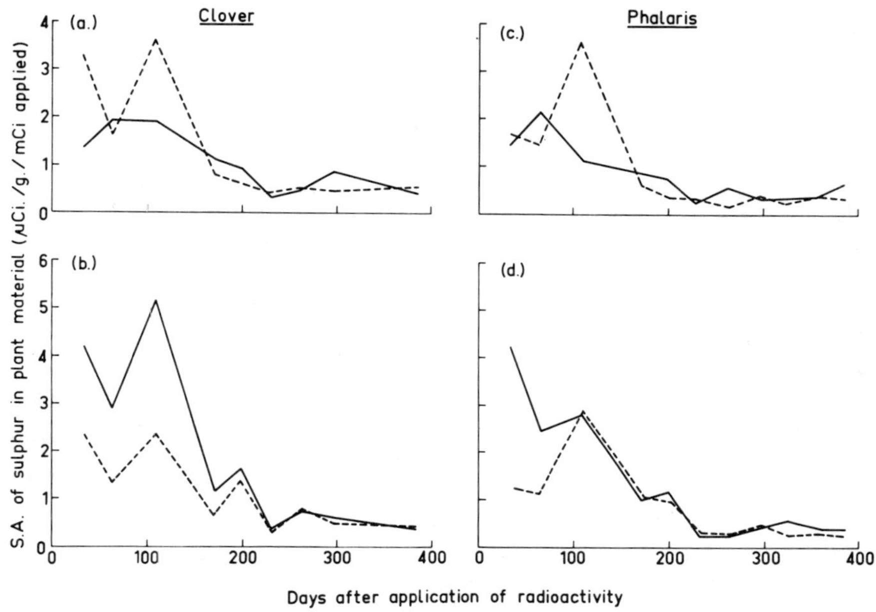


FIG. 12.4. SPECIFIC ACTIVITIES OF PLANT SAMPLES FROM STRIP-LABELED PADDOCK

Sulfur applications of 130 mg per hectare (-----) and 7.8 kg per hectare (—). (a.) and (c.) are solid applications, (b.) and (d.) liquid applications of Sulfur-35. The different masses and methods of application influenced the early stages of the plant sulfur specific activity transients for both phalaris and clover but in about 200 days similar plateaus were reached.

the amount of "available" sulfur, as estimated by some extraction procedure, and indicates the importance of the cycling sulfur pool (a large proportion of which must be organic sulfur in the soil) in determining the productivity of the system.

The application of a dose of sulfur-35 to a very small percentage of the total area of the paddock has produced wool specific activity transients similar to those from other experiments (Fig. 12.5). This experiment, which will be reported in detail elsewhere, provided data for measurement of the distribution of radioactivity down the soil profile and extent of sulfur translocation. After 254 days the areas initially labeled were excluded from grazing and some of the sheep replaced by nonradioactive animals. The replacement animals began growing radioactive wool which must have come from sulfur-35 that had been translocated and subsequently cycled through the pasture plants. The specific activities at the plateaus (Fig. 12.5) show that about 50% of the applied sulfur-35 had been translocated.

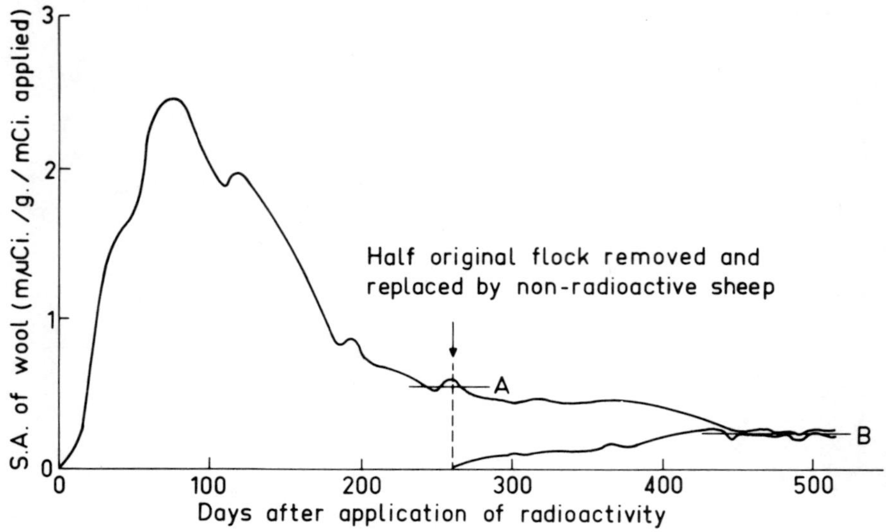


FIG. 12.5. WOOL SPECIFIC ACTIVITY TRANSIENTS FOLLOWING "HOT SPOT" LABELING OF PASTURE

The second group of sheep had access only to pasture labeled by translocation of Sulfur-35 by the first group of sheep. The wool specific activities at A and B were used to estimate the proportion of the dose that had been translocated. Arrow shows when "hot spots" were excluded from grazing.

The curves in Fig. 12.6 are an example of the use of an analog computer to simulate the activity transients of the system. These were derived from a very simple treatment of the cycle in Fig. 12.1 in which first order processes were assumed and no allowance made for environmental influences. Detailed computer studies of the system are to be presented elsewhere.

TABLE 12.2

PARAMETER VALUES FROM SIMPLE TREATMENT (TR20) AND DETAILED STUDY (EAI 8800)

| | Dose in Pools (%) | | | Estimated $T^1/2$ (Days) | |
|---|---|---|---|---|---|
| Pools | TR20 | EAI8800 | Transports[1] | TR20 | EAI8800 |
| P | 10 | 4 | D → S | 140 | 156 (73–408)[4] |
| A | 0.7 | 0.5 | D → P | 230 | 387 |
| R | 80 | 81 | S → P[3] | 35 | 125 (60–340)[4] |
| S | 9 | 12 | P → R | 38 | 35 |
| W | ...[2] | 2 | R → S | 335 | 274 (128–716)[4] |

[1]The remaining transports could not be compared due to differences in control and programming of the models.
[2]There were insufficient computing elements to represent this pool in the model.
[3]These transports are not directly comparable as in the more complex model an initial dose of radioactivity was added to the plant pool to represent the small input of high specific activity material that would arise from contamination. Thus the S → P transport had to be artificially high to represent this input.
[4]The figures in parentheses show the range of the $T^1/2$'s when a weather function was included.

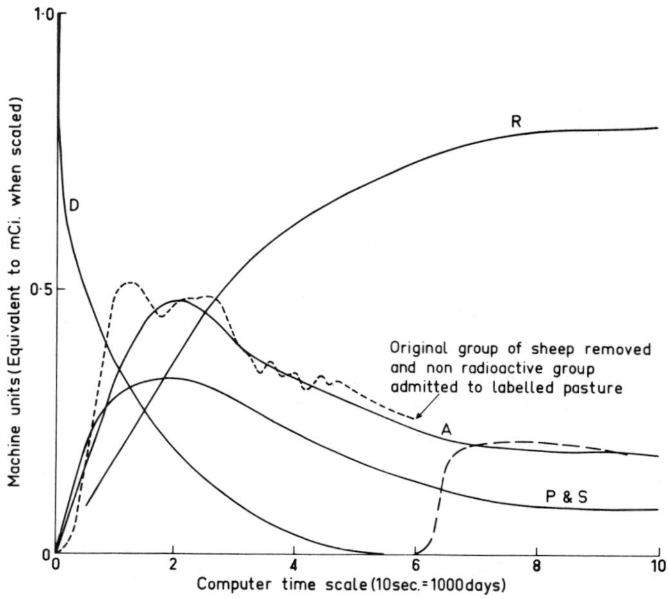


FIG. 12.6. RESPONSE CURVES FROM SIMULATION OF GRAZED PASTURE SYSTEM ON ANALOG COMPUTER

Experimental curve (-----). The animal pool (A) is scaled to represent wool. Limitations of the computer precluded the simulation of two separate wool curves. The pools to which the transients belong are represented by symbols which have the meanings defined in Fig. 12.1. The curves D, R, S, and P represent the radioactivity transients that would be expected and indicate the probable behavior of the pools. S and P were very similar on this scaling.

The transport rates estimated from the simple analog are presented in Table 12.2 together with values obtained from the more detailed analog.

## DISCUSSION

May *et al.* (1968) have demonstrated the feasibility of applying radiotracer techniques to studies of the grazed pasture system. A simple model of the sulfur cycle (Till and May 1970) and network analysis have been used to obtain quantitative information on the cycling of sulfur.

The advantages of the approach include the minimal degree of disturbance it introduces to the normal running of the system studied, and the possibility of inferring the behavior of parts of the cycling system not accessible to direct measurement.

The investigations have shown that a closed system recycling process was capable of satisfactorily accounting for observations on the specific activity of wool following the application of sulfur-35 to the pasture. However, direct observations on other parts of the system are needed to confirm the applicability of the model proposed and to provide further information for estimates of model parameters.

Inconsistent results from attempts to relate sulfur supplying capacity of the soil to measurements of total soil sulfur or various sulfur fractions in the soil are probably derived from the failure of the measurements to account for the rates of transport between the sulfur pools. For example, if we consider the soil pool in Fig. 12.1 it is obvious from the number of inputs and outputs that measurements of the transport rates would be more important than an estimate of pool size.

The isotope dilution experiments show that a large proportion of the total sulfur is in the cycling sulfur pool (Table 12.1) and the majority of this will be in the form of organic residues in the soil (Fig. 12.1b). Thus it appears that modification of the transport rates associated with the residues pool offer the greatest potential for obtaining increased plant and animal production. Should normal fertilizer treatments not make significant changes in the transport rate constants, then a large amount of sulfur would have to be added to make a sufficiently large increase in the amount mobilized from the residues pool. This amount would be many times greater than the estimate of "available" sulfur by extraction processes and is probably the reason for the need to build up "banks" of fertilizer.

The animal pool is very small compared with the other pools in the system but this provides a good example of the importance of considering the transports in the system. At times a large proportion of the plant material will pass through the animal and this provides a pathway for some of the sulfur to return directly to the soil pool. Sulfur passing along this alternative pathway will avoid the slow turnover of the residues pool, and the net effect should be equivalent to speeding up the transport rate from residues to soil.

The computer simulation provides a guide to the responses in the system that may be expected following certain manipulations and can be of considerable help in determining the limits of future experiments. It also allows the behavior of parts of a system to be inferred even though direct measurements may be impossible.

The curves from the simple analog are presented only to show that even such a simple representation can be used to illustrate the basic behavior of other pools and their relative importance.

Direct measurements of dynamic aspects of sulfur cycling in grazed pastures, in which the integrated system is observed with negligible disturbance, are much more meaningful than measurements possible by conventional techniques. From the indications of the present investigations of the method, no apparent theoretical problems are associated with its application to different situations or adaptation to other nutrients.

## ACKNOWLEDGEMENT

The authors wish to thank Mr. M. J. Cumming of the CSIRO Chemical Research Laboratories for permission to use results from the detailed simulation of the system.

## BIBLIOGRAPHY

CHAPMAN, H. D., and PRATT, P. F. 1961. Sulphur. *In* Methods of Analysis for Soils, Plants and Waters. Div. Agr. Sci., Univ. California.

ENSMINGER, L. E., and FRENEY, J. R. 1966. Diagnostic techniques for determining sulphur deficiencies in crops and soils. Soil Sci. *101*, 283.

GARRIDO, M. L. 1964. Determination of sulphur in plant material. Analyst (London) *89*, 61.

MAY, P. F., TILL, A. R., and DOWNES, A. M. 1968. Nutrient cycling in grazed pastures. I. A preliminary investigation of the use of [$^{35}$S] gypsum. Australian J. Agr. Res. *19*, 531.

TILL, A. R., and MAY, P. F. 1970. Nutrient cycling in grazed pastures. II. Further observations using [$^{35}$S] gypsum. Australian J. Agr. Res. *21*, No. 2. (In press)

CHAPTER 13

H. J. Almquist

# Sulfur Nutrition of Nonruminant Species

## INTRODUCTION

The relative nutritional effectiveness of different forms of methionine activity have for some time been a matter of conjecture.

To start out it is well to consider the general functions of this amino acid particularly those which involve interactions with other molecular species and their bearing on methionine requirement. This I have attempted to illustrate in Fig. 13.1. The processes indicated are based upon demonstrated nutritional replacements but do not necessarily mean that the biological mechanisms are identical with or limited to those indicated. In fact, previous discussions at this Symposium have delineated these relations more thoroughly.

For convenience in further discussion, I shall for the most part, represent the methionine isomers as simply L, D or DL. and the hydroxy analog (DL-$\alpha$-hydroxy-$\gamma$-methylmercaptobutryic acid) as A. SAA will designate

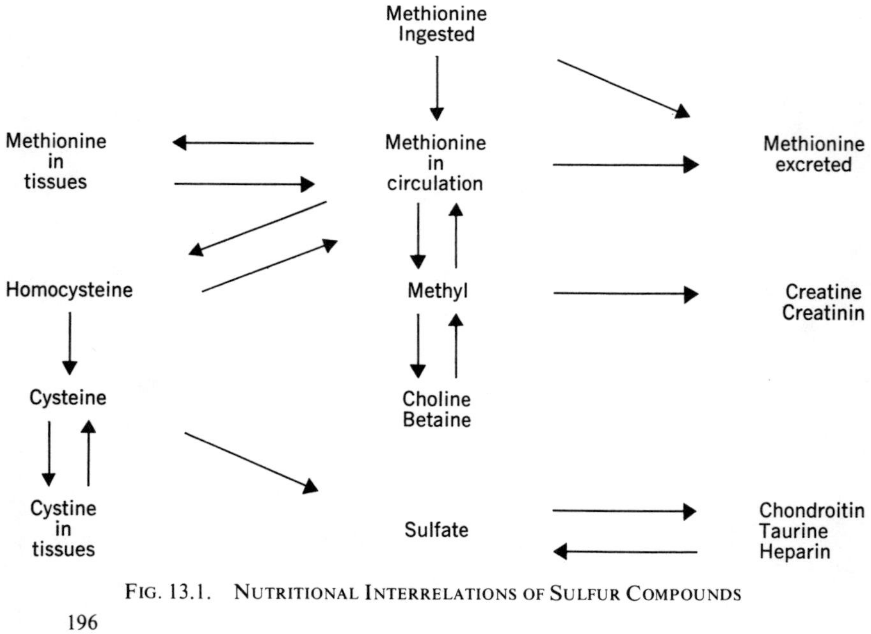


FIG. 13.1. NUTRITIONAL INTERRELATIONS OF SULFUR COMPOUNDS

total sulfur-containing amino acids, including cystine and analog precursors.

## RATS

Utilization of D and L by the young rat was studied using an amino acids diet containing 0.2% cystine. At a 0.2 level L was superior for growth in each of 2 trials. Growth became optimal at about 0.3 added L and 0.6 added D. At the 1.4 level the L was depressing to growth while the D was not (Wretlind and Rose 1950).

This latter finding has been reported several times in respect to extravagant levels of these isomers, and seems to be an established fact (Van Pilsum and Berg 1950). We may be dealing here with a rate factor. Due to time required for inversion the D may be a less intense source of the L and, therefore, show less of the inhibitory effects of the L when fed at higher than optimal levels.

Despite the differences in results the authors concluded that the D was utilized as effectively as the L because growth differences were hardly significant statistically. I submit that the true response curves may have been as indicated in Fig. 13.2. If correct, this figure indicates that, depending upon levels fed, D might appear inferior to L, equal to L or superior to L.

Whether the 0.2% level of the cystine was sufficient for maximal sparing action on methionine at the higher levels of supplementation may be ques-

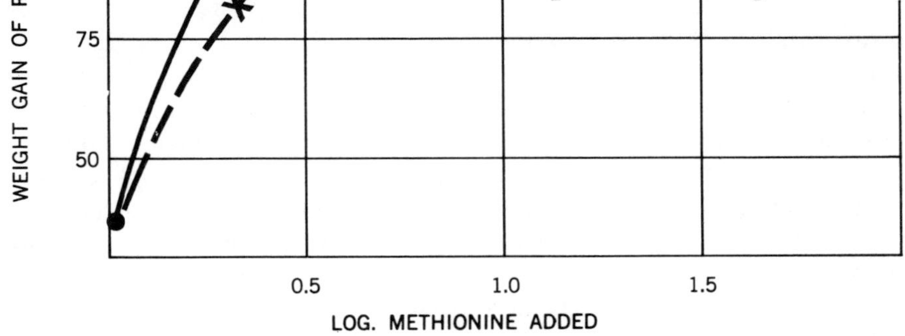


FIG. 13.2. RELATION OF RAT WEIGHT GAIN TO THE LOG OF L- OR D-METHIONINE ADDED TO THE DIET
(Grams per kilogram of diet)

tioned. At the higher levels the conversion to cystine for tissue growth could have caused an additional diversion of the methionine. The growth curves appear to be a compound type, indicating that some further demand for methionine became effective at the higher intake levels.

In studies using rats and amino acids diets it was observed that suboptimal intakes of D-methionine produced inferior growth as compared to L, if at the same time other essential amino acids were supplied in DL forms. If five of the other essential amino acids were supplied as L forms then the D and L methionines appeared to be closer in growth promotion; however, the D remained consistently lower as compared to the L. A usual type of statistical test indicated no difference (Wretlind 1952).

Rats fed purified L-amino acids diets in which D-amino acids replaced some of the L, showed a growth reduction with D-histidine which was accentuated by further substitutions of D forms of lysine, threonine, isoleucine, leucine, and valine. The growth impairment appeared related to no single amino acid but was cumulative. D-amino acids which were not readily utilized exerted an antagonistic action on those D-amino acids, such as methionine, which may be readily inverted (Kamath and Berg 1964).

Studies on utilization of methionine from foods by the rat have been conducted with a diet based primarily on oxidized casein. In this peroxide-treated casein practically all methionine is destroyed. Cystine, tryptophan, and tyrosine were added to the diet. All amino acids except methionine were calculated to exceed the estimated requirements for the protein-depleted rat. A response curve was established by results from L-methionine additions to the diet. The data follow the diminishing returns principle quite closely with both weanling and protein-depleted rats. The latter appear to require less methionine for optimal gains. Of particular interest in this case is the report that both types of rats utilized DL approximately 95% as well as the standard L (Schweigert and Guthneck 1954).

The effect of isomers of methionine on nitrogen balance in the adult rat was studied with a diet program involving an amino acid mixture containing no methionine or cystine, but ample choline. L and DL were fully equivalent at nitrogen balance, but the latter was progressively less effective as the nitrogen balance became more negative. It was suggested that the enzyme systems for inversion of the D isomer were less effective under negative nitrogen balance (Nasset and Anderson 1951).

## CHICKS

Chicks fed first a practical diet and then a purified diet based on isolated soybean protein showed equivalent growth responses to L or DL. This was true in the presence or in the absence of added cystine and choline. The basal diet provided about half the total methionine requirements in natural sources (Grau and Almquist 1943).

Growth of chicks was tested with all-amino acids basal diet containing no methionine or cystine. Adding D, L, or DL at 0.3% level led to little or no net gain in any case. At 0.5% supplementation the L was significantly better for gain than D or DL. At 0.8% supplementation all forms were equivalent for gain but there was a tendency for poorer gain-feed ratio with the D form. A level of 1.2% DL supported very little better gain (Marrett *et al.* 1964).

The diet contained 3.8% D-amino acids which the authors point out may have interfered somewhat with utilization of D-methionine. It is apparent from several studies that the D-amino acid oxidase system can be overloaded.

Further trials were made with a basal diet containing only L-amino acids. L or D-methionine additions supported gains not significantly different; however, the data are not sufficiently extensive to test the validity of the differences (Marrett *et al.* 1964).

A chick diet of natural ingredients based primarily on corn and peanut meal contained by calculation 0.28% methionine and 0.36% cystine, or total SAA, 0.64%. This indicated a deficiency of about 0.25%. Addition of 0.2% D or L stimulated chick growth equally well, up to a normal rate for practical diets. Feed efficiencies were identical. Measurement of D-amino acid oxidase in kidney and liver tissues indicated no difference due to the form of methionine added (Bauriedel 1963). It seems possible that on this latter point not enough D-amino acid was used in this natural foods diet to provide a measurable stress on the D-amino acid oxidase. The work demonstrates that a level of D-methionine about four times that which would be used in practical diets had no apparent effect on the oxidase system.

Other work has indicated that DL was less effective than L for growth of chicks. A log plot of the data does in fact show that DL was 71% as active as L for chick growth as observed with a simplified type diet; however, both products were equally effective for prevention of nutritional muscular dystrophy. L-cystine was found to be a specific sulfur amino acid for dystrophy prevention (Scott and Calvert 1962). This suggests that D is as readily converted as the L into L-cystine.

Adult male chickens were fed synthetic diets based on amino acids mixtures. Although the N-balance data appear somewhat erratic the results do suggest excellent utilization of the D isomer (Leveille *et al.* 1960).

Both the calcium salt of A and its acid amide were compared to DL for growth of chicks fed a basal simplified diet containing primarily peanut meal plus added lysine. This basal diet would appear to contain by calculation 0.2% L and 0.3% cystine. A 0.1% or equivalent, supplements of the methionine active products markedly improved growth and feed conversion in each case. All supplements reached their maximal effect at 0.3%; however, the general rating for growth promotion would indicate the amide to be inferior to the other two. The A calcium salt was evidently about 77% active compared to DL (Bird 1952).

The effect of dietary levels of vitamin $B_6$ on the utilization of L, DL, D, and A was studied with chicks over the period to three weeks age. The basal purified diet contained 0.21% methionine and 1.3 mg vitamin $B_6$ per kg. At a suboptimal level of 0.25 mg added Vitamin $B_6$ and 0.05% levels of added L, D, DL or A (molecular equivalent basis), the D and DL appeared slightly inferior for growth promotions. At 0.10% levels only the D was slightly inferior. At 0.15% levels all forms were equal. When 1.6 mg vitamin $B_6$ was added no differences appeared among the supplements. This work indicates that vitamin $B_6$ may be concerned with the utilization of D (Fuller and Hill 1964).

A chick diet composed largely of sucrose and isolated soybean protein at total protein levels of 12 to 14% supported faster gains when supplemented with 0.336% DL than with an equivalent level of A. At 12% protein level in the diet added urea or diammonium citrate improved response to the A but not to the DL. Results suggested that the extra forms of nitrogen were needed to promote amination of the A (Sullivan and Bird 1957).

In another laboratory isolated soybean protein in a chick diet similar to that mentioned above was supplemented by either 0.288% DL or A, and with either of these supplements plus 0.226% of diammonium citrate. In further tests, diets of 12 to 14% protein were supplemented with 0.3 to 0.35% levels of DL or A and also diammonium citrate or urea. Equivalent gains were registered throughout (Machlin and Gordon 1959). Apparently, extra sources of ammonia nitrogen did not, in this case, improve growth results from A.

A novel feeding technique was used in comparing DL and A for chick growth. The chicks, a male cross-bred type, were fed in individual cages throughout the studies. They were selected for uniform growth during a four-week pretest period on an otherwise complete diet based upon raw soybean meal, and then assigned into uniform lots.

During the test period of 3 weeks they received a basal diet of 40% crude protein and 2650 cal of M.E. per kilo with vitamin supplements. This diet also contained added threonine, valine, and lysine to be sure these amino acids were not deficient. The chicks were given 13 gm of this diet daily. The chicks also had unlimited access to a nitrogen-free diet based on dextrose, starch, corn oil, fiber, and minerals, and containing 3465 cal M.E. per kilo. The purpose of this arrangement was to control nitrogen intake but at the same time allow the chick to select all the energy it required.

Addition of DL or A to the basal protein diet at a 0.3% level, or the equivalent, promoted marked but equivalent improvement in weight gains and feed utilization. The data indicate that A, as the calcium salt, was about 80% as active as DL for both weight gain and feed conversion. One percent addition of DL led to further improved gains and feed utilization, showing that the 0.3% test level was in the submaximal range, as expected (Calet and Baratov 1960). It is interesting to note that consumption of the high calorie nitrogen-free part of the diet increased with increased efficiency of the basal protein diet.

A methionine-deficient purified diet based on isolated soy protein was fed to chicks from the 3rd to 17th day of life. The diet received two suboptimal levels of supplements of L, D, DL, and A (molecular equivalent basis). The latter contained 90% calcium salt. Relative potencies for growth promotion were found as follows: L—100, D—82, DL—97, and A—98. The potency of A was 78% on the product basis (Brueggemann *et al.* 1962).

Studies on relative efficacies for chick growth to four weeks of L, DL, and A have been made with a purified type diet based on isolated soy protein, starch, glucose, corn oil, and necessary vitamin and mineral supplements. The basal diet was fed alone and supplemented with 0.05, 0.1 and 0.2% of methionine, or the analog calcium salt (80% active constituent). Results were compared to a standard response curve based on L. From feed intake data the consumption of the isomers and analog was calculated. From this the efficiencies of the supplements added to the diet were derived (Bishop 1964):

| | 0.05% of diet | 0.10% of diet | 0.20% of diet |
|---|---|---|---|
| L as standard | 100 | 100 | 100 |
| DL | 89 | 81 | 71 |
| A (calcium salt) | 89 | 72 | 54 |

A review of the data by another method leads to different conclusions. In Fig. 13.3 are shown plots of the growth data obtained with the L, DL, and A supplements. Response lines fell on closely parallel slopes. This parallelism indicates essentially constant ratios of activities between the different supplements over the growth range covered. If the potency ratio, R, of any two forms is expressed as $R = A \div B$, at any chosen growth level, then $\log R = \log A - \log B$. The horizontal spreads of these lines measure the logs of the potency ratios, thus DL was 76% as active as L, whereas the A was 54% as active for growth promotion.

A repetition of the work (Bishop 1964) using a soybean meal instead of the isolated protein in the same diet showed the DL response to be closer to that with L, but the A was again clearly less efficient. In comparison to L the DL was 91% and the A was 69% as effective.

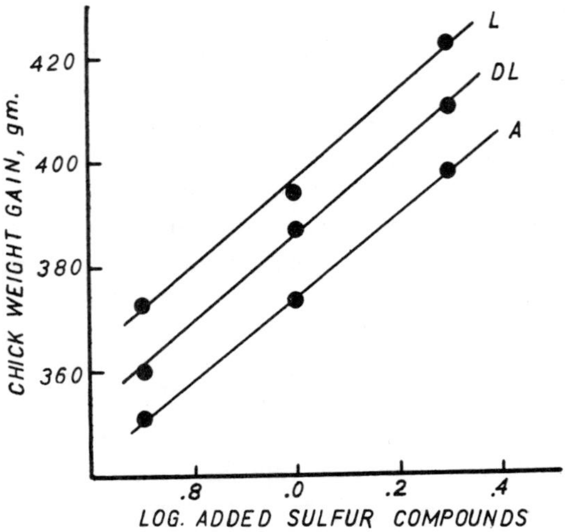


FIG. 13.3. RELATION OF CHICK WEIGHT GAIN TO THE LOG OF SULFUR-CONTAINING AMINO ACIDS AND ANALOGUE ADDED TO THE DIET (Grams per kilogram of diet, characteristic omitted)

In comparison to DL the A was 76% as effective. This figure is fairly close to the amount (80%) of active material stated to be present in the commercial analog product. From this it would be concluded that the A, on a molecular basis, is approximately as efficient as the DL, but on the usual product basis about 80% as active, for promotion of growth rate.

In another study of these relative activities an essentially uncooked soybean meal was used in a basal diet to magnify growth responses to methionine (Guttridge and Lewis 1964). The authors mention that response lines did not deviate significantly from linearity and parallelism but these responses certainly appear to be both nonlinear and nonparallel as presented. However, when placed upon a log plot the data line up as shown in Fig. 13.4. In this work there was no apparent difference in growth responses to L and DL. On the weight basis A appeared to be about 80% as active, with a suggestion that relative activity decreased slightly at higher levels of intake. In this case also, the growth activity is in agreement with the expected percentage of active material in the analog product.

Chick weight at 28 days was used to construct response curves to 28-day total SAA intake, as varied by supplements of either DL or A. Fitting of the curves to quadratic equations or by the log method gave closely similar results. In a purified diet the relative efficacy of the analog was 67%, and in a semipractical diet 84% (Halloran and Bishop 1966).

It is plain that complications intervene in the utilization of the D. Among these may be recognized the level of the D ingested, presence of relatively

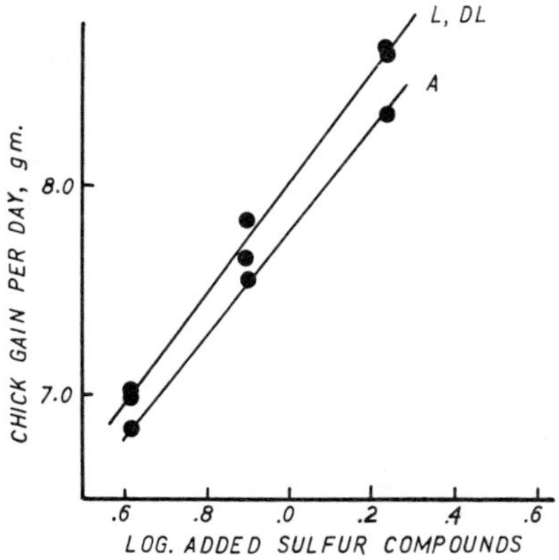


FIG. 13.4. RELATION OF CHICK WEIGHT GAIN TO THE LOG OF SULFUR-CONTAINING AMINO ACIDS AND ANALOGUE
(Same as for Fig. 13.3)

large amounts of D forms of other amino acids, the general plane of nutrition, the rate of inversion of the D, and the relative rate of excretion of the D.

It may be commented that lowered efficiency of the DL as compared to L shows up for growth results with purified or simplified types of diets and is seen rarely with more practical diets. We face the probability, in evaluating comparative data on L, D, and A forms, that the character of the basal test diet has a large influence on results, especially on the activity of the D.

Cysteine appears to be the functional sulfur compound in one pathway involved in the prevention of nutritional muscular dystrophy in the chick, while vitamin E may be concerned in another pathway. Adequate levels of sulfur amino acids and of vitamin E will prevent this disease under commercial conditions (Hathcock *et al.* 1968). It is reported that methionine has a specific antithyrotoxic effect in the chick (Kano *et al.* 1968).

In addition to the amino acids methionine and cystine, sulfur is found throughout the animal as various compounds of the sulfate and ester form.

Prominent among these are:

*Chondroitin* is an important structural component of cartilage, bone, tendons, and walls of the blood vessels. Chondroitin is a muco-protein which can be split into: (1) protein, (2) N-acetyl chondrosamine, (3) glucuronic acid, and (4) sulfuric acid. Chondrosamine is 2-amino-D-galactose. Glucosamine is reported also present as a part of chondroitin.

*Taurine* is found throughout the body and as a conjugate in a bile acid, taurocholic acid. Taurine is 2-amino-ethane-sulfonic acid.

*Heparin* is a polysaccharide composed of hexosamine and hexuronic acid groups containing sulfuric acid ester groups. It functions as a blood anticoagulant.

*Sulfate* is used in the detoxification of many substances.

It is hardly necessary to point out the important role of chondroitin in the development of bones and supportive tissues, that without normal synthesis of chondroitin the young animal, particularly, could suffer from impaired formation of such tissues, although significant effects on weight gains and feed efficiency might not be seen.

If sulfate is inadequately supplied in the diet, then obviously the animal must make it from other sources, which would be most probably the sulfur-containing amino acids.

Nevertheless, in most studies on requirements and functions of the sulfur-containing amino acids, the influence of the sulfate intake has been disregarded. Since it is a common practice to include compounds such as manganese sulfate, zinc sulfate, etc., in mineral mixtures perhaps the sulfate has been provided fortuitously. Some feed ingredients, in the preparation of which sulfuric acid has been used (such as some solubles products and molasses), may contain quite large amounts of sulfate. I have wondered how often sulfate has been the actual identity of somebody's "unidentified growth factor."

Balance studies have indicated some retention of dietary inorganic sulfur in the chick (Evans and McGinnis 1946). It has been shown in several studies with the radioisotope $^{35}S$ that sulfate taken orally or injected can be found in other forms and widely distributed in the animal (Berry *et al.* 1969). It takes part in the enzymatic exchange of sulfate in chondroitin (Bostrom and Mannson 1952).

Radioactive sulfate was found to be distributed into the cystine of feathers of chicks and in the lining of the gizzard. Analyses of the lining showed quite appreciable amounts of methionine, cystine, and sulfate (Machlin *et al.* 1954). The feathers contained a further organic radio-sulfur compound which was identified as taurine (Machlin and Pearson 1957). In the body of the bird 23% of the ingested sulfate $^{35}S$ was incorporated into tissues, small amounts into cystine and methionine, and larger amounts into taurine and sulfate compounds.

Growth responses on adding sulfate to cystine-low diets, free of inorganic sulfate, showed a "sparing action" of the latter on the requirement of the chick for the sulfur-containing amino acids (Machlin and Pearson 1957; Gordon and Sizer 1955; Jukes 1956). Rats fed sulfate-free diets based on casein showed growth responses to sodium sulfate in the range of 0.25 to 1.0% of the diet (Martin and Patrick 1960).

Increased vitamin $B_{12}$ in the diet enhanced retention of $^{35}S$ in the skeletal tissues of the chick (Binet and Wellers 1955). Attention could be called to two papers which showed that isolated chondroitin had a growth-promoting action when placed in chick diets (Robinson *et al.* 1939; Almquist *et al.* 1940).

A very complete set of data obtained with chicks fed a simplified sulfate-free diet was kindly made available by Dr. T. H. Jukes. The work was done at the Lederle Laboratories (Jukes 1956).

I have expressed the results on a graph using a log plot of the sulfur compound additions according to the diminishing returns principle (Fig. 13.5). These results show a linear growth response when methionine and cystine were added to the sulfate-free diet. The response terminated in a definite plateau. The growth response line corresponding to the same additions to a diet containing 0.5% sodium sulfate, including the sulfate in the total sulfur added, was displaced to the right but parallel to the first mentioned response line. In the latter case, however, growth reached a definitely higher peak and then declined. Such a decline is often seen in amino acid work and is most probably evidence of an imbalance caused by too much

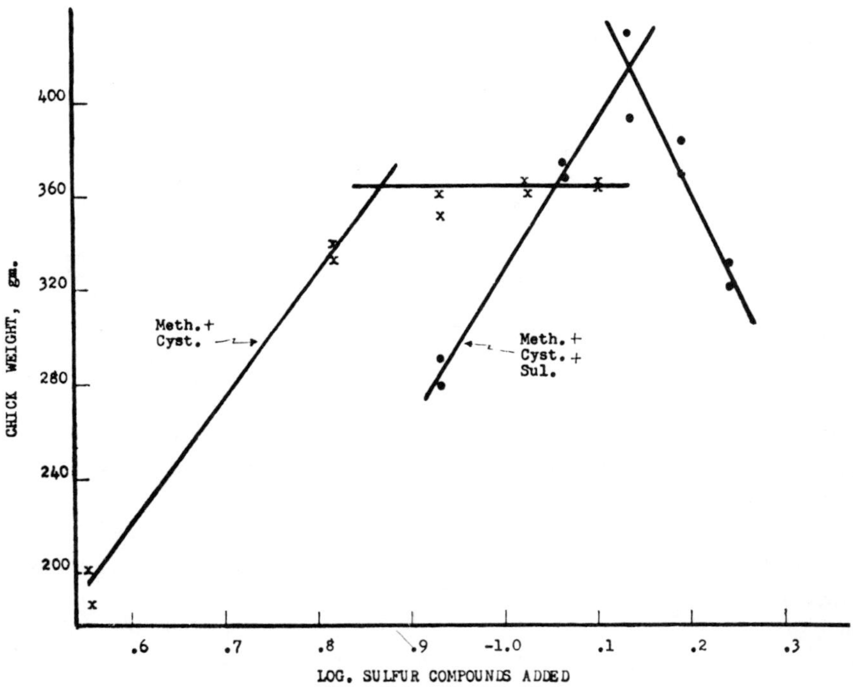


FIG. 13.5. RELATION OF CHICK WEIGHT TO THE LOG OF SULFUR-CONTAINING AMINO ACIDS AND SULFATE ADDED TO THE DIET
(Percent of diet)

instead of too little. The fact that sulfate helps the animal to achieve a greater gain seems to imply a need for sulfate as such. In absence of dietary sulfate the chick does not make it fast enough from the sulfur-containing amino acids, for the maximal rate of gain.

The horizontal displacement of the two log response lines can be interpreted to mean in this case that 0.21% amino acid was replaced by the 0.50% sodium sulfate, or that the latter was about 40% efficient in sparing action.

The above mentioned work with monogastric animals was conducted with simplified diets. Being aware of the possible significance of sulfate in the diet, we ran some tests with practical chick diets and obtained the results shown in Table 13.1. These results show consistently favorable responses from sodium sulfate addition. They also show that the relatively insoluble calcium sulfate was of no value.

TABLE 13.1

TRIALS WITH PRACTICAL BROILER DIETS WITH AND WITHOUT METHIONINE OR SULFATE SUPPLEMENTATION

| Experiment | Diet | Groups | Average Live Weight Lb | Average Feed Conversion |
|---|---|---|---|---|
| A | Control | 2 | 3.21 | 2.88 |
| | C. + 0.06% methionine | 2 | 3.39 | 2.69 |
| | C. + 0.10% methionine | 2 | 3.39 | 2.71 |
| | C. + 0.12% sodium sulfate | 2 | 3.44 | 2.67 |
| B | Control | 2 | 3.27 | 2.56 |
| | C. + 0.05% methionine | 2 | 3.30 | 2.41 |
| | C. + 0.10% sodium sulfate | 2 | 3.22 | 2.48 |
| C | Control | 2 | 2.71 | 2.80 |
| | C. + 0.05% methionine | 2 | 2.83 | 2.64 |
| | C. + 0.10% sodium sulfate | 2 | 2.82 | 2.57 |
| | C. + 0.10% calcium sulfate | 2 | 2.70 | 2.78 |

It is evident that sulfate can exert an appreciable sparing action on the sulfur-containing amino acids, at least to the extent of minimizing the destruction of these amino acids for the formation of sulfate according to body needs.

One may speculate on the possible other advantages of insuring adequate sulfate in the diet. Among these could be included possibly improved skeletal development, less bone deformities, possibly improved condition of blood vessels, etc. I have deliberately called this a speculation, because that is all it can be at present, but it is something to think about.

## BIBLIOGRAPHY

ALMQUIST, H. J., STOKSTAD, E. L. R., MECCHI, E., and MANNING, P. V. D. 1940. Identification of the rice factor. J. Biol. Chem. *134*, 213–216.

BAURIEDEL, W. R. 1963. The effect of feeding D-methionine on the D-amino oxidase activity of chick tissues. Poultry Sci. *42*, 214–217.

BERRY, R. K., HANSARD, S. L., ISMAIL, R. J., and WYSOCKI, A. A. 1969. Absorption, deposition, and placental transfer of sulfate sulfur by gilts. J. Nutr. *97*, 399–408.

BINET, L., and WELLERS, G. 1955. Physiological value of the mineral sulfates. J. Physiol. (Paris) *47*, 883–892.

BIRD, F. H. 1952. A comparison of methionine and two of its analogues in the nutrition of the chick. Poultry Sci. *31*, 1095–1096.

BISHOP, R. B. 1964. Poultry and Livestock Comment, E. I. du Pont de Nemours & Co., Wilmington, Del.

BOSTROM, H., and MANNSON, B. 1952. On the enzymatic exchange of the sulfate group of chondroitin sulfuric acid in slices of cartilage. J. Biol. Chem. *196*, 483–488.

BRUEGGEMANN, J., DREPPER, K., and ZUCKER, H. 1962. Quantitative determination of utilization of D-, L-, and DL-methionine and Ca DL-2-hydroxy-4-(methylthio) butyrate by chicken. Naturwissenschaften *14*, 334–340.

CALET, C., and BARATOV, J. 1960. Comparison of growth efficiency for chicks of methionine and the hydroxy analog. Rech. Avicoles, CNR2, *4*, 261–267. (French)

EVANS, R. J., and MCGINNIS, J. 1946. The influence of autoclaving soybean oil meal on the availability of cystine and methionine for chicks. J. Nutr. *31*, 449–461.

FULLER, H. J., and HILL, R. L. 1964. The relationship of vitamin $B_6$ and the utilization of the various forms of methionine by the chick. Poultry Sci. *43*, 1320. Abstr. (Data provided by Dr. Fuller.)

GORDON, R. S., and SIZER, I. W. 1955. The biological equivalence of methionine hydroxy analogue. Poultry Sci. *34*, 1198.

GRAU, C. R., and ALMQUIST, J. J. 1943. Utilization of the sulfur amino acids by the chick. J. Nutr. *26*, 631–640.

GUTTRIDGE, D. G. A., and LEWIS, D. 1964. Chick bioassay of methionine and cystine. Brit. Poultry Sci. *5*, 193–200.

HALLORAN, H. R., and BISHOP, R. B. 1966. A new approach to evaluating the relative efficacy of DL-methionine and methionine hydroxy analogue calcium in practical nutrition. Poultry and Livestock Comment, E. I. du Pont de Nemours & Co., Wilmington, Del.

HATHCOCK, J. N., HULL, S. G., and SCOTT, M. L. 1968. Derivatives and analogs of cysteine and selected sulfhydryl compounds in nutritional muscular dystrophy in chicks. J. Nutr. *94*, 147–150.

JUKES, T. H. 1956. The growth effects of methionine and sulfate additions to a sulfate-free diet for chicks. Private communication.

KAMATH, S. H., and BERG, C. P. 1964. Antagonism of the D-forms of the essential amino acids toward the promotion of growth by D-histidine. J. Nutr. *82*, 243–248.

KANO, A. K., HOUGHAM, D. F., and CHARKEY, L. W. 1968. Effects of dietary DL-methionine on tissue levels of glutathione in hypothyroid chicks. J. Nutr. *94*, 233–236.

LEVEILLE, G. A., SHAPIRO, R., and FISHER, J. 1960. Amino acid requirements for maintenance in the adult rooster. J. Nutr. *72*, 8–15.

MACHLIN, L. J., and GORDON, R. S. 1959. Equivalence of methionine hydroxy analog and methionine for chickens fed low-protein diets. Poultry Sci. *38*, 650–652.

MACHLIN, L. J., JACKSON, J. T., LANKENAU, A. H., and PEARSON, P. B. 1954. Uptake of $^{35}S$ in the feathers, gizzard lining, liver, and muscle of young chickens after injection with radioactive methionine or sulfate. Poultry Sci. *33*, 234–238.

MACHLIN, L. J., and PEARSON, P. B. 1957. Metabolism of taurine in the growing chicken. Arch. Biochem. Biophys. *70*, 35–42.

MARRETT, L., BIRD, H. R., and SUNDE, M. L. 1964. The effects of different isomers of methionine on growth of chicks fed amino acid diets. Poultry Sci. *43*, 1113–1118.

MARTIN, W. G., and PATRICK, H. 1960. The effect of certain dietary constituents, including protein and vitamin $B_{12}$, on the metabolism of inorganic $^{35}S$ sulfate in the chick. Poultry Sci. *39*, 1501–1510.

NASSETT, E. S., and ANDERSON, J. T. 1951. Nitrogen balance in the adult rat as affected by diets low in L- and DL-methionine. J. Nutr. *44*, 237–247.

ROBINSON, H. E., GRAY, R. E., CHESLEY, F. F., and CRANDALL, L. A. 1939. Chondroitin sulfuric acid as a growth factor. J. Nutr. *17*, 227–233.

SCHWEIGERT, B. S., and GUTHNECK, B. T. 1954. Utilization of amino acids from foods by the rat. J. Nutr. *54*, 333–343.

SCOTT, M. L., and CALVERT, C. C. 1962. Evidence of a specific effect of cystine in the prevention of nutritional muscular dystrophy in vitamin E-deficient chicks. J. Nutr. *77*, 105–108.

SULLIVAN, T. W., and BIRD, H. R. 1957. Effect of quantity and source of dietary nitrogen on the utilization of the hydroxy analogues of methionine and glycine by chicks. J. Nutr. *62*, 143–150.

VAN PILSUM, J. E., and BERG, C. P. 1950. The comparative availabilities of the mixtures of the L and D modifications of the essential amino acids for growth in the rat. J. Biol. Chem. *183*, 279–285.

WRETLIND, K. A. J. 1952. The effect of D-amino acids on the stereonaturalization of D-methionine. Acta Physiol. Scand. *25*, 267–275.

WRETLIND, K. A. J., and ROSE, W. C. 1950. Methionine requirement for growth and utilization of its optical isomers. J. Biol. Chem. *187*, 697–703.

SECTION V

# Sulfur in Human Nutrition

CHAPTER 14

**Marian E. Swendseid**
**Marian Wang**[1]

## Sulfur Requirements of the Human

### INTRODUCTION

Compounds containing sulfur are found in all body cells and they represent a highly diversified group of substances performing a number of vital functions (Young and Maw 1958). It seems quite remarkable that in addition to two sulfur-containing vitamins, the dietary needs for sulfur can be met by a single amino acid, namely methionine. There is in ordinary food, however, another sulfur-containing amino acid, cystine, which either as such or in its reduced form can partially replace the requirements for methionine. Inorganic sulfur in the diet does not appear to contribute substantially to sulfur nutrition since in studies (Papadopoulou 1957) with labeled sulfate given to normal subjects it was found that from 80 to 90% of the radioactivity was excreted in the inorganic sulfate fraction of the urine and 5% was excreted in conjugation with phenolic derivatives.

Although qualitatively the sulfur needs are quite well-established, many questions remain with regard to the quantitative requirements of the amino acids. This review will be limited to a discussion of these compounds since they represent the chief source of all metabolizable sulfur compounds.

There are several reasons why the nutritional requirements for the sulfur-containing amino acids are of particular interest. In the first place, there is evidence that for many plant proteins, methionine is the first limiting amino acid. Hence in populations subsisting on vegetable proteins such as the legumes, the addition of methionine may improve the nutritive quality of the food. There is also evidence, mainly in experimental

[1]Supported in part by NIH training grant AMO5557.

animals, that methionine is the most toxic of all amino acids, and there may be reason to be concerned about too much as well as too little of this amino acid in low protein diets. Then since methionine is the principal methyl donor compound in the body, this amino acid assumes an extremely important role in metabolism.

There have been three main approaches used in the study of sulfur nutrition. The first of these involved a sulfur balance study which became in practice a measurement of the nitrogen to sulfur ratio (N:S ratio). Then, the determination of nitrogen balance was carried out with diets containing varying amounts of the sulfur-containing amino acids. More recently, plasma levels of methionine and cystine have been measured in different nutritive and metabolic states.

### SULFUR BALANCE AND THE N:S RATIO

Perhaps the first definitive study was reported by Wilson (1925). He observed that the urinary sulfur excretion, similar to the nitrogen excretion, varied with the amount of sulfur in the diet. When a protein meal was fed, the amount of sulfur excreted in the urine reached a peak value before nitrogen. However, in terms of 24 hr periods, the N:S ratio in the urine varied with this ratio in dietary protein. It was also observed that subjects could store proteins of varying N:S ratios and that they first retained labile products relatively high in sulfur content. The studies of Wilson led to the suggestion of Miller and Naismith (1958) that the urinary N:S ratio might be an index to the quality of protein consumed since it paralleled the N:S ratio of the diet and many diets were shown to be low in methionine. This approach has not yet been adequately studied. However, Waterlow (1968) has observed that when the protein intake of infants is reduced to low levels, the N to S ratio in urine is decreased. He attempted to explain this finding by noting that if the protein originated in liver, the N to S ratio of this protein is very low.

The N to S ratio has also been studied during fasting by several investigators (Lusk 1931) who observed high ratios at least during the first few days of the fast. We have had the opportunity to correlate N and S excretion in fasting obese subjects over prolonged time periods and the results are shown in Table 14.1. A reduction occurs in both total and inorganic sulfur excretion that progresses as fasting continues. The inorganic sulfur fraction tends to be decreased more than the organic and since the N:S ratio is greater than for muscle tissue it appears that sulfur is being conserved by the body to a greater extent than nitrogen. Figure 14.1 shows the effect of fasting on the urinary nitrogen and sulfur excretion of a single subject as determined in five day pools. It can be seen that the excretion of total sulfur tends to reach a plateau more rapidly than nitrogen

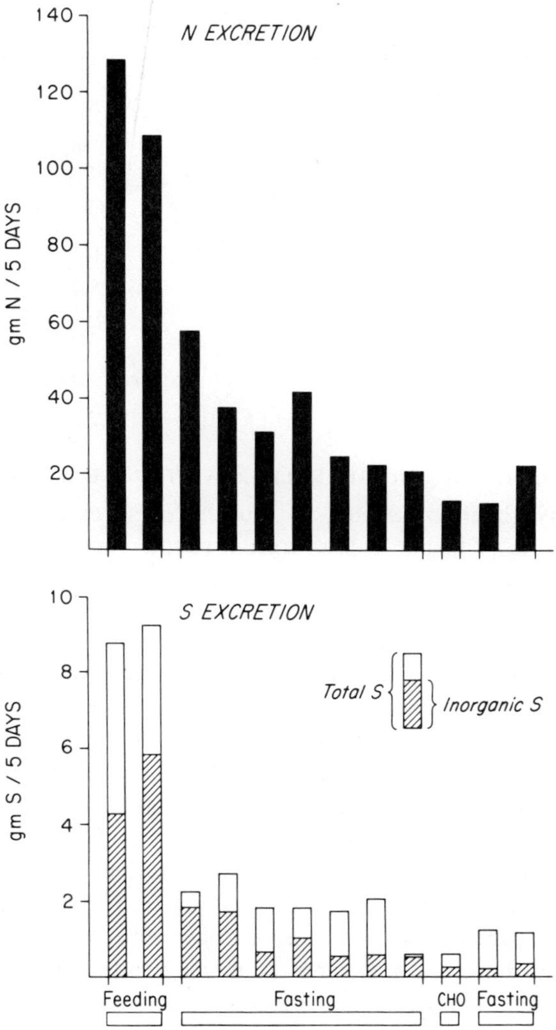


FIG. 14.1. URINARY NITROGEN AND SULFUR EXCRETION IN A SUBJECT FASTING FOR 7- to 5-DAY PERIODS

but that the inorganic sulfur continues to decrease. A further reduction in inorganic sulfur occurs after 35 days of fasting when 100 gm of glucose is given per day for a 5 day period. This has also been observed in two other fasting subjects when carbohydrate alone is administered. When carbohydrate is given, the nitrogen content of the urine is also decreased. However, the sulfur is decreased proportionately more than the nitrogen so that N:S ratios of ten and lower are observed. Apparently then during refeeding

TABLE 14.1

URINARY S AND N EXCRETION IN FOUR FASTING OBESE SUBJECTS[1]

| Fast Days | Mean Daily N Excretion Gm | Mean Daily S Excretion Total Mg | Inorganic Mg | N:S |
|---|---|---|---|---|
| 1–15 | 8.8 | 550 | 366 | 16.2 |
| 16–30 | 5.5 | 333 | 195 | 16.4 |
| 31–45 | 2.7 | 280 | 142 | 17.5 |

[1]Normal values reported by Freyberg *et al.* (1940): N excretion, 8.3 gm; total S, 682 mg; inorganic S, 518 mg; and N:S, 12.1.

the nitrogenous compounds that are being spared are relatively rich in methionine.

## REQUIREMENTS FOR THE SULFUR-CONTAINING AMINO ACIDS

Studies have been conducted in subjects of several age groups using amino acid diets or in one study, protein supplemented with amino acids, to determine the minimum quantity of methionine and cystine needed to maintain nitrogen equilibrium or to induce nitrogen retention. The first of these studies was carried out by Rose *et al.* (1955) with college-age men as subjects. Subsequently, studies in young women (Swendseid *et al.* 1956; Reynolds *et al.* 1958), in men over fifty (Tuttle *et al.* 1965), with infants (Holt and Snyderman 1961) and with 11-yr-old boys (Nakagowa *et al.* 1961) were reported. The results are summarized in Table 14.2 with methionine requirements being defined for growing subjects as those amounts necessary to promote nitrogen retention. For adult subjects, the methionine requirements are the amounts needed to maintain nitrogen equilibrium. For subjects in the older age group, the requirements should be stated more specifically for a daily nitrogen intake of 7 gm since there is some evidence that amino acid needs in this group vary with the protein content of the diet (Tuttle *et al.* 1965). There appears to be a wide range of requirements for the various age groups and also between men and women. It should be pointed out that the values reported from each investigation are the highest obtained for any of the subjects studied. While greater

TABLE 14.2

DAILY METHIONINE REQUIREMENT

| Subjects | No. | Daily Requirement |
|---|---|---|
| Infants | 6 | 45 mg/kg |
| 11-Yr-old boys | 4 | 800 mg (27 mg/kg) |
| Young men | 6 | 1100 mg |
| Young women | 27 | 550 mg |

requirements for infants and growing boys as compared with those of adults are to be expected, the basis for the differences between men and women and between men of the two age groups remains unexplained. Are there variations in experimental techniques which account for the different requirement values?

One independent observer, Hegsted (1963) has attempted to answer this question for the requirements of men and women. He calls attention to the fact that the criteria used to determine requirements were somewhat different for the two sexes. For young men, Rose *et al.* (1955) attempted to determine amounts that would promote "distinctly positive balance" in each subject and selected as the estimated minimal requirement the value for the subject with the highest requirement, a procedure that would lead to a relatively high estimate. In studies on young women, the amount of amino acid which would induce nitrogen equilibrium in all subjects was selected with equilibrium being "a zone in which difference in intake and excretion does not exceed ±5%." Since theoretically, no subjects would have to reach the point of equilibrium, this definition may yield a low estimate. Hegsted attempted to define absolute equilibrium in the young women series by carrying out a regression analysis with the data. Using this procedure the requirement values he obtained for the nitrogen equilibrium point were not appreciably different from those in the original articles. He concluded that the accuracy of estimates available for women was better than for men since more subjects were utilized and the procedures were standardized, but that differences in methods used to determine nitrogen equilibrium could not account for the variations found in men and women. It should be noted also that there is no overlap in the values for methionine requirements obtained in the two series of studies. Rose's lowest requirement value was 800 mg; for women, 3 subjects were in positive nitrogen balance (by Rose criteria) with a 300 mg intake of sulfur-containing amino acids.

There is little evidence that the difference in body mass is the basis for the explanation of differing methionine needs. With one exception, that of lysine (Clark *et al.* 1960), no amino acid requirement value has been shown to be correlated with body size. With the recognition that substrate and hormonal concentrations in body tissues can markedly influence amino acid dissimilation and with the variety of metabolic pathways for sulfur, the hypothesis is plausible that the metabolic state rather than the amount of protein synthesized in the adult is the major factor in determining amino acid needs. It is possible that in women the utilization of methionine for a variety of body processes is more efficiently carried out.

It should be noted further that the methionine requirements for men over 50 are much greater than those of college-age men. Again there is the question as to whether these differences are real or are the result of variations in experimental technique. For men over 50, the studies were conducted so that a dietary period when ordinary food was given was interposed between every period of amino acid diets, with all diets being isonitrogenous. Hence the body reserves of protein would remain at a relatively constant level throughout the study. In the series with women, one 64-yr-old woman was included (Reynolds *et al.* 1958). She did not attain nitrogen equilibrium when receiving 850 mg of methionine and 10 mg of cystine. She was not given a larger quantity because all younger persons in the study achieved nitrogen balance with smaller quantities. The results with this one subject thus tend to confirm the observations that larger quantities of methionine are required for nitrogen balance in older individuals. Again it would appear plausible to seek an explanation for this finding in differences in metabolic utilization of sulfur-containing acids. Since the older men were given only methionine in their diet, it is possible that the increased needs of methionine for this age group are not due to an increased requirement for the sulfur-containing amino acids *per se* but to a decreased efficiency in the conversion of methionine to cystine. Other possibilities include alterations in activities of some of the enzymes concerned with sulfur metabolism. We have recently observed that in old rats, the activity of the t-RNA methylating enzyme as measured *in vitro* is much less than in younger rats (Umezawa 1969). This could conceivably alter the requirement for methionine in these experimental animals.

Rose and Wixom (1955) have also investigated the extent to which cystine can substitute for methionine in their experimental diets. They have found that approximately 80% of the methionine requirement can be replaced by cystine. Young women have been shown to require between 150 and 350 mg of methionine with 200 mg of cystine in the diet (Swendseid *et al.* 1956). Reynolds *et al.* (1958) observed that with 500 mg of cystine the addition of 150 to 180 mg of methionine to the diet resulted in either nitrogen storage or nitrogen equilibrium. She also found that a period of time was needed to adjust to an effective utilization when the dietary source of the sulfur-containing amino acids was changed.

Studies have also indicated that alone of the amino acids that have been investigated in humans, the D-form of methionine is as well utilized as the L-form. Rose *et al.* (1955) found that in two subjects the requirement for D-methionine was similar to the DL form. Again the metabolic explanation for this finding is not known with certainty.

There are indications in animals that methionine is a toxic substance and when added to low protein diets will cause a reduction in growth rate, anemia, and gross pathologic changes in liver tissue (Klavins and Johansen

1965). In children, there is one report by Bressani *et al.* (1960) where supplementation of wheat protein with methionine to conform to the suggested FAO reference pattern of amino acids caused a reduction in nitrogen retention. This observation suggests the possibility that excess dietary methionine may also be harmful in humans. In one other study, Block and coworkers (1965) fed 9 gm of methionine daily for 4 days to two subjects when they were receiving 54 and 104 gm of protein in their diet. In these subjects at both levels of protein intake between 80 and 90% of the methionine sulfur was excreted as inorganic sulfate and the subjects were always in nitrogen balance. Since this amount of methionine was five times greater than the content of the basal diet, adult subjects on adequate protein intakes apparently have the ability to metabolize large quantities of methionine without affecting nitrogen balance at least.

The determination of optimum methionine requirements for subjects of all ages assumes particular significance not only because of the possibility of toxicity of excesses given with low protein diets but also because of reports that this amino acid may be limiting in many vegetable protein diets. However, the evidence for this is based for the most part on administering these diets to rats and noting the beneficial effects of methionine supplementation on improving protein quality (Miller and Naismith 1958). However, the methionine requirement for the rat may be relatively greater than in humans because of the high cystine content of hair. A preliminary note by de Oliveria (1969) indicates that a soya milk marketed in Brazil which was methionine-deficient for rats, when fed to malnourished children had the same nutritive value as cow's milk.

## SULFUR-CONTAINING AMINO ACIDS IN PLASMA AND NUTRITIVE STATE

Studies have indicated that in both man and animals the levels of amino acids in plasma particularly during the period of postabsorption are influenced by the nutritive state. However, it is the opinion of some investigators that with low protein diets the limiting dietary amino acid is not necessarily the amino acid that is most reduced in the plasma. For example, in kwashiorkor, valine among the plasma amino acids has been found to be decreased to the greatest extent although there is no evidence that this amino acid is limiting in the food. Since methionine has been implicated as one of the amino acids most likely to be deficient in the diets of populations subsisting on vegetable protein, it is of interest to determine whether the low protein diets do indeed affect the levels of the sulfur-containing amino acids in plasma.

The studies on kwashiorkor, beginning with the original one of Holt *et al.* (1960) have shown an alteration in the ratio of cystine to methionine in plasma due primarily to a reduction in the level of cystine. It should be noted that in similar fashion, the ratio of tyrosine to phenylalanine is also altered due to a reduction in the level of tyrosine. Hence, apparently when low protein diets are fed, metabolic adjustments result in reductions of the semiessential amino acids in the plasma with sparing of their obligate metabolic precursors.

We have recently conducted a study of children with kwashiorkor residing in Guadalajara, Mexico (Padilla *et al.* 1969). When levels of amino acids in plasma of the patients were compared with those of healthy well-nourished children, it was found that there were significant reductions in all of the essential amino acids except threonine, lysine, and also histidine, which was slightly elevated. The decrease in the methionine level was much less than for some of the other essential amino acids, including the semiessential amino acids tyrosine and cystine. The data for the sulfur-containing amino acids are shown in Table 14.3. It would appear,

TABLE 14.3

PLASMA LEVELS OF S-CONTAINING AMINO ACIDS IN CHILDREN

| Subjects | Methionine μmoles/Ml | Cystine-SH μmoles/Ml |
|---|---|---|
| Normal | 1.6 | 7.7 |
| Kwashiorkor | 1.0 | 1.4 |
| Phenylketonuria | 1.8 | 5.6 |

then, on the basis of the much greater reduction in cystine than in methionine, that a decrease in cystine or a decrease in the ratio of cystine to methionine would be possible indicators of a protein deficiency state.

We have one other result obtained in children relating to sulfur-containing amino acid levels, and this in children with phenylketonuria. Many children with this genetic disorder have now been living and growing for several years while receiving diets containing free amino acids as their chief source of nitrogen. They are probably unique examples of human subjects who have subsisted for prolonged periods of time on amino acid diets. Their plasma essential amino acid levels show little variation from normal healthy children except for phenylalanine and tyrosine. However, the cystine level is somewhat reduced. Since the source of amino acids is a hydrolysate of casein supplemented with methionine, the cystine content of the diet is probably very low and the low plasma values for cystine may result from the limited capacity of body tissues to form cystine from methionine.

There are also data available on the plasma levels of the sulfur-amino acids for adults on controlled diets of either high or low protein content. In one study (Swendseid *et al.* 1966) postabsorption plasma amino acid levels were measured at intervals during a 5 week period in 6 men aged 55 yr or older who were receiving a diet containing 3.5 gm of nitrogen per day. The subjects were in negative nitrogen balance throughout this period and the plasma levels of the total essential amino acids were also decreasing. Valine showed a significant reduction after subjects had received the low-protein diet for one week. The effects on the sulfur-containing amino acids are shown in Table 14.4. After 3 weeks on the low

TABLE 14.4

PLASMA LEVELS OF S-CONTAINING AMINO ACIDS IN ADULT SUBJECTS

| Subjects | Dietary or Metabolic State | Methionine μmoles/Ml | Cystine-SH μmoles/Ml |
|---|---|---|---|
| Young adults | 14 gm N diet | 2.4 | 7.8 |
| | 14 gm N diet, lo CHO, 1 wk | 2.5 | 7.9 |
| | 3.8 gm N diet, 1 wk | 3.0 | 8.2 |
| | fasting, 2 wk | 3.3 | 12.0 |
| | fasting, 6 wk | 2.4 | 8.5 |
| | fasting and refed, lo Cal diets, 1 mo | 2.0 | 9.2 |
| Men over 50 | 14 gm N diet, 1 wk | 1.8 | 9.9 |
| | 3.5 gm N diet, 3 wk | 2.3 | 10.1 |
| | 3.5 gm N diet, 5 wk | 2.0 | 9.6 |

protein diet, the cystine level remained essentially constant, but the methionine was significantly increased and remained thus for the 5 week study period. This seemingly anomalous result is the more surprising when it is recognized that the content of the sulfur-containing amino acids in the diet is reduced to a proportionately greater extent than the total nitrogen. The amount of these amino acids in the diet containing 3.5 gm of nitrogen is calculated as 20% of that found in the diet with 14 gm of nitrogen. These results have been confirmed in another study with young men to whom the same diets were administered (Swendseid *et al.* 1968). The data are as shown in Table 14.4 and indicate that again with cystine remaining relatively constant, there is a significant elevation in methionine when the diet containing 3.5 gm of nitrogen is substituted for the 14 gm nitrogen diet for a single week.

We have also had the opportunity to measure plasma amino acids in obese subjects during prolonged periods of total fasting continuing for five weeks or longer (Swendseid *et al.* 1969). In general, the total of essential amino acids in plasma for these subjects does not show a reduction until fasting has been continuing for five weeks. In the interim, some of the essential amino acids have tended to be increased. Methionine is one of these although a significant elevation from the prefasting mean value of 2.7 μmoles/100 ml could not be demonstrated. In the discussion on N to

S ratios it was reported that during fasting relatively more sulfur than nitrogen was retained by the body and these results are consonant with that observation. What is the explanation for these findings? One possibility is that the body tissue protein being degraded has a relatively higher content of methionine than the ordinary dietary protein. Alternatively, the tissue need for methionine in processes such as transmethylation may be greatly decreased with these dietary conditions. There is also the possibility that tissue enzyme systems respond in such a fashion that methionine is conserved. All of these factors may be operating.

When refeeding is undertaken in the fasting subjects by administering a diet having an energy value of 500 Cal and a content of 8.3 gm of nitrogen, the methionine level in plasma falls and remains significantly low for periods of 1 month or longer. It may be that in addition to a relatively small requirement for this amino acid when nitrogen is being lost from the body, there is a relatively greater requirement when nitrogen is being retained.

We have been dealing here with a variety of protein deficiency states in children and adults, some of them having associated energy deficiencies. There have been different responses in regard to the sulfur-containing amino acids in plasma. In kwashiorkor the level of methionine stayed relatively constant while the cystine level fell. With low protein diets and with fasting the methionine level was enhanced and the cystine level remained stationery. Is the first event in protein deficiency an increase in plasma methionine levels followed in a second phase by a reduction in methionine and a greater fall in cystine with the end result being the levels obtained in kwashiorkor? Or do the plasma levels in kwashiorkor reflect the relative inability of young children to convert methionine to cystine? We are not certain, of course, that in any of these studies methionine was a limiting amino acid in the diet.

To answer some of these questions we have been conducting experiments in rats of various ages force-fed diets free of methionine but nutritionally adequate in other respects. We have also fed diets containing excess amounts of methionine. In one series of experiments (Sanchez and Swendseid 1969), 200 gm rats were force-fed diets containing amino acid mixtures proportioned as in casein (21% of diet) but with varying amounts of methionine for a 3 day period. One group received a complete amino acid mixture. Other groups were given this same diet either with methionine and cystine removed or with 4% of methionine added as a supplement. Plasma, liver and gastrocnemius muscle of the three groups of animals were analyzed for amino acids. With the diet free from the sulfur-containing amino acids as compared to the complete amino acid diet, it was found that the methionine content of the tissues was similar but the cystine level in plasma was greatly decreased and some other amino acids were increased as, for example, threonine and arginine. With the 4% methionine-supplemented diet as compared to the complete amino acid diet, methionine was elevated in all tissues but a number of other amino acids were decreased including ornithine, arginine, glycine, serine, and glutamic acid. It is apparent that methionine excesses affect amino acid homeostasis to a greater degree than methionine deficiency states. That the decreased tissue levels for some of the amino acids at least affect metabolic activity is shown by the findings that serine and glycine supplementation overcomes the toxic effect of methionine (Hardin and Hove 1951). Methionine excess then, appears in some fashion to induce a metabolic deficiency of a nonessential amino acid. An increased activity of serine-threonine dehydratase has been shown in this condition and may be the explanation for the serine depletion that results. It should be noted that in the experiments of Block *et al.* (1965), when methionine was fed to subjects for four days there was a decrease in the urinary excretion of glycine, evidence that similar metabolic effects take place in humans.

It has been found that the plasma levels of the sulfur-containing amino acids in rats on diets deficient in methionine and those in kwashiorkor are similar. The methionine level in both instances appears to be relatively constant while cystine is reduced. These results suggest that an alteration in the cystine to methionine ratio is an indicator not only of protein deficiency but also of a deficiency of sulfur-containing amino acids. The preservation of the methionine level at a constant value when the dietary intake is low suggests that a remarkable homeostatic mechanism is operating, very possibly through alterations in enzyme activity. In the studies with rats, however, the methionine-activating enzyme and cystathionase as well showed no significant differences when diets devoid of methionine were fed.

It is becoming more apparent in nutrition that dietary needs and metabolism are closely interrelated through feedback mechanisms affecting enzyme activity. It is hoped that further studies with sulfur-containing amino acids in relation to dietary intake, tissue levels, and enzyme activities will be carried out to determine how plasma levels of methionine are maintained in deficiencies and to investigate the metabolic basis of the seeming differences in methionine requirements for men and women.

## BIBLIOGRAPHY

BLOCK, W. D., MARKOVS, M. E., and STEELE, B. F. 1965. Effect of protein and of free L-methionine intake on amino acid excretion by human subjects. J. Nutr. *86*, 256–264.

BRESSANI, R., WILSON, D. L., BEHAR, M., and SCRIMSHAW, N. S. 1960. Supplementation of cereal proteins with amino acids. III Wheat flour as measured by nitrogen retention of young children. J. Nutr. *70*, 176–186.

CLARK, H. E., YANG, S. P., WALTON, W., and MERTZ, E. T. 1960. Amino acid requirements of men and women. II. Relation of lysine requirement to sex, body size, basal caloric expenditure and creatine excretion. J. Nutr. *71*, 229–234.

FREYBERG, R. H., BLOCK, W. D., and FROMER, M. F. 1940. A study of S metabolism and the effect of S administration in chronic arthritis. J. Clin. Invest. *19*, 423–435.

HARDIN, J. O., and HOVE, E. L. 1951. Prevention of DL-methionine toxicity in rats by vitamin E, $B_{12}$, folacin, glycine and arginine. Proc. Soc. Exptl. Biol. Med. *78*, 728–731.

HEGSTED, D. M. 1963. Variation in requirements of nutrients-amino acids. Federation Proc. *22*, 1424–1430.

HOLT, L. *et al.* 1960. Protein and Amino Acid Requirements in Early Life. New York Univ. Press, New York.

HOLT, L. E., and SNYDERMAN, S. E. 1961. The amino acid requirements of infants. J. Am. Med. Assoc. *175*, 100–103.

KLAVINS, J. V., and JOHANSEN, P. V. 1965. Pathology of amino acid excess. Arch. Path. *79*, 600–614.

LUSK, G. 1931. The Elements of the Science of Nutrition. W. B. Saunders Co., Philadelphia.

MILLER, D. S., and NAISMITH, D. J. 1958. Correlation between sulphur content and net dietary-protein value. Nature *182*, 1786–1787.

NAKAGOWA, I., TAKAHASHI, T., and SUZUKI, T. 1961. Amino acid requirements of children: minimal needs of lysine and methionine based on nitrogen balance method. J. Nutr. *74*, 401–407.

DE OLIVERIA, D. 1969. Quoted in Bulletin of League for International Food Education. Aug., 1969, p. 3.

PADILLA, H. *et al.* 1969. Plasma amino acid patterns in children from Guadalajara with kwashiorkor. Unpublished.

PAPADOPOULOU, D. B. 1957. Urinary sulfur partition in normal men and in cancer patients. Clin. Chem. *3*, 257–262.

REYNOLDS, M. S., STEEL, D. L., JONES, E. M., and BAUMAN, C. A. 1958. Nitrogen balances of women maintained on various levels of methionine and cystine. J. Nutr. *64*, 99–111.

ROSE, W. C., COON, M. J., LOCKHART, H. B., and LAMBERT, G. F. 1955. The amino acid requirements of man. XI. The threonine and methionine requirements. J. Biol. Chem. *215*, 101–110.

ROSE, W. C., and WIXOM, R. L. 1955. The amino acid requirements of man. XIII. The sparing effect of cystine on the methionine requirement. J. Biol. Chem. *216*, 763–773.

SANCHEZ, A., and SWENDSEID, M. E. 1969. Amino acid levels and enzyme activity in tissues of rats force-fed diets differing in methionine content. J. Nutr. *99*, 145–151.

SWENDSEID, M. E. *et al.* 1966. Plasma amino acid levels of men fed diets differing in protein content. Some observations with valine-deficient diets. J. Nutr. *88*, 239–248.

SWENDSEID, M. E., UMEZAWA, C. Y., and DRENICK, E. 1969. Plasma amino acid levels in obese subjects before, during, and after starvation. Am. J. Clin. Nutr. *22*, 740–743.

SWENDSEID, M. E., WILLIAMS, I., and DUNN, M. S. 1956. Amino acid requirements of young women based on nitrogen balance data. J. Nutr. *58*, 495–505.

SWENDSEID, M. E., YAMADA, C., VINYARD, E., and FIGUEROA, W. G. 1968. Plasma amino acid levels in young subjects receiving diets containing 14 or 3.5 gm nitrogen per day. Am. J. Clin. Nutr. *21*, 1381–1383.

TUTTLE, S. G. *et al.* 1965. Further observations on the amino acid requirements of older men. I. Effects of nonessential nitrogen supplements fed with different amounts of essential amino acids. II. Methionine and lysine. Am. J. Clin. Nutr. *16*, 225–231.

UMEZAWA, C. Y. 1969. Unpublished results.

WATERLOW, J. C. 1968. The adaptation of protein metabolism to low protein intakes. *In* Calorie Deficiencies and Protein Deficiencies, R. A. McCance, and E. M. Widdowson (Editors). Little, Brown and Co., Boston.

WILSON, H. E. C. 1925. XLIX. The relation between sulphur and nitrogen metabolism. Biochem. J. *19*, 322–337.

YOUNG, L., and MAW, G. A. 1958. The Metabolism of Sulphur Compounds. John Wiley & Sons, New York.

CHAPTER 15

S. Harvey Mudd

# Errors of Sulfur Metabolism

Studies of microbial biochemical mutants have contributed to our understanding of metabolic processes, the nature of the gene, and regulation of gene action in ways familiar to all. Similar insights may be expected from the study of biochemical mutants in higher organisms. However, to date, the labor of producing and selecting such mutants has been too great, and biochemical mutants of higher organisms have not become readily available. A major exception lies in the field of the inborn errors of human metabolism. These genetic diseases, which Garrod in 1908 called "Nature's experiments" (Garrod 1908A), are being defined at an increasing rate as biochemical methodology is applied to clinical medicine. Although limited by the restrictions inherent in working with human beings, and often forced to base conclusions on study of a very small number of cases, investigations into these diseases have yielded and will surely continue to yield significant information about normal molecular events in the mammal. In this presentation I will discuss the known errors of sulfur metabolism, and attempt to point out some of the contributions studies of these conditions have made to our understanding of the biochemistry of sulfur-containing compounds.

At the time of writing there are nine diseases which are known to affect in a major way the handling of sulfur-containing compounds. It would be difficult in the available time to cover the details of what is known about almost any one of these diseases. Clearly, it is impossible to discuss all nine in depth. I will say only a few words about most of these derangements, emphasizing the major lessons we have learned, or may hope to learn, from these natural experiments. In the latter part of the talk I will go into more detail about homocystinuria, studies of which have gone on in our laboratory for the past several years.

Data about the nine known errors of sulfur metabolism are shown in Table 15.1. The diseases are oriented according to the area of metabolism affected, beginning with the ingestion of methionine and ending with the excretion of inorganic sulfate. Listed in the first column are the major chemical aberrations which characterize the disease. In the second column are the publications in which the disease was first recognized as a specific entity, separate from other diseases. In the third column is a brief statement of the current concept of the etiology of each disease.

TABLE 15.1

ERRORS OF SULFUR METABOLISM (1969)

| Chemical Abnormalities | Recognition | Causation |
|---|---|---|
| α-Hydroxybutyric aciduria<br>Methionine, methionine sulfoxide, α-hydroxybutyrate, and α-aminobutyrate increased in faeces | Smith and Strang (1958)<br>Hooft *et al.* (1965) | Intestinal absorption of methionine impaired (Hooft *et al.* 1965) |
| Hypermethioninemia<br>Questionable association with tyrosinemia | Perry *et al.* (1965)<br>Partington *et al.* (1967) | Unknown |
| Homocystinuria<br>Homocystine and methionine raised in plasma | Carson and Neill (1962)<br>Gerritsen *et al.* (1962) | Cystathionine synthase activity deficient (Mudd *et al.* 1964) |
| Homocystinuria<br>Cystathionine high, methionine low in plasma | Levy *et al.* (1969) | $B_{12}$ metabolism deranged (Mudd *et al.* 1969) |
| Cystathioninuria | Harris *et al.* (1959) | Cystathionase activity deficient (Frimpter 1965; Finkelstein *et al.* 1966) |
| Cystinuria<br>Dibasic amino aciduria | Wollaston (1810)<br>Dent and Harris (1951) | Intestinal absorption of cystine and dibasic amino acids defective (Thier *et al.* 1964, 1965)<br>Renal transport of dibasic amino acids defective (Fox *et al.* 1964) |
| Cystinosis—excessive cystine in tissues | Dent and Harris (1951) | Cystine storage in lysosomes (Patrick and Lake 1968; Schulman 1969) |
| β-Mercaptolactate-cysteine disulfide in urine | Crawhall *et al.* (1968) | Unknown |
| Sulfite, thiosulfate, and S-sulfocysteine increased in urine<br>Inorganic sulfate decreased in urine | Irreverre *et al.* (1967) | Sulfite oxidase activity deficient (Mudd *et al.* 1967) |

The first disease listed is characterized chiefly by excessive excretion of α-hydroxybutyric acid in the urine. This is a rare condition, described to date in two patients only. Hooft *et al.* (1965, 1968) found not only α-hydroxybutyric acid in the urine of their patient, but also the presence in the faeces of abnormally great amounts of methionine, methionine sulfoxide, α-hydroxybutyrate and α-aminobutyrate. It was postulated that the underlying defect is a failure to absorb methionine normally from the intestine.[1] The abnormal compounds in the stool are thought to be formed from methionine by the bacteria of the colon. Although some puzzling

---

[1]Active transport of methionine by intestinal tissues from experimental animals has been demonstrated (Wiseman 1954).

observations remain to be explained in these cases,[2] it seems likely that patients with $\alpha$-hydroxybutyric aciduria may have an isolated defect in the active transport of methionine and that study of such patients may elucidate the early events in the physiological handling of this amino acid.

The second disease listed is hypermethioninemia, a condition in which methionine accumulates in plasma and tissues and spills over into the urine. Other sulfur amino acids are relatively normal, although an increased amount of the keto analogue of methionine, $\alpha$-keto-$\gamma$-methiolbutyric acid, is also present in the urine (Perry *et al.* 1965). This is the least understood of the diseases to be discussed. Indeed it is not clearly established whether this is a distinct entity or whether the accumulation of methionine is a secondary result of a disturbance in tyrosine metabolism or some other lesion. The references listed discuss this question thoroughly and I will not do so here. Current concepts of methionine metabolism in the mammal (see Fig. 5.1) suggest that abnormal accumulation of methionine unaccompanied by abnormalities of other sulfur amino acids might plausibly result from a decreased flow of methionine through the step catalyzed by S-adenosylmethionine synthetase. This decreased utilization of methionine could be due either to failure of methionine to penetrate the cell, to lack of S-adenosylmethionine synthetase, to decreased affinity of this enzyme for one of its substrates or to inhibition by an abnormal metabolite. I have had an opportunity to study liver tissue obtained at postmortem examination of the hypermethioninemic patient reported by Perry *et al.* (1965). Some results from these studies are shown in Table 15.2. They provide strong data against all the possibilities just mentioned. Not only was there a normal activity of S-adenosylmethionine synthetase, but this enzyme must have been functioning well *in vivo*, since the liver contained a greatly increased concentration of S-adenosylmethionine. Therefore, other causes for the methionine accumulation in this patient and similar patients must be sought.[3]

---

[2]Hooft and co-workers made two observations after administration of an oral dose of methionine to their patient which are not readily reconciled with the postulate of impaired intestinal absorption: (a) The patient excreted an amount of methionine in the urine 3 to 4 times that excreted by normal subjects (Hooft *et al.* 1968). (The patient of Strang and Smith also showed increased methioninuria under basal conditions.) These findings suggest that possibly the active transport of methionine is defective in such patients not only in intestine, but also in kidney. (b) The plasma concentration of methionine rose normally after the oral load. This observation, if representative, is difficult to explain, since both intestinal and renal transport deficiencies would tend to blunt the normal rise of methionine. Caution is indicated in interpretation of these findings because homocystinuria was known to occur in the family of the patient of Hooft *et al.* (1968) and it is possible the patient himself was heterozygous for cystathionine synthase deficiency.

[3]Decreases in the activities of both methionine activating enzyme and cystathionine synthase were observed in extracts made from liver biopsies of two hypermethionemic patients (Gaull *et al.* 1968A). S-adenosylmethionine concentrations were not reported.

TABLE 15.2

S-ADENOSYLMETHIONINE CONCENTRATION AND ENZYME SPECIFIC ACTIVITIES IN EXTRACTS OF LIVER FROM AN INFANT WITH HYPERMETHIONINEMIA[1]

| | Control Subjects | | | Patient |
|---|---|---|---|---|
| | Number | Range | Mean | (R. L.) |
| S-adenosylmethionine μg/gm wet liver | 6 | 1.2–47.5 | 14.2 | 321 |
| Methionine activating enzyme mμm/mg protein/ 60 min | 15 | 0.3–5.7 | 1.5 | 2.9 |
| Cystathionine synthase mμm/mg protein/ 135 min | 15 | 1–333 | 132 | 27 |
| Betaine-homocysteine methyltransferase mμm/mg protein/45 min | 15 | 2.1–15.5 | 7.2 | 1.2 |
| $N^5$-methyltetrahydrofolate methyltransferase mμm/mg protein/60 min | 11 | 0.5–1.7 | 1.0 | 2.2 |

[1]Patient R. L. was a hypermethioninemic infant who died at 12 wk of age. The case has been reported by Perry and co-workers (1965). Dr. Perry kindly sent a specimen of the liver obtained at postmortem examination. The 15 control subjects were under 4-yr at death. They died of diseases without known amino acid abnormalities. Values from only those subjects less than 1 yr of age are presented for $N^5$-methyltetrahydrofolate methyltransferase since the specific activity of this enzyme in postmortem human liver decreases with age. Enzyme assays were performed as described (Mudd *et al.* 1965; Mudd *et al.* 1969). S-adenosylmethionine concentration was determined according to Baldessarini and Kopin (1963).

The next two conditions share the common end result of producing excess homocystine in the urine. I would like to discuss these two situations in more detail after completion of the remainder of our survey of the errors of sulfur metabolism.

The fifth disease listed is usually called primary cystathioninuria to distinguish it from cystathioninuria secondary to lack of vitamin $B_6$ or to the presence of certain tumors. The initial case was reported by Harris and coworkers in 1959. Only a few additional cases have been discovered since that time. The clinical condition of these patients has varied widely. Primary cystathioninuria may exist in children who enjoy good physical and mental health (Perry *et al.* 1968B). Several groups have demonstrated a deficient activity of γ-cystathionase in this condition (Frimpter 1965; Finkelstein *et al.* 1966; Tada *et al.* 1968).[4] As a result of this enzyme deficit, cystathionine accumulates in the tissues (Harris *et al.* 1959; Brenton *et al.* 1965) and is excreted in excessive amounts in the urine. Reported cystathionine excretions have ranged up to 5 to 6 millimoles/day. N-acetylcystathionine is also excreted in substantial amounts (Perry *et al.* 1968A).

To evaluate the role of cystathionase in normal metabolism, it would be important to assess the balance between methionine intake and excretion

[4]Among the four cases in which enzyme assays have been performed there were differences noted in regard to the *in vitro* and *in vivo* response to added pyridoxal phosphate (Frimpter 1965; Finkelstein *et al.* 1966; Tada *et al.* 1968). Addition of this coenzyme restored the γ-cystathionase activity of one extract to normal (Frimpter 1965), but did not do so in the other three cases. Interpretations of these facts have been presented elsewhere (Frimpter 1965; Finkelstein *et al.* 1968; Tada *et al.* 1968).

of cystathionine (and other abnormal metabolites such as N-acetylcystathionine). In general, such balance studies have not been reported. One cystathioninuric patient given an oral load of methionine, excreted the equivalent of only 13% of the administered methionine as cystathionine (Shaw *et al.* 1967). This finding, if representative, could be explained by either of two general hypotheses. Either (a) the cystathionine pathway is not the major pathway for methionine degradation or (b) the residual activity of $\gamma$-cystathionase in such patients is sufficient to allow metabolism of a major portion of cystathionine. Further evidence bearing on this question will be presented in the discussion of cystathionine synthase deficiency.

Cystinuria, the next disorder on our list, was the earliest disease of sulfur metabolism to be recognized and is much the most common. Cystinuria may be considered to have been discovered at the same time as cystine itself by Wollaston, who in 1810 examined what he described as "a new species of urinary calculus" and found these stones to be composed of an unknown substance to which he gave the name cystic oxide because of its supposed origin in the urinary bladder. These stones were assuredly taken from cystinuric patients and were composed largely of cystine.

Cystinuria was one of the four diseases discussed by Garrod in his Croonian lectures of 1908 (Garrod 1908B) in which he formulated the concept of inborn errors of metabolism and provided the beginnings of the one gene, one enzyme hypothesis.

Cystinuria is not an uncommon condition. In one study of more than 10,000 apparently healthy college students, cystinuria was found in 1 in 600 (Lewis 1932). Clinical difficulties in this condition are almost exclusively due to the formation of cystine stones in the urinary tract, an event which apparently occurs in a minority of cystinurics.

Cystinuria has served as the stimulus for a great deal of investigative work and a great deal of debate as to the etiology of the disease. Following the demonstration by Yeh *et al.* (1947) that, in addition to cystine, lysine and arginine were also present in abnormal amounts in the urine of cystinuric patients, Dent and Rose (1951) found that none of these amino acids were elevated in plasma (see also Dent *et al.* 1954A, 1954B) and suggested the unifying concept that cystine and the dibasic amino acids share a common renal transport mechanism and that it is this mechanism which is defective in cystinuria.

A decade later, Milne, Asatoor and associates first clearly demonstrated by loading experiments that cystinuric subjects do not absorb lysine, ornithine (Milne *et al.* 1961), or arginine (Asatoor *et al.* 1962) normally from the intestine. *In vitro* studies of intestinal transport of cystine and the dibasic amino acids have fully supported the postulate of a common intestinal absorption mechanism for these substances which is defective in cystinuria (Thier *et al.* 1964; Thier *et al.* 1965).

Paradoxically, the situation with respect to renal transport is far less clear. The postulate of Dent and Rose (1951) of a common renal transport mechanism was supported by the infusion studies of Robson and Rose (1957), but has not been sustained by *in vitro* measurements of transport in kidney slices (Fox *et al.* 1964). The latter studies suggest that cystine is handled by a transport system different from that which handles the dibasic amino acids. The excessive excretion of cystine in the urine of human cystinurics remains unexplained (Segal and Smith 1969). Recently attention has turned to the possibility that there is a common renal transport system for lysine, ornithine, arginine, and cysteine (Schwartzman *et al.* 1966). It seems reasonable to hope that these continuing investigations of the lesion in cystinuria will ultimately give us more insight not only into active transport mechanisms, but also into the metabolic relationships between cysteine and cystine.

The next disorder to be discussed, cystinosis, or cystine storage disease, is an uncommon condition which was often confused with cystinuria until the clear clinical, biochemical, and genetic differences were pointed out by Dent and Harris (1951). Patients with cystinosis deposit crystals of cystine in many tissues including the eye, bone marrow, peripheral white cells, and kidney. Clinical onset usually occurs around six months of age and the patients suffer progressive renal failure with eventual early death.

In spite of intensive investigations the cause of this disease remains obscure. A number of enzymes involved in degradation of cystine or cysteine are not deficient (Patrick 1962), nor has any abnormality in reduction of cystine to cysteine been demonstrated (Seegmiller *et al.* 1968). Recently an important lead has been developed by Schneider and co-workers (1967A, 1967B) who demonstrated greatly increased cystine contents of peripheral leukocytes and of fibroblasts cultured from patients with cystinosis. Although the fibroblasts may have as much as 100 times the normal content of cystine, they can not mobilize this cystine fast enough to prevent cellular death when cystine is removed from the external medium (Schneider *et al.* 1967B). Following this lead Patrick and Lake (1968) have published indications, and Schulman, Wong, Schneider, Bradley and Seegmiller (Schulman 1969) have obtained extensive evidence (now in press) that the abnormal store of cystine is located in lysosomes. It appears possible that the defect in cystinosis, rather than being a low activity of an enzyme involved specifically in cysteine or cystine metabolism, may involve the general functioning of lysosomes. It would be disappointing if further investigations of this disease do not give us a great deal more understanding of normal lysosomal function and of the intracellular fate of cysteine and cystine.

The eighth disease on our list was discovered by Crawhall *et al.* (1968) in the course of screening mentally retarded patients in Massachusetts. Only a single case has been reported. The distinguishing feature was urinary excretion of a new disulfide, the structure of which is shown in Fig. 15.1. No

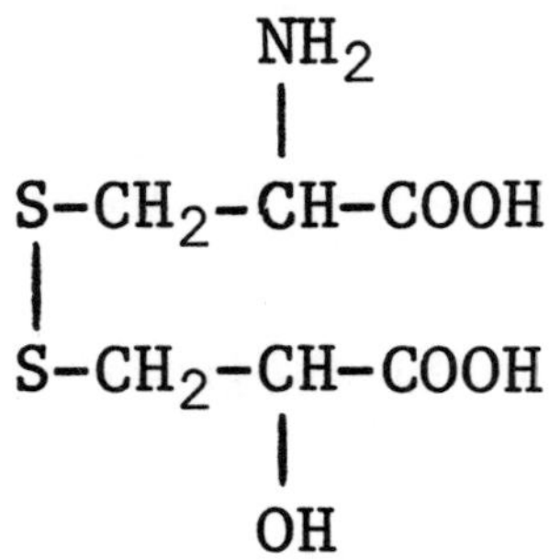

Fig. 15.1. β-MERCAPTOLACTATE-CYSTEINE DISULFIDE

reports have yet appeared as to the etiology of this disease, although the structure of the unusual metabolite suggests that a step in cysteine degradation is affected.

The final disease to be discussed is sulfite oxidase deficiency. This also is a very recently discovered condition reported in only a single case. The patient in this instance was a child with severe neurological disfunction who died in his third year of life. It is still too early for definition of the clinical picture of this disease or its incidence. The urine of this child contained greatly increased amounts of sulfite, thiosulfate and S-sulfocysteine and virtually no inorganic sulfate (Irreverre *et al.* 1967). No sulfite oxidase activity could be demonstrated in liver, kidney, or brain extracts from this patient although sulfite oxidase was present in the same tissues of all control subjects tested (Mudd *et al.* 1967). The enzyme deficiency accounts for the lack of inorganic sulfate and for the accumulation of sulfite. Accumulated sulfite could react by known pathways to form thiosulfate (Sorbo 1957) and S-sulfocysteine (Clark 1932).

$$\underset{\text{cystine}}{R{-}S{-}S{-}R} + \underset{\text{sulfite}}{HSO_3^{\ominus}} \rightleftharpoons R{-}S{-}SO_3^{\ominus} + HSR$$

$$\text{where } R = {-}CH_2\underset{\displaystyle NH_2}{\underset{|}{C}}HCOOH$$

Alternatively, S-sulfocysteine could be formed from thiosulfate and serine, as it is in the mold, *Aspergillus nidulans* (Nakamura and Sato 1963).

$$\underset{\text{serine}}{R{-}OH} + \underset{\text{thiosulfate}}{HS_2O_3^{\ominus}} \rightarrow R{-}S{-}SO_3^{\ominus} + H_2O$$

The discovery of this condition appears to establish the vital physiological role of the enzyme sulfite oxidase (Mudd *et al.* 1967), a point which had been in some doubt due to the propensity of sulfite to undergo autoxidation (MacLeod *et al.* 1961).

Having briefly surveyed the known errors of sulfur metabolism, I would like to turn now to a more detailed discussion of the conditions leading to homocystinuria. Homocystine was first discovered in human urine in 1962 independently by groups in both Ireland and Wisconsin (Carson and Neill 1962; Gerritsen *et al.* 1962). Since then, about 100 similar cases have been discovered and reported. The clinical syndrome caused by cystathionine synthase deficiency may range from a rather benign condition marked perhaps only by dislocated ocular lenses to severe disease with mental retardation, bony and connective tissue changes, and early death, usually from thromboembolic episodes.

The early descriptions of homocystinuric patients reported that in addition to the abnormal excretion of homocystine in the urine, there was an abnormal accumulation of methionine in plasma and urine (Gerritsen *et al.* 1962; Carson *et al.* 1963; Gerritsen and Waisman 1964). Examination of Fig. 5.1 will show that this was a most important finding because it helped to localize the block. Homocysteine is at a metabolic branch-point. It may be metabolized to cystathionine or it may be remethylated to form methionine. A block in metabolism of homocysteine along either of these pathways might theoretically lead to abnormal accumulation of homocysteine (and, secondarily, homocystine). The fact that methionine accumulated abnormally allowed a choice between these two possible lesions and suggested that the conversion of homocysteine to cystathionine was blocked and that the activity of cystathionine synthase was deficient.[5] In collaboration with Drs. Finkelstein, Irreverre, and Laster at the National Institutes of Health, we devised an assay for cystathionine synthase sensitive enough to permit quantitation of this enzyme activity in milligram amounts of control human liver. It was necessary to study liver since more accessible tissues such as skin, gut mucosa, or red or white blood cells had no detectable activity. The cumulative results of these studies are presented in Fig. 15.2, a scatter diagram, in which our studies of cystathionine

[5]A more complete explanation of this reasoning has previously been published (Laster *et al.* 1965B).

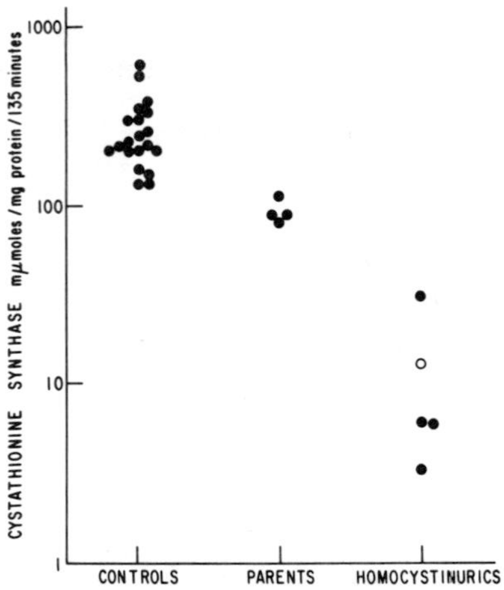


Fig. 15.2. SPECIFIC ACTIVITIES OF CYSTATHIONINE SYNTHASE IN EXTRACTS OF FRESH HUMAN LIVER

Assays were performed as described (Mudd *et al.* 1965). Closed circles indicate measured specific activities. Open circles indicate that no cystathionine synthase activity was detected. The position of the open circle indicates the sensitivity of the assay.

synthase in fresh human liver are summarized. The specific enzyme activity in each patient is represented by a single point. Note that the vertical axis is plotted on a logarithmic scale. These studies confirmed our hypothesis. In each case tested, humans excreting homocystine and methionine have had abnormally low hepatic cystathionine synthase activity.

Although the cystathionine synthase activity is low, it is not entirely missing. The specific activity in 3 instances was 1 to 2% of the mean control value. One exceptional homocystinuric patient had 12% of the mean control value.

This figure shows also that parents of homocystinuric patients had specific activities of cystathionine synthase intermediate between the values found in the patients and the values in the control subjects (Finkelstein *et al.* 1964). There was no overlap between the three groups. This finding is taken as direct proof that the parents are heterozygous for the enzyme deficiency, an interpretation fully compatible with numerous pedigree studies of these families.

Subsequently, it was shown that cystathionine synthase activity is lacking not only in liver, but also in brain tissue of homocystinuric patients (Mudd *et al.* 1966) (Table 15.3).

Recently, in collaboration with Dr. William Uhlendorf, we have investigated the presence of cystathionine synthase in fibroblasts grown in tissue culture (Uhlendorf and Mudd 1968). A very small piece of skin is removed

TABLE 15.3

ENZYME ACTIVITIES IN EXTRACTS OF BRAIN TISSUE FROM CONTROL AND HOMOCYSTINURIC SUBJECTS[1]

| | Methionine Activating Enzyme Mμm/Mg/30 Min | Cystathionine Synthase Mμm/Mg/135 Min | Cystathionase Mμm/Mg/60 Min |
|---|---|---|---|
| Control subjects (5) | | | |
| Range | 0.35–1.30 | 19–27 | 0.17–0.51 |
| Homocystinurics | | | |
| R. S. | 0.22 | 0.07 | 0.21 |
| M. L. | 0.88 | <0.08 | 0.19 |

[1]Data taken from Mudd *et al.* (1966).

from the patient and cells are allowed to grow from it. Such cultures may be propagated for many generations, producing substantial amounts of fresh cells from a single small biopsy. At any time after outgrowth the cells may be frozen in suitable media and stored indefinitely, then revived for further growth. The results of our studies are summarized in Fig. 15.3. Because skin biopsy involves no risk and very little discomfort for the patient, we have been able to study cells from a larger number of individuals. In all instances, patients excreting homocystine have had low

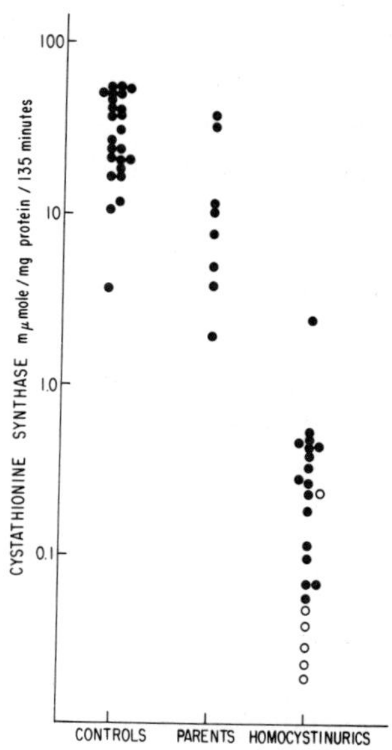


Fig. 15.3. SPECIFIC ACTIVITIES OF CYSTATHIONINE SYNTHASE IN EXTRACTS OF FIBROBLASTS GROWN IN TISSUE CULTURE

Cells were grown as described (Uhlendorf and Mudd 1968). Closed and open circles as in Fig. 15.2.

activities of cystathionine synthase in their fibroblasts grown in tissue culture. With this system also, in many instances there is residual enzyme activity of as much as 2% of the mean control value. The cells from one patient were exceptional in that their specific activity was close to the lower limit observed in a series of cells obtained from control subjects. Cells from parents of homocystinuric patients tend to show low values of cystathionine synthase activity, although again there are two exceptions, the parents of the girl with the unusually high activity. The availability of these many mutant cell lines obviously provides a powerful tool for further investigations. From these cells one may hope to obtain insight into questions such as a possible correlation between severity of clinical disease and the level of residual enzyme activity, and whether the defect in a given instance is in a structural or regulatory gene.

The availability of patients with proven deficiency of cystathionine synthase has made possible studies designed to assess the role of this enzyme in normal physiology (Laster *et al.* 1965A). When normal control subjects are given oral doses of L-methionine, an amount of sulfur equivalent to 60 to 80% of the administered dose is excreted in the urine as inorganic sulfate within the next 24 hr. In contrast to normal subjects, cystathionine synthase deficient patients were unable to convert methionine to inorganic sulfate at a rapid rate. These patients were able to convert the sulfur of orally administered cysteine to sulfate normally. These experiments, which have now been confirmed and extended in other cystathionine synthase deficient patients, indicate that the pathway through cystathionine is quantitatively the dominant one by which normal humans convert the sulfur of a large dose of methionine to inorganic sulfate.

These studies of sulfur balance during methionine loads led to several other conclusions. Figure 15.4 was drawn on the basis of data presented by Laster and co-workers (Laster *et al.* 1965A). It shows during a chronic methionine load given to a cystathionine synthase deficient patient, the balance between oral intake of cystine and methionine and the urinary excretion of sulfur-containing metabolites. It should be noted that the balance was not complete since sulfur loss in stool, respiratory air, and sweat were not taken into account. Nevertheless, the great amount by which sulfur intake exceeded sulfur excretion indicates that during this study the patient was in strongly positive sulfur balance. This conclusion is supported by the fact that methionine accumulated in the plasma of the patient. In spite of the proven inability of this patient to convert the sulfur of methionine to inorganic sulfate at a normal rate, inorganic sulfate was the single most important sulfur-containing urinary metabolite. The inorganic sulfate excretion exceeded the cystine intake. Some methionine was being converted to sulfate. It has been calculated that this conversion

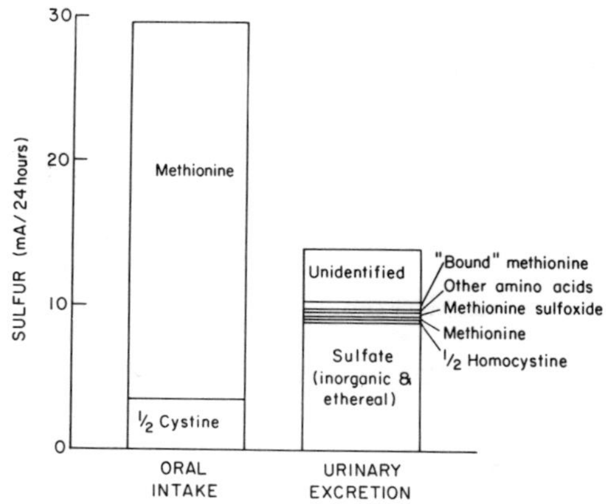


Fig. 15.4. Balance Between Oral Sulfur Intake and Urinary Sulfur Excretion During a Chronic Methionine Load Given a Cystathionine Synthase Deficient Patient

Drawn after data from Laster *et al.* (1965A).

could be accounted for by the residual cystathionine synthase activity and need not indicate the existence of an alternative minor pathway from methionine to sulfate (Laster *et al.* 1965A).

Although cystathionine synthase deficiency was initially recognized by the abnormal homocystine excretion found in this condition and although the disease continues sometimes to be called homocystinuria, homocystine accounted for only a minor portion of the abnormal urinary metabolites shown to be present. Amounts comparable to homocystine were excreted of methionine, methionine sulfoxide, and material which gave rise to methionine after acid hydrolysis, called here "bound" methionine. The known metabolites, however, failed to account for all the urinary sulfur, and in these studies the major portion of organic sulfur metabolites was not identified. Following this lead Perry and co-workers have succeeded in isolating five additional abnormal metabolites from the urine of cystathionine synthase deficient patients (Perry *et al.* 1966A, B, 1967). The structures of three of these compounds have been proven and are shown in Fig. 15.5. Together, these studies make it quite apparent that, although homocysteine is the compound which primarily accumulates as a result of the block in cystathionine synthase, a major portion of this amino acid is metabolized back to methionine, and both homocysteine and methionine are further converted to a variety of other compounds which are excreted in abnormal amounts. A good deal of insight into minor metabolic pathways for sulfur amino acids may be expected from continued study of these compounds and their modes of formation.

Brenton and his co-workers (1966) have utilized a cystathionine syn-

```
     NH2                     NH2
     |                       |
HOOC CH CH2CH2 - S - CH2CH COOH
```

Cystathionine

```
     NH2                           NH2
     |                             |
HOOC CH CH2CH2 - S - CH2CH2CH COOH
```

Homolanthionine

Perry, et al., 152, 1750 (1966).

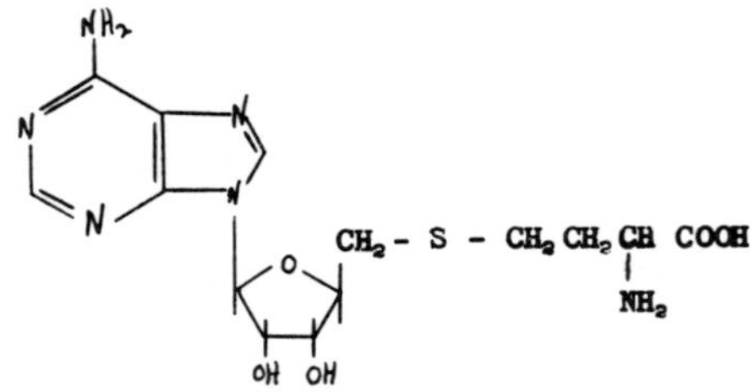


S-Adenosylhomocysteine

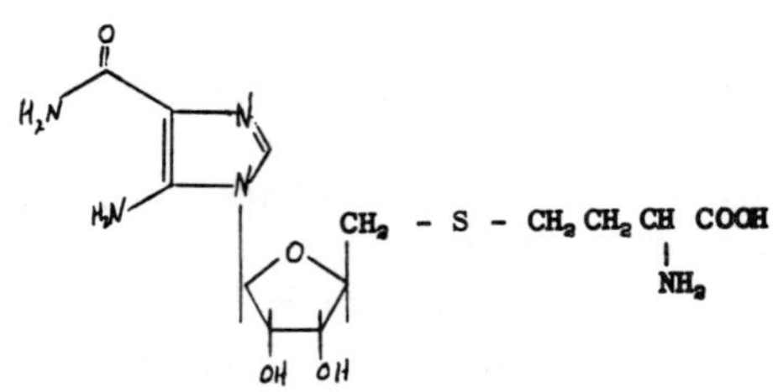


5-Amino-4-imidazole-carboxamide-
5'-S-homocysteinylriboside

*Perry et al. (1966A)*

Fig. 15.5. Some Abnormal Metabolites Excreted by Cystathionine Synthase Deficient Patients

The structure of cystathionine is shown for comparison only.

thase deficient patient to demonstrate that if the cystathionine pathway is missing, cystine becomes an essential amino acid. As has been known, cystine in the diet may be removed and replaced by methionine without sending a normal subject into negative nitrogen balance. However, when cystine was removed from the diet of a cystathionine synthase deficient subject, he went into immediate, marked, negative nitrogen balance. In addition to their implications for therapy, these experiments serve to confirm the quantitative importance of the cystathionine pathway for cystine biosynthesis in the mammal. Crude extracts of several tissues of vertebrates are quite active in the synthesis of cysteine from serine and sulfide, a reaction catalyzed by serine sulfhydrase (Bruggemann *et al.* 1962). Apparently serine sulfhydrase is not capable of supplying the normal cystine requirement of man. Perhaps, as suggested by Waldschmidt (1962), insufficient quantities of sulfide are normally present.[6]

The role of pyridoxine and related $B_6$ compounds has recently attracted a good deal of experimental attention in relation to cystathionine synthase deficiency. Miss Conerly and I have been studying the properties of cystathionine synthase partially purified from human liver. Human cystathionine synthase completely dependent for activity upon the addition of pyridoxal phosphate can be prepared by dialysis of the enzyme against hydroxylamine and removal of hydroxylamine by gel filtration. A representative experiment is shown in Fig. 15.6. Pyridoxal phosphate is quite specific in restoration of maximal activity. These, and other studies, indicate

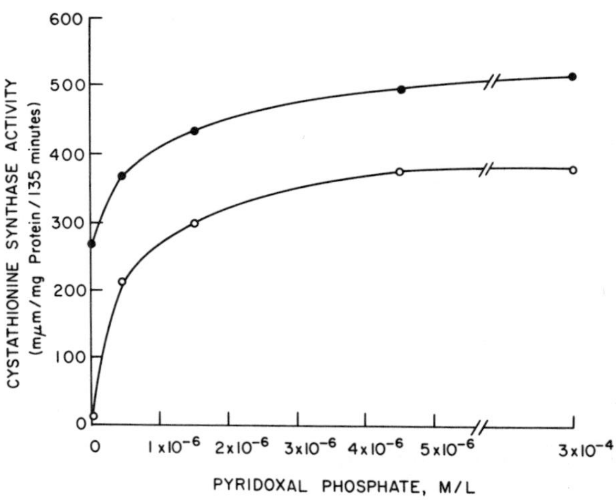


Fig. 15.6. EFFECT OF PYRIDOXAL PHOSPHATE ON CYSTATHIONINE SYNTHASE PARTIALLY PURIFIED FROM HUMAN LIVER

The top line represents the activity of an enzyme partially resolved during routine protein fractionation procedures. The bottom line represents the activity of the same enzyme preparation after treatment with hydroxylamine as described in the text.

[6]Very recent evidence has raised the possibility that the serine sulfhydrase and the cystathionine synthase activities of chicken liver are catalyzed by the same enzyme protein (Braunstein *et al.* 1969). If this possibility applies to the human enzyme, cystathionine synthase deficient patients may well lack serine sulfhydrase activity and the reasoning of this paragraph with respect to serine sulfhydrase would be inapplicable.

that pyridoxal phosphate is the cofactor for mammalian cystathione synthase.[7]

The *in vitro* assays of liver, brain, and fibroblast extracts from homocystinuric patients which I have discussed above were all carried out in the presence of pyridoxal phosphate added to the incubation medium. The abnormally low enzyme activity is not due to lack of cofactor. Therefore, it was somewhat surprising when Barber and Spaeth (1967) reported that administration of pyridoxine in large doses to cystathionine synthase deficient patients led to elimination of homocystine from the urine and plasma and decrease in the plasma methionine concentration. These findings have since been confirmed in other patients (Hooft *et al.* 1967; Turner 1967; Kelly and Copeland 1968; Yoshida *et al.* 1968; Hollowell *et al.* 1968). Some cystathionine synthase deficient patients fail to respond in this manner (Chase *et al.* 1967; Hooft *et al.* 1967).

The mechanism whereby pyridoxine exerts this effect is of interest for both theoretical and therapeutic reasons. In collaboration with Mr. Edwards and Drs. Loeb, Brown, and Laster, we have recently carried out some studies designed to investigate this mechanism. Table 15.4 summarizes some of our observations. Pyridoxine was administered to two patients who had previously been shown by liver biopsy to be cystathionine synthase deficient. Each responded with a marked drop in urinary homocystine. Other metabolites arising proximal to the cystathionine synthase step also decreased. (These data are not shown.) During pyridoxine response second liver biopsies were obtained and cystathionine synthase measured. As is

TABLE 15.4

EFFECT OF PYRIDOXINE TREATMENT ON HOMOCYSTINE EXCRETION AND CYSTATHIONINE SYNTHASE ACTIVITY OF TWO PATIENTS

| Pyridoxine | K. B. | J. H. |
|---|---|---|
| | Homocystine, μm/24 Hr[1] (approximate mean values) | |
| – | 500 | 450 |
| + | 50 | 10 |
| | Cystathionine synthase | |
| – | Less than 4% of mean control value | |
| + | | |

[1]The values for homocystine excretion are approximations of the mean daily excretions. Taken from unpublished observations of S. H. Mudd, W. A. Edwards, P. M. Loeb, M. S. Brown, and L. Laster (1969).

[7]Although some studies of the enzyme from rat liver have led to different conclusions (Brown *et al.* 1968), other findings indicate that pyridoxal phosphate is a cofactor for cystathionine synthase from this source also (Nakagawa and Kimura 1968; Kashiwamata and Greenberg 1969).

shown, even during response, the specific activity of cystathionine synthase did not rise above 4% of the mean control activity. This was true regardless of *in vitro* supplementation of the extracts with pyridoxal phosphate. Similar findings have recently been reported by other groups (Hollowell *et al.* 1968; Gaull *et al.* 1968B).[8]

These experiments make it clear that pyridoxine may lead to a dramatic change in the excretion of homocystine without stimulation of cystathionine synthase activity to a level even close to normal. Further work is now in progress to ascertain whether an alternate metabolic pathway is opened for homocysteine disposal following pyridoxine administration. Another hypothesis meriting very serious consideration is that pyridoxine does lead to a small stimulation of cystathionine synthase activity, for example, from 1 to 2% of normal up to 3 to 4% of normal, and that this small increment in enzyme activity is sufficient to allow disposal of most of the normal metabolic load of homocysteine. This possibility is also under investigation. If proven, it may illustrate the feasibility of treating inborn errors by specific stimulation of the deficient enzyme activity and show that beneficial results are attainable even though the enzyme activity is only minimally restored.

The final subject to be discussed is a recently discovered disease which may be an alternate cause of excessive homocystine excretion. Levy and co-workers discovered that an infant who failed to thrive and died during his seventh week had a combination of excessive homocystine and cystathionine in plasma and urine and less than normal methionine in plasma and tissues (Levy *et al.* 1969). This unusual pattern of sulfur amino acid abnormalities suggests that the child was unable to methylate homocysteine to methionine at a normal rate. As may be deduced from Fig. 5.1, the expected consequences would be primary accumulation of homocysteine (and homocystine), with possible secondary failure to maintain normal methionine concentrations and enhanced formation of cystathionine. Assays of enzyme activities in several tissues obtained from the child at postmortem examination confirmed this hypothesis (Mudd *et al.* 1969). Some of these data are reproduced in Table 15.5 where specific activities of the two enzymes known to be involved in mammalian methylation of homocysteine are shown. The $N^5$-methyltetrahydrofolate-dependent enzyme activity was significantly low in both liver and kidney. In other studies the activity of this enzyme has been shown to be decreased also in brain (Levy *et al.* 1969). In contrast the betaine-dependent enzyme was normal, or close to normal, in both liver and kidney.

---

[8]Somewhat different results were obtained by Yoshida *et al.* (1968) in their pyridoxine responsive patient.

TABLE 15.5

ENZYME SPECIFIC ACTIVITIES IN EXTRACTS OF TISSUES FROM THE PATIENT AND CONTROL SUBJECTS[1]

| Enzyme and Source | Patient | Control Subjects Mean ± SE[2] | Control Subjects Range |
|---|---|---|---|
| Methylfolate-$H_4$ methyltransferase | | | |
| Liver | 0.25 | 1.10 ± 0.14 | 0.52–1.62 |
| Kidney | 0.10 | 2.34 ± 0.46 | 0.80–4.56 |
| Betaine methyltransferase | | | |
| Liver | 4.66 | 6.22 ± 0.98 | 2.1 –12.5 |
| Kidney | 4.05 | 10.4 ± 1.1 | 7.33–15.4 |

*Reproduced with permission of Academic Press, New York*

[1]Taken from Mudd *et al.* (1969). For the control series, each enzyme was assayed in an extract of liver from each of 10 control subjects. Similar studies were performed in extracts of kidneys from 8 control subjects (not necessarily the same subjects that were used for the studies of liver). Specific activities, all per mg protein, are given in the following units: methylfolate-$H_4$ methyltransferase, mμmoles methionine/60 min; betaine-homocysteine methyltransferase, mμmoles methionine/45 min.

[2]Mean ± standard error.

$N^5$-methyltetrahydrofolate homocysteine methyltransferase is one of the two mammalian enzymes known to utilize $B_{12}$ derivatives as coenzymes. The exact form in which $B_{12}$ is active in this system is not known, but it is known to be a derivative other than deoxyadenosyl-$B_{12}$, the coenzyme for methylmalonyl CoA isomerase, the other $B_{12}$-dependent mammalian enzyme.

To investigate the possibility that the patient had a low activity of $N^5$-methyltetrahydrofolate-homocysteine methyltransferase due to lack of suitable coenzyme rather than lack of apoenzyme, his urine was examined for methylmalonic acid, a compound known to be excreted in elevated amounts when methylmalonyl-CoA isomerase activity is low (Cox and White 1962; Oberholzer *et al.* 1967). The child did indeed have abnormally high methylmalonic aciduria, thus indicating that both known $B_{12}$-dependent enzyme activities were depressed.

These findings suggested a difficulty in accumulation of coenzymatically active derivatives of $B_{12}$, a hypothesis which was confirmed by examination of the tissues of the child for total $B_{12}$ and for deoxyadenosyl-$B_{12}$. Total $B_{12}$ was within the control range but deoxyadenosyl-$B_{12}$ was low in both liver and kidney (Mudd *et al.* 1969) (Table 15.6). Although there are several possible sites for the exact metabolic lesion in this patient, it now seems most plausible that there is at least one common step in the transformation of vitamin $B_{12}$ to deoxyadenosyl-$B_{12}$ and to the $B_{12}$ derivative active in methyl transfer. It would be in such a step that the patient was blocked.

A culture of fibroblasts was initiated from a skin biopsy from this patient. In collaboration with Drs. Uhlendorf and Levy, we have recently been studying these cells. A few of our unpublished results will be sum-

TABLE 15.6

$B_{12}$ DERIVATIVES IN SERUM AND LIVER OF THE PATIENT AND CONTROL SUBJECTS[1]

| | Patient | Control Subjects Mean ± SE | Range |
|---|---|---|---|
| Serum, total $B_{12}$ (day 20), mμg/ml | 1.75 | – | 0.2–0.5 |
| Liver, mμg/gm wet tissue | | | |
| Total $B_{12}$ | 56 | 104 ± 22 | 52–150 |
| Deoxyadenosyl-$B_{12}$ | 13 | 160 ± 22 | 67–240 |
| Kidney, mμg/gm wet tissue | | | |
| Deoxyadenosyl-$B_{12}$ | <2 | – | 9–14 |

*Reproduced with permission of Academic Press, New York*

[1]Taken from Mudd *et al.* (1969). Total $B_{12}$ was measured in livers from 4 control subjects; deoxyadenosyl-$B_{12}$, in liver from 7 and kidney from 2 control subjects. The patient's tissues were frozen within 1 hr of death and assayed for deoxyadenosyl-$B_{12}$ 6 months later. Control tissues were obtained more than 1 hr after death and assayed after storage from 3 months to 3 yr. The values for liver are discordant in that the content of deoxyadenosyl-$B_{12}$ is apparently higher than total $B_{12}$. That this discrepancy results from a systematic difference in the 2 assay systems employed, rather than random error, is indicated by the fact that for the control patients the 2 sets of values correlate well ($r = 0.96$, $p < 0.05$).

marized here. The fibroblasts show metabolic lesions in both areas that the patient did: (a) When incubated with radioactive propionate the cells do not form radioactive $CO_2$ at a normal rate, but rather accumulate unusually large amounts of radioactivity in methylmalonate. (b) When methionine in the culture medium is replaced by homocystine the mutant cells from the patient are unable to grow under conditions in which control fibroblasts grow normally unless the medium is supplemented with relatively high concentrations of $B_{12a}$. Apparently, at limiting $B_{12}$ concentrations, the mutant cells are unable to form coenzyme for homocysteine methylation.

Perhaps the chief interest of this new disease for those concerned with sulfur metabolism is that a unique system is available in these patients to examine questions such as the metabolic importance of remethylation of homocysteine, the normal balance between routing of homocysteine to cystathionine or back to methionine, and the relative contributions of methylation from methyltetrahydrofolate or from betaine. Clearly, under some conditions, methylation by use of methyltetrahydrofolate may be of great physiological importance. In view of these findings it may also be worthwhile to examine more thoroughly sulfur amino acid metabolism in vitamin $B_{12}$ and folic acid deficiency states since important abnormalities may well be present in these conditions.

In summary, I have discussed nine inborn errors of human metabolism which affect sulfur biochemistry in major ways. Although this subject goes back a long way in biochemical history, it appears still to be in its infancy. None of the diseases is completely understood. Seven of them have been defined within the last decade, and 4 of the diseases are exemplified by 1 or 2 patients only. Additional patients will surely be found, and almost as surely other inborn errors remain to be discovered. I have tried to emphasize the contributions already made by the study of these conditions and the reasons why they are worthy of continued general interest.

## BIBLIOGRAPHY


ASATOOR, A. M., LACEY, B. W., LENDER, D. R., and MILNE, M. D. 1962. Amino acid metabolism in cystinuria. Clin. Sci. *23*, 285–304.

BALDESSARINI, R. J., and KOPIN, I. J. 1963. Assay of tissue levels of S-adenosylmethionine. Anal. Biochem. *6*, 289–292.

BARBER, G. W., and SPAETH, G. L. 1967. Pyridoxine therapy in homocystinuria. Lancet *1*, 337.

BRAUNSTEIN, A. E., GORYACHENKOVA, E. V., and LAC, N. D. 1969. Reactions catalysed by serine sulfhydrase from chicken liver. Biochim. Biophys. Acta *171*, 366–368.

BRENTON, D. P., CUSWORTH, D. C., DENT, C. E., and JONES, E. E. 1966. Homocystinuria. Clinical and dietary studies. Quart. J. Med. *35*, 325–346.

BRENTON, D. P., CUSWORTH, D. C., and GAULL, G. E. 1965. Homocystinuria. Biochemical studies of tissues including a comparison with cystathioninuria. Pediatrics *35*, 50–56.

BROWN, C., BRENNAN, C., and GORDON, P. H. 1968. Cystathionine synthase. Federation Proc. *27*, 782.

BRUGGEMANN, J., SCHLOSSMANN, K., MERKENSCHLAGER, M., and WALDSCHMIDT, M. 1962. The occurrence of serine sulfhydrase. Biochem. Z. *335*, 392–399. (German)

CARSON, N. A. J., and NEILL, D. W. 1962. Metabolic abnormalities detected in a survey of mentally backward individuals in Northern Ireland. Arch. Disease Childhood *37*, 505–513.

CARSON N. A. J. *et al.* 1963. Homocystinuria: A new inborn error of metabolism associated with mental deficiency. Arch. Disease Childhood *38*, 425–436.

CHASE, H. P., GOODMAN, S. I., and O'BRIEN, D. 1967. Treatment of homocystinuria. Arch. Disease Childhood *42*, 514–519.

CLARK, H. T. 1932. The action of sulfite upon cystine. J. Biol. Chem. *97*, 235–248.

COX, E. V., and WHITE, A. M. 1962. Methylmalonic acid excretion: An index vitamin $B_{12}$ deficiency. Lancet *2*, 853–856.

CRAWHALL, J. C. *et al.* 1968. Beta mercaptolactate-cysteine disulfide: Analog of cystine in the urine of a mentally retarded patient. Science *160*, 419–420.

DENT, C. E., and HARRIS, H. 1951. Genetics of cystinuria. Ann. Eugen. (London) *16*, 60–87.

DENT, C. E., HEATHCOTE, J. G., and JORON, G. E. 1954A. The pathogenesis of cystinuria. I. Chromatographic and microbiological studies of the metabolism of sulphur-containing amino acids. J. Clin. Invest. *33*, 1210–1215.

DENT, C. E., and ROSE, G. A. 1951. Amino acid metabolism in cystinuria. Quart. J. Med. *20*, 205–219.

DENT, C. E., SENIOR, B., and WALSHE, J. M. 1954B. The pathogenesis of cystinuria. II. Polarographic studies of the metabolism of sulphur-containing amino-acids. J. Clin. Invest. *33*, 1216–1226.

FINKELSTEIN, J. D., MUDD, S. H., IRREVERRE, F., and LASTER, L. 1964. Homocystinuria due to cystathionine synthetase deficiency: The mode of inheritance. Science *146*, 785–787.

FINKELSTEIN, J. D., MUDD, S. H., IRREVERRE, F., and LASTER, L. 1966. Deficiences of cystathionase and homoserine dehydratase activities in cystathioninuria. Proc. Natl. Acad. Sci. *55*, 865–872.

FINKELSTEIN, J. D., MUDD, S. H., IRREVERRE, F., and LASTER, L. 1968. Cystathioninuria. Am. J. Diseases Children *115*, 388–390.

FOX, M. *et al.* 1964. Evidence against a single renal transport defect in cystinuria. New Engl. J. Med. *270*, 556–561.

FRIMPTER, G. W. 1965. Cystathioninuria: Nature of the defect. Science *149*, 1095–1096.

GARROD, A. E. 1908A. The Croonian lectures on inborn errors of metabolism. Lancet *2*, 1–7.

GARROD, A. E. 1908B. The Croonian lectures on inborn errors of metabolism. Lancet *2*, 142–148, 214–220.

GAULL, G. E., RASSIN, D. K., and STURMAN, J. A. 1968A. Significance of hypermethionaemia in acute tyrosinosis. Lancet *1*, 1318–1319.

GAULL, G. E., RASSIN, D. K., and STURMAN, J. A. 1968B. Pyridoxine-dependency in homocystinuria. Lancet *2*, 1302.

GERRITSEN, T., VAUGHN, J. G., and WAISMAN, H. A. 1962. The identification of homocystine in the urine. Biochem. Biophys. Res. Commun. *9*, 493–496.

GERRITSEN, T., and WAISMAN, H. 1964. Homocystinuria, an error in the metabolism of methionine. Pediatrics *33*, 413–420.

HARRIS, H., PENROSE, L. S., and THOMAS, D. H. H. 1959. Cystathioninuria. Ann. Human Genet. *23*, 442–453.

HOLLOWELL, J. G., JR. *et al.* 1968. Homocystinuria as affected by pyridoxine, folic acid, and vitamin $B_{12}$. Proc. Soc. Exptl. Biol. Med. *129*, 327–333.

HOOFT, C., CARTON, D., and SAMYN, W. 1967. Pyridoxine treatment in homocystinuria. Lancet *1*, 1384.

HOOFT, C. *et al.* 1965. Methionine malabsorption syndrome. Ann. Paediat. *205*, 73–104.

HOOFT, C. *et al.* 1968. Further investigations in the methionine malabsorption syndrome. Helv. Paediat. Acta *23*, 334–349.

IRREVERRE, F., MUDD, S. H., HEIZER, W. D., and LASTER, L. 1967. Sulfite oxidase deficiency: Studies of a patient with mental retardation, dislocated ocular lenses, and abnormal urinary excretion of S-sulfo-L-cysteine, sulfite, and thiosulfate. Biochem. Med. *1*, 187–217.

KASHIWAMATA, S., and GREENBERG, D. M. 1969. Highly purified cystathionine synthetase of rat liver. Federation Proc. *28*, 668.

KELLY, S., and COPELAND, W. 1968. A hypothesis on the homocystinuric's response to pyridoxine. Metabolism *17*, 794–795.

LASTER, L., MUDD, S. H., FINKELSTEIN, J. D., and IRREVERRE, F. 1965A. Cystathionine synthase deficiency: The metabolism of L-methionine. J. Clin. Invest. *44*, 1708–1719.

LASTER, L., SPAETH, G. L., MUDD, S. H., and FINKELSTEIN, J. D. 1965B. Homocystinuria due to cystathionine synthase deficiency. Ann. Internal Med. *63*, 1117–11142.

LEVY, H. L. *et al.* 1969. A derangement in $B_{12}$ metabolism associated with homocystinemia, cystathioninemia, hypomethioninemia, and methylmalonic aciduria. Am. J. Med. (in press)

LEWIS, H. B. 1932. The occurrence of cystinuria in healthy young men and women. Ann. Internal Med. *6*, 183–192.

MacLeod, R. M., Farkas, W., Fridovich, I., and Handler, P. 1961. Purification and properties of hepatic sulfite oxidase. J. Biol. Chem. *236*, 1841–1846.

Milne, M. D., Asatoor, A. M., Edwards, K. D. G., and Loughridge, L. W. 1961. The intestinal absorption defect in cystinuria. Gut. *2*, 323–337.

Mudd, S. H., Finkelstein, J. D., Irreverre, F., and Laster, L. 1964. Homocystinuria: An enzymatic defect. Science *143*, 1443–1445.

Mudd, S. H., Finkelstein, J. D., Irreverre, F., and Laster, L. 1965. Transsulfuration in mammals: Microassays and tissue distributions of three enzymes of the pathway. J. Biol. Chem. *240*, 4382–4392.

Mudd, S. H., Irreverre, F., and Laster, L. 1967. Sulfite oxidase deficiency in man: Demonstration of the enzymatic defect. Science *156*, 1599–1602.

Mudd, S. H., Laster, L., Finkelstein, J. D., and Irreverre, F. 1966. Studies on homocystinuria. *In* Amines and Schizophrenia, H. E. Himwich, S. S. Kety, and J. R. Smythies (Editors). Pergamon Press, Oxford and New York.

Mudd, S. H., Levy, H. L., and Abeles, R. H. 1969. A derangement in $B_{12}$ metabolism leading to homocystinemia, cystathioninemia and methylmalonic aciduria. Biochem. Biophys. Res. Commun. *35*, 121–126.

Nakagawa, H., and Kimura, H. 1968. Purification and properties of cystathionine synthetase from rat liver: Separation of cystathionine synthetase from serine dehydratase. Biochem. Biophys. Res. Commun. *32*, 208–214.

Nakamura, T., and Sato, R. 1963. Enzymatic synthesis of S-sulphocysteine from thiosulphate and serine. Nature *198*, 1198.

Oberholzer, V. G., Levin, B., Burgess, E. A., and Young, W. F. 1967. Methylmalonic aciduria: An inborn error of metabolism leading to chronic metabolic acidosis. Arch. Disease Childhood *42*, 492–504.

Partington, M., Scriver, C. R., and Sass-Kortsak, A. (Editors) 1967. Conference on Hereditary Tyrosenemia. Can. Med. Assoc. J. *97*, 1045–1101.

Patrick, A. D. 1962. The degradative metabolism of L-cysteine and L-cystine *in vitro* by liver in cystinosis. Biochem. J. *83*, 248–256.

Patrick, A. D., and Lake, B. D. 1968. Cystinosis: Electron microscopic evidence of lysosomal storage of cystine in lymph node. J. Clin. Pathol. *21*, 571–575.

Perry, T. L., Hansen, S., Bar, H.-P., and MacDougall, L. 1966A. Homocystinuria: Excretion of a new sulfur-containing amino acid in urine. Science *152*, 776–778.

Perry, T. L., Hansen, S., Love, D., and Finch, C. A. 1968A. N-acetylcystathionine: A new urinary amino-acid in congenital cystathioninuria. Nature *219*, 178–179.

Perry, T. L., Hansen, S., and MacDougall, L. 1966B. Homolanthionine excretion in homocystinuria. Science *152*, 1750–1752.

Perry, T. L., Hansen, S., MacDougall, L., and Warrington, P. D. 1967. Sulfur-containing amino acids in the plasma and urine of homocystinurics. Clin. Chim. Acta *15*, 409–420.

Perry, T. L. *et al.* 1965. Hypermethioninemia: A metabolic disorder associated with cirrhosis, islet cell hyperplasia, and renal tubular degeneration. Pediatrics *36*, 236–250.

Perry, T. L. *et al.* 1968B. Cystathioninuria in two healthy siblings. New Engl. J. Med. *278*, 590–592.

Robson, E. B., and Rose, G. A. 1957. The effect of intravenous lysine on the renal clearances of cystine, arginine and ornithine in normal subjects, in patients with cystinuria and Fanconi syndrome and in their relatives. Clin. Sci. *16*, 75–93.

SCHNEIDER, J. A., BRADLEY, K., and SEEGMILLER, J. E. 1967A. Increased cystine in leukocytes from individuals homozygous and heterozygous for cystinosis. Science *157*, 1321–1322.

SCHNEIDER, J. A., ROSENBLOOM, F. M., BRADLEY, K. H., and SEEGMILLER, J. E. 1967B. Increased free-cystine content of fibroblasts cultured from patients with cystinosis. Biochem. Biophys. Res. Commun. *29*, 527–531.

SCHULMAN, J. D. 1969. Personal communication. Bethesda, Md.

SCHWARTZMAN, L., BLAIR, A., and SEGAL, S. 1966. A common renal transport system for lysine, ornithine, arginine and cysteine. Biochem. Biophys. Res. Commun. *23*, 220–226.

SEEGMILLER, J. E. *et al.* 1968. Cystinosis. Ann. Internal Med. *68*, 883–905.

SEGAL, S., and SMITH, I. 1969. Delineation of separate transport systems in rat kidney cortex for L-lysine and L-cystine by developmental patterns. Biochem. Biophys. Res. Commun. *35*, 771–777.

SHAW, K. N. F., LIBERMAN, E., KOCH, R., and DONNELL, G. N. 1967. Cystathioninuria. Am. J. Diseases Children *113*, 119–128.

SMITH, A. J., and STRANG, L. B. 1958. An inborn error of metabolism with the urinary excretion of $\alpha$-hydroxybutyric acid and phenylpyruvic acid. Arch. Disease Childhood *33*, 109–113.

SORBO, B. 1957. Enzymic transfer of sulfur from mercaptopyruvate to sulfite or sulfinates. Biochim. Biophys. Acta *24*, 324–329.

TADA, K. *et al.* 1968. Cystathioninuria not associated with vitamin $B_6$ dependency: A probably new type of cystathioninuria. Tohoku J. Exptl. Med. *95*, 235–242.

THIER, S., FOX, M., and SEGAL, S. 1964. Cystinuria: *In vitro* demonstration of an intestinal transport defect. Science *143*, 482–484.

THIER, S. *et al.* 1965. Cystinuria: Defective intestinal transport of dibasic amino acids and cystine. J. Clin. Invest. *44*, 442–448.

TURNER, B. 1967. Pyridoxine treatment in homocystinuria. Lancet *2*, 1151.

UHLENDORF, B. W., and MUDD, S. H. 1968. Cystathionine synthase in tissue culture derived from human skin: Enzyme defect in homocystinuria. Science *160*, 1007–1009.

WALDSCHMIDT, M. 1962. Comparison of the incorporation of sulfide-$S^{35}$ and sulfate-$S^{35}$ into the body protein of rats. Biochem. Z. *335*, 400–407. (German)

WISEMAN, G. 1954. Preferential transference of amino-acids from amino-acid mixtures by sacs of everted small intestine of the golden hamster (*Mesocricetus auratus*). J. Physiol. *124*, 414–422.

WOLLASTON, W. H. 1810. On cystic oxide, a new species of urinary calculus. Phil. Trans. Roy. Soc. London *100*, 223–230.

YEH, H. L. *et al.* 1947. The urinary excretion of amino acids by a cystinuric patient. Am. J. Med. Sci. *214*, 507.

YOSHIDA, T., TADA, K., YOKOYAMA, Y., and ARAKAWA, T. 1968. Homocystinuria of vitamin $B_6$ dependent type. Tohoku J. Exptl. Med. *96*, 235–242.

SECTION VI

# Conclusion

CHAPTER 16

E. J. Underwood

## Summation and Discussion

I imagine that at this late hour you are not looking for a long dissertation from me, and I can assure you immediately that you are not going to get one. I would like to make one or two general comments before making any comments on the papers individually, and that is to mention what to me were the two most striking aspects of this Symposium. The first is how clearly it was planned and organized and how very ably and enthusiastically the individual contributors performed their tasks. In fact, I don't think I've ever attended a conference in which the quality of the papers was so well-sustained from beginning to end, and also, the attendance listened to the papers from the beginning to the end. The other striking aspect of the Symposium which I think has some significance in relation to any thoughts on future symposiums on sulfur is the breadth of coverage which has been maintained without any significant loss of depth of consideration of the individual contributions. It surely has been an extremely refreshing thing for all of us to hear the points of view from individuals so widely divergent in their background and their experience. I imagine this is one of the most satisfactory aspects of the Symposium from the point of view of its sponsors, because all of us must go away with some feeling of challenge in error in their particular area and of some addition or opportunities opened up by the contributions that we've heard in the papers and in the discussions such as we've had just a few moments ago.

Now the stage for the whole Symposium was very capably set for us by Dr. Allaway in his introductory paper. He traced the movement of sulfur through the whole biological cycle and reminded us simply, but forcibly, that without sulfur in the enzyme world there is no life. And it was quite appropriate, too, that he should tell us at this early stage that at one end of the scale a shortage of sulfur-containing amino acids is one of the most crucial, critical, nutritional problems facing the human race. In the next to last paper Dr. Swendseid discussed in detail some of the subjects introduced by Dr. Stracher. I found this a particularly revealing paper, particularly because of the remarkable age and sex differences in the requirements. And I found also rather comforting the little aside she gave of the value of the soybean preparation despite its low methionine content in improving the condition of cases of kwashiorkor or protein malnutrition. Dr. Allaway also mentioned that at the other end of the spectrum, contamination with sulfur from industrial sources can pose serious problems of human health in some circumstances, so that we had the importance of too little and of too much sulfur emerging as problems within the first era of the Symposium.

It was appropriate, too, that the very first morning should be followed by a series of papers on the basic chemistry and biochemistry of sulfur and its compounds and that the first of these should be given by Dr. Kice, the physical organic chemist who gave, I felt, a dynamic account of the chemistry of the sulfur-sulfur bond. Most of us here, I think, were familiar with the ubiquitous occurrence of this bond in key biological compounds, despite the little digs concerning the ignorance of the biochemist that he put in. But very few of us, I suspect, were familiar with the physical-chemical factors affecting the cleavage of this bond and of the enormous changes in the disulfide reactivity brought about by the displacement of hydrogen by methyl groups and by other steric differences which he made very clear. It was stimulating, and perhaps surprising, or should I say surprising to me, to learn from Dr. Kice that neutrophilic substitution reactions of the kind he described are very much greater for sulfur than they are for carbon.

Following this fundamental approach to the biology of sulfur, we were treated to a succession of papers of a strictly biochemical nature. Dr. Stracher told us of the three different classes of sulfhydryl groups present in the reactions relating to the acting of muscle and shared very revealingly how the subunit structure of niacin merged with the sulfhydryl problem. Dr. Parks dealt with the transmethylation reactions in the tissues, pinpointed the role of S-adenosylmethionine in the complex process of methionine biosynthesis, and rather startled us by pointing out the number of transmethylations which are now known and which have arisen, I think he said, from about 40 just a few years ago, to some 70 at the present time.

This was followed by an illuminating and quite entertaining, high-velocity paper from Dr. Finkelstein on the enzymic processes involved in methionine utilization and regeneration. In fact, during this address I was reminded of a story which some of you may have heard before, and I felt like the man in the large American car who drove up to the garage to get the tank filled and after a few minutes the garage attendant put his head in the window and said, "Shut the engine off, she's gaining on me." Well, that's exactly how I felt at that moment. However, Dr. Finkelstein did go on to tell us of the effects of dietary changes upon the individual enzymes in rat tissues and to spot the wealth of data now available in this area. He pointed out to us that the molecular mechanisms of sulfur amino acids homeostasis in the animal body is still quite unknown.

In the paper on sulfur requirements in metabolism of microorganisms by Dr. Peck, the unique contribution of microorganisms to the sulfur cycle in nature was made clear to us and the stimulating suggestion arose, also rather incidentally, that it was through the metabolism of sulfur, perhaps that life originated.

Dr. Thompson in the next paper pointed out that cysteine and methionine are quantitatively and qualitatively the most important sulfur compounds in plants, and that they constitute the real reason for both the sulfur requirements of plants and for the sulfur supply to nonruminant animals. He shared with us, too, that array of other sulfur compounds occurring in plants which do not appear in his own words to do anything vital either for the plants or for the animals consuming the plants. Obviously a great deal more needs to be learned of the significance of these particular sulfur compounds. He pointed out the important differences between plants and nonruminant animals in the metabolism of sulfur and emphasized further that although important differences exist, there are also important resemblances. And I was pleased that he made special mention of the need of animals for vitamin $B_{12}$ for the conversion of homocysteine to methionine, whereas plants do not.

Turning to this morning, the second day, Dr. Dziewiatkowski demonstrated to us the wide range of sulfate esters produced in animals, some of which are quite transient while others, like the sulfate of polysaccharides, serve as relatively stable tissue components and appear to play a role in the transition of cartilage into bone. Sulfated polysaccharides were also shown to be present in other tissues including the brain, as he demonstrated with some slides, but their functions in this important tissue remain obscure.

The rest of this morning, I imagine you can all cast your minds back that far, was taken up by three excellent papers by Drs. Garrigus, Whanger, and Moir on aspects of the nutritional physiology of sulfur in ruminant animals. Dr. Garrigus, I thought, gave an extremely scholarly account of a very wide field, and it was good that he should, at this state in the proceedings, remind us of the enormous and overwhelming quantitative significance of these ruminant animals to the supply of the world's human needs of high quality protein. The nutritional uniqueness of the ruminant with its vast microbial rumen population intervening between the third ingesters and the host tissues was emphasized by all three speakers. Dr. Whanger gave a particularly illuminating account of the early sequence of events, in the rumen of the sulfur-deficient sheep and showed how this resulted in an accumulation of D-lactate which is not available to these animals as a source of energy. Dr. Moir told us of a central role played by sulfide in sulfur metabolism in the rumen in a manner analogous to the role played by ammonia in nitrogen metabolism in the rumen, and showed further that because of the differential recycling of sulfur and nitrogen to the rumen sulfur can become a limiting factor in nonprotein nitrogen utilization by the species.

This afternoon's performance began with a paper by Dr. Almquist giving a detailed and a very authoritative account of sulfur nutrition in non-ruminant species, using the rat and the chick as examples. Of course he paid particular attention to the requirements of cysteine and methionine and their isomers, but an extremely interesting feature of this paper to me was the evidence presented for the limited sparing effect by dietary inorganic sulfate. We then had what I felt was a fascinating account from Dr. Mudd, on errors of sulfur metabolism, the so-called nature's mistakes. Not only did we get a quick outline of the nine diseases, as you just heard, known to involve errors in sulfur metabolism, but we got a detailed account, including the results of some of his own outstanding work, on the metabolic defects and the enzyme changes involved in homocystinuria. This illustrated again the very wide range of coverage that this Symposium has managed to achieve and the way in which some superficially unsuspectant areas can contribute so significantly to our thinking in quite other areas. Toward the end of his presentation he clearly outlined these errors of metabolism. These can be used as a basis for further studies of the sulfur-amino acid metabolism in normal physiological states as well as relating these to the diseased conditions themselves and to their inheritance.

There are, to my mind, only two notable omissions from the coverage of this Symposium. One was mentioned by Dr. Finkelstein during a discussion period. That is regarding the transport functions or activities of methionine. The other one is of special interest to me with my feeling for trace elements, and that is the absence of any consideration of membrane transport mechanisms of sulfite. A lot of the extreme significance of sulfate to the metabolism of molybdenum and copper was mentioned by several speakers, and although little is known other than theory as to the mechanism by which these remarkable effects on molybdenum excretion and copper retention are mediated by sulfate, there was no serious attention given to this subject in this Symposium. Perhaps this could be one of the subjects either for a future symposium or for some presentation at a later meeting. However, there is no doubt that we can agree without any equivocation, that the Symposium has, in its coverage, achieved its purpose, as I said at the beginning of these few remarks. It has achieved its purpose by widening our interests and our knowledge of this wide range of research in sulfur in plant and animal nutrition, and it has given many of us the opportunity in informal discussions to meet others and to hear their point of view. I can think of no more satisfactory way in which one could comment upon a symposium, whether it is on sulfur or whether it is on any other element one could nominate.

A good deal of the success of the Symposium clearly goes to the speakers, but I think I would be in error, and I'm sure you wouldn't approve, if I failed to mention also the extraordinarily adequate and skillful planning carried out by Dr. Oldfield and his colleagues in their choice of speakers, apart from the speaker summarizing the proceedings of course, and in the way in which the sequence of treatment of this subject was planned and carried out. I think it would be very appropriate in these concluding remarks if we as a gathering here, expressed our appreciation in the usual manner to Dr. Oldfield and his colleagues.

# Index